to the memory of Carlo Perone Pacifico

The Common Agricultural Policy after the Fischler Reform

National Implementations, Impact Assessment and the Agenda for Future Reforms

Edited by

ALESSANDRO SORRENTINO
Università della Tuscia, Italy

ROBERTO HENKE
National Institute of Agricultural Economics, Italy

SIMONE SEVERINI
Università della Tuscia, Italy

Routledge
Taylor & Francis Group

LONDON AND NEW YORK

First published 2011 by Ashgate Publishing

Published 2016 by Routledge
2 Park Square, Milton Park, Abingdon, Oxon OX14 4RN
711 Third Avenue, New York, NY 10017, USA

Routledge is an imprint of the Taylor & Francis Group, an informa business

British Library Cataloguing in Publication Data
European Association of Agricultural Economists. Seminar (109th : 2008 : Viterbo, Italy)
    The Common Agricultural Policy after the Fischler Reform: National
    Implementations, Impact Assessment and the Agenda for Future Reforms.
    1. Agriculture and state – European Union countries – Congresses. 2. Agricultural
    laws and legislation--European Union countries – Congresses
    I. Title II. Sorrentino, Alessandro. III. Henke, Robert, 1955- IV. Severini, Simone.
    338.1'84–dc22

Library of Congress Cataloging-in-Publication Data
The Common Agricultural Policy after the Fischler Reform: National Implementations,
    Impact Assessment and the Agenda for Future Reforms / [edited] by Alessandro
    Sorrentino, Roberto Henke and Simone Severini.
        p.   cm.
    Includes bibliographical references and index.
    1. Agriculture and state – European Economic Community countries.
    I. Sorrentino, Alessandro. II. Henke, Roberto. III. Severini, Simone.
    HD1920.5.C682 2011
    338.1'84--dc22   2011004226

ISBN 9781409421948 (hbk)

Contents

List of Figures

List of Tables

Notes on Contributors

Arfini, Filippo – Università degli Studi di Parma, Parma, Italy.
Bartolini, Fabio – University of Bologna, Bologna, Italy.
Belhouchette, Hatem – IAMM Montpellier, France.
Bergmann, Holger – University of Goettingen, Goettingen, Germany.
Bertoni, Danilo – University of Milan, Milan, Italy.
Bulgheroni, Claudia – University of Milan, Milan, Italy.
Canali, Gabriele – Università Cattolica del S. Cuore, Piacenza, Italy.
Carbone, Anna – Università degli Studi della Tuscia, Viterbo, Italy
Cavicchioli, Daniele – University of Milan, Milan, Italy.
Chatellier, Vincent – INRA UR1134, LERECO, Nantes, France.
Daniel, Karine – INRA UR1134, LERECO, Nantes, France.
Daugbjerg, Carsten – University of Aarhus, Aarhus, Denmark.
Davidova, Sophia – University of Kent, Kent, United Kingdom.
Dax, Thomas – Bundesanstalt für Bergbauernfragen, Vienna, Austria
de Andrés, Rosario – Instituto di Economìa, Geografìa y Demografìa, Centro de Ciencias Humanas y Sociales, Madrid, Spain
de Diego, Jose Luis Miguel – Escuela Tecnica Superior de Ingenieros Agronomos, Madrid, Spain.
de Miguel, Jose Maria – Escuela Tecnica Superior de Ingenieros Agronomos, Madrid, Spain.
Del Giudice, Teresa – University of Naples Federico II, Naples, Italy.
Donati, Michele – Università degli Studi di Parma, Parma, Italy.
Esposti, Roberto – Università Politecnica delle Marche, Ancona, Italy.
Fischler, Franz – Former European Commissioner, Absam, Austria.
Flichman Guillermo – IAMM Montpellier, France.
Gallerani, Vittorio – University of Bologna, Bologna, Italy
Genius, Margarita – University of Crete, Crete, Greece.
Giacchè, Giulia – University of Perugia, Perugia, Italy.
Haniotis, Tassos – European Commission, Brussels, Belgium.
Henke, Roberto – Istituto Nazionale di Economia Agraria (INEA), Rome, Italy.
Hočevar, Vida – University of Ljubljana, Ljubljana, Slovenia.
Hovorka, Gerhard – Bundesanstalt für Bergbauernfragen, Vienna, Austria.
Iagatti, Matteo – Università degli Studi della Tuscia, Viterbo, Italy.
Ibáñez, Miguel – Escuela Tecnica Superior de Ingenieros Agronomos, Madrid, Spain.
Judez, Lucinio – Escuela Tecnica Superior de Ingenieros Agronomos, Madrid, Spain.
Juvančič, Luka – University of Ljubljana, Ljubljana, Slovenia.

Kröger, Melanie – Institut für Ländliche Strukturforschung, Frankfurt am Main, Germany.
Leat, Philip – Scottish Agricultural College, Edinburgh, United Kingdom
Lelyon, Baptiste – INRA LERCO (UR1134), Nantes, France.
Lobianco, Antonello – Università Politecnica delle Marche, Ancona, Italy.
Louhichi, Kamel – INRA, UMR Economie Publique, Thiverval-Grignon, France.
Malak-Rawlikowska, Agata – Warsaw University of Life Sciences, Warszawa, Poland
Moran, Dominic – Scottish Agricultural College, Edinburgh, United Kingdom.
Mouratiadou, Ioanna – Scottish Agricultural College, Edinburgh, United Kingdom.
Musotti, Francesco – University of Perugia, Perugia, Italy.
Nieberg, Hiltrud – Federal Research Institute for Rural, Forestry and Fisheries, Braunschweig, Germany.
Offermann, Frank – Federal Research Institute for Rural, Forestry and Fisheries, Braunschweig, Germany.
Olper, Alessandro – University of Milan, Milan, Italy.
Panico, Teresa – University of Naples Federico II, Naples, Italy.
Pascucci, Stefano – University of Naples Federico II, Naples, Italy.
Petriccione, Gaetana – Istituto Nazionale di Economia Agraria (INEA), Rome, Italy.
Pitlik, Hans – Austrian Institute of Economic Research, Vienna, Austria.
Pretolani, Roberto – University of Milan, Milan, Italy.
Pupo D'Andrea, Maria Rosaria – INEA – Istituto Nazionale di Economia Agraria (INEA), Rome, Italy.
Raggi, Meri – University of Bologna, Bologna, Italy.
Renwick, Alan – Scottish Agricultural College, Edinburgh, United Kingdom.
Revoredo-Giha, Cesar – Scottish Agricultural College, Edinburgh, United Kingdom.
Russell, Graham – University of Edinburgh, Edinburgh, United Kingdom.
Sali, Guido – University of Milan, Milan, Italy.
Sanders, Juern – Federal Research Institute for Rural, Forestry and Fisheries, Braunschweig, Germany.
Sardone, Roberta – Istituto Nazionale di Economia Agraria (INEA), Rome, Italy.
Schmid, Erwin – University of Natural Resources and Applied Life Sciences Vienna, Austria.
Severini, Simone – Università degli Studi della Tuscia, Viterbo, Italy.
Sinabell, Franz – Austrian Institute of Economic Research, Vienna, Austria.
Solazzo, Roberto – Univerità degli Studi di Parma, Parma, Italy.
Sorrentino, Alessandro – Università degli Studi della Tuscia, Viterbo, Italy.
Subioli, Giovanna – Università degli Studi della Tuscia, Viterbo, Italy.
Swinbank, Alan – University of Reading, Reading, United Kingdom.
Swinnen, Johan – Katholieke Universiteit Leuven, Leuven, Belgium.

Taglioni, Chiara – University of Perugia, Perugia, Italy.
Therond, Olivier – INRA- UMR AGIR, Tolosan, France.
Thomson, Kenneth – University of Aberdeen, Scotland, United Kingdom.
Topp, Cairistiona – Scottish Agricultural College, Edinburgh, United Kingdom.
Torquati, Bianca Maria – University of Perugia, Perugia, Italy.
Tzouvelekas, Vangelis – University of Crete, Crete, Greece.
Urzainqui, Elvira – Instituto di Economìa, Geografìa y Demografìa, Madrid, Spain.
Valle, Stefano – Università degli Studi della Tuscia, Viterbo, Italy.
Velazquez, Beatriz – European Commission, Brussels, Belgium.
Viaggi, Davide – University of Bologna, Bologna, Italy.
Was, Adam – Warsaw University of Life Sciences, Warszawa, Poland.
Wery, Jacques – SupAgro, UMR System, Montpellier, France.

Foreword
A Common Agricultural Policy for the Twenty-first Century

Although the CAP has been reformed fundamentally commencing with the McSharry – Reform and continuing with the Agenda 2000 followed by the 2003 – Reforms, there is still a need to adapt the agricultural policy to the challenges we are already facing but even more to the changes coming up internally and globally.

The more and more challenging question is: "How to feed a world with meanwhile more than a Billion people starving from hunger, with an increase of 80 Million people every year and 1 Billion people moving from the rural areas to the urban zones?"

At the same time we are forced to fight against climate change, knowing that we cannot avoid it totally but that we can mitigate weather disasters and adapt to the changed conditions.

At the same time we see additional demand coming up of animal protein, biomass, biofuels, green chemicals, and more. We are confronted with ageing problems of the rural population and a progressive thinning out of the countryside.

In short: food, environmental and rural security is at the risk of becoming a market failure.

Without sustainable food security, a proper protection of the environment and a vibrant countryside the looming shortages will cause growing human suffering and civil strike not only in the developing world but also in the many parts of the European Union.

Much has been made of the concept of multifunctional agriculture, but the discussion so far has ignored reality checks of this concept. We cannot simply leave it to the market and to technology to find a balance between various, partly contradictory aims for our agricultural and food sectors. The invisible hand of the market needs sometimes guidance as to what kind of objectives it should be directed.

There is no lack of criticism about European agricultural support programmes and especially the level of CAP expenditure, which accounts for less than on per cent of total public expenditure. However, scrapping the CAP, as a growing number of people advocates, would actually mean throwing out the baby with the bath water.

The risks include production intensification in the plains with increased pollution of soil and groundwater and new damages to the remaining biotops, land abandonment, rural desertification, reduced farm output in the less favoured areas, accelerated urbanization with additional infrastructural and environmental costs.

However, the status quo is not an option! Therefore the CAP reform process must continue! Before the next reform steps are taken it would be useful to launch a discussion about the future objectives of the CAP. The more so, because Member States were not able to incorporate a modern set of policy objectives in the new Lisbon treaty.

It is unacceptable that Europe continues to answer the questions of the twenty-first Century with the policy instruments of Post-War Europe.

Just as important is further development of the existing and the introduction of new policy instruments. How can we steer "know-how intensification"? How should we deal with growing market volatility? How can we strengthen the link between support and environmental responsibility? What are sustainable contributions of the agricultural sector to mitigate climate change? How can we reach a better social balance amongst farmers? What are the right instruments to strengthen the food chains? How should we strengthen a rural development policy involving all people living in our villages? What is the right approach to development cooperation? These are only a few questions we must give a reasonable answer to if the CAP is to be more appreciated in the European society and on the international scene.

FRANZ FISCHLER

Acknowledgements

The idea of this book has its origins in a European Association of Agricultural Economists (EAAE) seminar concerning future developments of the Common Agricultural Policy (CAP) after the 2003 reform. The seminar took place on the 20 and 21 November, 2008, at Tuscia University (Viterbo, Italy). This book includes a selection of papers which were presented on that occasion. The selection of papers followed a two-step process of peer reviewing. The first step was completed by the chairmen of the seminar sessions while the second step was completed by referees who were selected on the basis of the content of each paper. The structure and the contents of the book were then supervised and accepted for publication by the scientific-editorial committee of the Ashgate Publishers.

Sadly, just the day before the EAAE Seminar took place, Professor Carlo Perone Pacifico passed away. He was the person who had inspired this scientific meeting and was an authoritative and renowned "Maestro" for Italian Agricultural Economists. As a symbol of our gratitude to him, this book is dedicated to his memory.

Many institutions and people have made valuable contributions to the EAAE Seminar and to the publication of this book. We would like to thank Tuscia University (Department of Agro-Forestry Economics and Rural Environment - DEAR, Department of Business, Technological and Quantitative Studies - DISTATEQ and Department of Ecology and Sustainable Economic Development - DECOS), the Italian National Institute of Agricultural Economics (INEA) and the Europe Direct Lazio for their precious organizational and financial contributions towards the creation of the seminar. Financial aid was also gratefully received from the Cassa di Risparmio di Viterbo Foundation (CARIVIT), the Viterbo Province Administration and the Chamber of Commerce of Viterbo (CCIAA).

Our genuine thanks go to the Scientific Committee of the EAAE Seminar and to all the people who took part in the seminar and who, with their scientific contribution and participation, stimulated the debate on the future of the CAP. They provided analytical points and critical comments which have been invaluable in the creation of this book. We are also grateful to those who contributed to the peer reviewing process of the papers which were presented during the seminar.

Last but not least, heartfelt thanks go to Franz Fischler, the former European Commissioner for Agriculture, who offered his authoritative experience to stimulate the debate during the seminar and wrote the foreword for this book.

Viterbo, October 2010
ALESSANDRO SORRENTINO, ROBERTO HENKE
and SIMONE SEVERINI

Chapter 1

From the Fischler Reform to the Future CAP

Roberto Henke, Simone Severini and Alessandro Sorrentino

Introduction

The chapters of this book cover several issues regarding the CAP reform: the political economy of the reform, the impact of the Fischler Reform on the EU agriculture, the environmental effects of the new CAP and the consequences of the new measures on the rural development policies. While most analyses look at what happened after the Fischler Reform, other chapters have a more prospective nature, looking at the future of the CAP in general and of specific instruments including market measures and direct payments.

This section analyses three broader issues: the achievements of Fischler Reform; the relative instability of the current CAP and the need for further reform; the new political, institutional and financial conditions in which the reform might occur.

While opinions regarding the Fischler Reform can be very articulated, there is no doubt that it has been a turning point for the evolution of the CAP. It has introduced radical changes in the way the EU provides support to the farm sector, has developed instruments that have actually become cornerstones of the new CAP, and has reached some of the important objectives that originally motivated it. However, despite these achievements, after less than ten years, it is becoming evident that the current structure of the CAP shows problems that make it rather unstable and call for further reform. Finally, now that the debate about the new CAP reform is mounting, it is especially important to look at current institutional and financial settings.

The achievements of the Fischler Reform

The achievements of the Fischler Reform should be analysed considering the particular institutional and political context in which it took place, because several factors have influenced the Reform's outcome (Moehler 2008; Swinnen 2008). One of these factors refers to the external pressures from the World Trade Organization (WTO) Doha development round that has particularly influenced the choices in terms of decoupling of the support. Other internal factors are the important EU budget constraints and the pressure from EU enlargement (Henning 2008). Furthermore, the Reform has been developed in a context of diminishing power of agricultural interests and of the active presence of a broader number of new actors such as consumers, environmentalists, animal rights activists and other social groups (Matthews 2010).

The main elements of Fishler's Reform that we would like to focus on here are: the decoupling of direct payments, the introduction of cross-compliance and the strengthening of rural development policies through modulation.

The decoupling of direct payments

The decoupling of direct payments, the main core of both the Single Payment Scheme (SPS) and the Single Area Payment Schemem (SAPS), have reduced production distortions caused by the coupling of direct payments to production level or input use. This move has been perceived as a more efficient way of providing support because coupled support is often characterised by limited transfer efficiency and relevant distributive leakages (OECD, 2000). Furthermore, the shift of financial resources from coupled to decoupled direct payments has been a strategy to maintain an adequate level of income support provided by the EU without incurring the limitations that the WTO obligations on internal support are going to pose.

However, it is important to look at this topic considering both pros and cons. On one hand, some efficiency has been gained especially in those cases where, after the decoupling process, farmers have quit production activities where direct payments made them artificially profitable. This has been the case of some specific Mediterranean crops such as, for example, durum wheat, tobacco and cotton. As a consequence, a drastic reduction of the supply of these products has been recorded, which has also had effects on the related agro-food chains, including the downstream industries that are now experiencing a drastic restructuring process. On the other hand, this shift has had implications in terms of the use of resources. In many cases, it has caused the release of part of the resources previously used in those activities, including labour, but also contract works and other inputs. Because of the structural constraints that characterise the farm sector in many of those areas, the adjustment process deriving from decoupling has been slow and has often resulted in the impossibility to find alternative uses for the released resources. Indeed, especially in the period of economic recession that has been experienced in the last few years, some of those resources have been left unused or, at least, underused. This phenomenon has been particular strong in the most internal and remote rural areas, especially in the South Mediterranean Member States where this has had negative consequences on their overall economic situation and rural development. In synthesis, where the released resources did not find a proper alternative use, it seems very likely that the negative consequences of decoupling on rural development have exceeded the gains of efficiency deriving from the decoupling.

The decision to allow Member States to choose how to apply the SPS (i.e. historical vs. regional model and the possibility of leaving some partially coupled payments) contributed to reaching a political consensus on the reform. However, this choice has had very different redistributive implications among farmers and production factors, particularly in those regions where farms differ in terms of production patterns.[1] Indeed, decoupling has also had consequences on the distribution of support among production factors. Especially with the historical model, farmers have often shifted to less intensive production activities or have even decided to leave some of the land idle. In these cases, the most negatively affected by the reform have been temporary workers and those supplying services to farmers. Furthermore, the fact that some direct payments have been left coupled or partially coupled has strongly reduced the extent of the simplification of the CAP that was one of the objectives of the reform.

The decoupling of direct payments also poses some questions regarding their justifications and policy objectives. Indeed, direct payments can compensate EU farmers for the high production costs due to, for example, health and hygiene compliance, traceability and origin requirements, environment protection, preservation of biodiversity and countryside management. However, it can be questioned if direct payments are actually structured in the best way to reach this goal given that their distribution is not tailored to compensate these costs.

Furthermore, direct payments can be seen as a basic income safety-net for EU farmers that allows them to cope more adequately with increasing price volatility. However, because direct payments are not specifically targeted to revenue or income stabilisation, a more targeted policy could, at least theoretically, reach the same goal more efficiently.

Finally, direct payments could support the structural adjustment of the sector. In the less competitive farms where the main strategy is to leave the sector and to find non-farm job opportunities, the opportunity to count on an annual *quasi* constant payment can be used to cover some of the costs farmers may incur in pursuing this strategy. However, this opportunity could also be used to finance some on-farm investments (maybe co-financing rural development measures) that may enable farmers to improve the economic efficiency of their farms and to reinforce their competitiveness.

Cross-compliance

Direct payments represent the basis for cross-compliance that requires EU farmers to fulfil statutory management requirements and to maintain land in good agricultural and environmental conditions. This instrument seems very much in line with what the EU citizens expect from EU agriculture: the provision of safe, healthy food and a sustainable use of natural resources. Therefore, it provides

---

1 While with the historical model the redistribution of the support provided by direct payments among farmers from the pre-reform to the post-reform can be considered generally very limited, this is not so in some of the cases where the regional model has been chosen.

one of the main justifications to the support provided to EU farmers through direct payments. However, several criticisms have also been raised regarding this measure.

The European Court of Auditors concludes that: '… at farm level many obligations are still only for form's sake and therefore have little chance of leading to the expected changes' (ECA, 2008: p. 1) also because the conditions required by cross-compliance merely ask for what is already legally binding according to existing EU legislation. Other criticisms originate from the perceived high administrative burden for farmers and public administrations. However, it should be analysed whether this burden is caused more by the complexity of EU legislation on the fields covered by cross-compliance or by the cross-compliance itself.

It is also important to look at cross-compliance in perspective and to pose the following question: is cross-compliance enough to provide a political justification for the direct payments? Indeed, it should be acknowledged that cross-compliance is a strong move towards the legitimacy of a concept: farmers have the right to receive direct payments only if they behave in a way that is consistent with the goals of society. Note that this is particularly important because this approach can be extended to other sectors and requirements (e.g. land and water management). However, the future of cross-compliance is very much linked to three important questions that are central in the debate on the future reform of the CAP. How will the extent of direct payments change? How are agro-environmental measures going to be modified? Will the future CAP Reform introduce new instruments (e.g. market-based) to improve the provision of public goods?

If, as expected, the level of direct payments will be reduced by future reforms, the scope for cross-compliance will also be reduced because this reduces the 'leverage' of cross-compliance. The same can happen if it will be decided to give more emphasis to agro-environmental measures[2] or if new instruments aimed at incentivising the provision of public goods referred to by cross-compliance are introduced.

Strengthening rural development policies through modulation

The last 15 years have seen a progressive increase of the range of the political objectives of the CAP and most of these would be better dealt with through rural development policies. However, financial resources available for the two pillars are still quite unevenly distributed in favour of the first pillar.

The introduction of modulation has been a way of correcting this imbalance and to strengthen rural development policies without altering either the two-pillar structure of the CAP, or the policy equilibrium that has permitted agreement on

---

2   In particular, any increase in the amount of the overall financial resources for these measures and in the level of payment rates, any enlargement of the fields of application and any increase in the level of the requirements attached to agro-environmental payments, could reduce the scope for cross-compliance.

the financial perspectives and the Fishler Reform. Even if the relatively small modulation rate and the 5,000 Euro per farm threshold have limited the amount of resources obtained by modulation so far, the amount of additional resources moved to rural development measures has not been negligible. Because of the strong pressure towards reduction of the CAP budget, modulation has been an effective way to avoid the reduction of resources assigned to the second pillar of the CAP that, in the planning period 2007-2013, decreased from the expected amount. Indeed, modulation is playing an important role given that in the EU-15, according to estimates by Henke and Sardone (in this book), about 20 per cent of additional resources are added to the RDP budget. Furthermore, the introduction of modulation has permitted the 2009 Health Check Reform to double the modulation rate and the resources moved to rural development policies and also to use these to meet the 'new challenges' identified by that reform.

Despite these achievements, modulation has also been criticised by some farmers' organizations because of the re-distributional effect it has had (Henke and Sardone, in this book). Additional criticisms come from those national governments that have problems adding the national resources required to fulfil the co-financing rules of the second pillar, or that have already not been able to spend all available resources because of the administrative problems they face.

The instability of the CAP after the Fischler Reform

In spite of the many important achievements of the 2003 Fischler Reform, after the Health Check the CAP is at a turning point. Indeed, the recent reforms have considerably changed the features of the CAP:

- The profile and features of the first pillar have been re-shaped, with a substantial increase in the direct payments that have been now shaped as a form of income support for farmers whose nature and goals are the main issues of the current debate about public support for agriculture.
- The second pillar of the CAP has been enhanced in terms of financial resources, but at the same time its scope and goals have been widened, so that the final balance is not necessarily positive. In spite of the continuous process of improvement of the second pillar, it keeps the double and somehow ambiguous nature of a sector-oriented and territory-oriented tool box. It does not lose its original feature of 'accompanying policy' to the first pillar, being unable to stand and be acknowledged as a fully independent pillar of the CAP.
- There is a clear attempt to justify the whole public support for agriculture as a form of incentives for the production of European public goods, but at the moment there seems to be a gap between the theoretical justification and the instruments implemented.
- The current structure of the CAP is the result of a reform process designed

for and led by the 15 older Members of the EU. The new Member States have had to adapt themselves to the new CAP, but in the near future, agenda of reforms will be highly influenced by the EU-12 and that will lead to a redistribution of roles and resources within the EU-27.
• Finally, the whole CAP Reform process is strictly dependent on the debate about the budget review and the future agreements on the financing system of the EU.

One of the main issues left open by the latest CAP Reform is that of the goals addressed by the CAP itself and the instruments provided for the fulfilment of those goals. Are the two pillars of the CAP the most adequate set of tools with which to address the new objectives of the CAP? To this end, the main issue is to find a new and valid justification for the high rate of expenditure of the CAP in the total EU budget (Swinnen 2009). The Fischler Reform has reduced trade distortions thanks to the decoupling process and has more vigorously linked direct payments to environmental, ethical and health issues, with the main goal and quote explicit goal to re-legitimize the CAP (Cooper et al. 2009). It is quite paradoxical that the more transparent and the less distortive the CAP becomes, the more difficult it is to legitimize it properly: the basic CAP is no longer a product-based support aiming at the market stabilization, but it looks more and more like a history-based income support that is not among the legitimate goals of a common agricultural policy. Moreover, since decoupled direct payments are considered non-distortive in international competition there is no reason, at least in theory, why they should be a 'common' policy[3]. The main justification for keeping them at European level is that they contribute to providing public goods that are better produced and consumed at the supra-national level. On the other hand, with the Fischler Reform and the subsequent Health Check, the whole issue of targeting the CAP support to the production of public goods is sort of half-way: it is still rather weak in itself and it also does not justify the existence of two pillars.

The other relevant aspect of the new CAP is its dependency on the whole process of past reforms (Iagatti and Sorrentino 2007, Kay 2003). This is, of course, very relevant in understanding the decision process of CAP reform within the EU and the problems of implementing measures that represent a net break with the past (such as regionalisation or flat rate). The so-called path dependency can create a paradoxical situation in which a support measure like direct payments does not find a proper theoretical and distributive justification and there is growing pressure for a radical solution to this problem. On the other hand, CAP has always been characterised by slow and limited changes, and this may be the main reason why even though in principle everybody thinks direct payments should be phased out,

---

3   Evidently a 'common' agricultural policy has the main goal of supporting farmers' income, but in that case direct payments based on a historical approach are not the best way to accomplish such a relevant goal.

it is very probable that, in the future, they will still play a relevant role in the distribution of resources.

With regards, more specifically, to the first pillar, the debate is at the moment mainly focused on two relevant aspects: 1) to what extent are direct payments able to improve the production of public goods in agriculture and 2) is the supranational level the best one to produce those public goods (subsidiarity)? Once again, the main problem addressed in this debate is twofold: one has to do with definitions (what are, today, the direct payments?) and the other has to do with justification (why should farmers receive such a substantial support from the EU budget)[4]?

It is not easy to provide effective answers to these questions; however, the current debate seems to converge on the idea of a radical change of direction, making direct payments more directly tied to the ability of the primary sector to produce public goods which are not producible at the local level (Swinbank and Zahrnt 2009). Other keywords in this new direction of public support are: competitiveness of agriculture, diversification of rural areas, stabilising of markets. Specifically, to this last point, the debate involves also the future of market policies, given the ongoing decline of the traditional distortive market policies of the first pillar. Today attention is focused on market stabilisation and food security as the main goals of market policies, and on an effective safety net that can substantially contribute to improving income stabilisation and reducing price volatility (Copa-Cogeca 2010).

All in all, it seems that the debate is drawing the attention to a twofold set of tools: on one hand highly targeted and multi-tiered direct payments that are designed to cover the needs and demands of society; on the other, a set of new – less distortive market policies, designed for the sector-specific features of the agriculture activity.

Moving to the second pillar, in this case most scholars share the view that it actually contributes to the production and enhancement of public goods (Zahrnt 2009). Moreover, since it is co-financed by the Member States, the subsidiarity issue is less relevant in this case. However, there are other crucial questions regarding the current features of the second pillar and its future after 2013:

- the heterogeneity of the intervention programmed under that umbrella (sector-based measures, environmental measures, territory measures);
- the need of a deeper area targeting;
- the delicate balance between the financial reinforcement of the second pillar and the expenditure capacity and effectiveness of the national and

---

4   It is worth remembering, to this end, the paper signed by a number of prominent European agricultural economists about the desirable CAP for the future that points out the relevant questions of the justification of direct payments and the need of the CAP to effectively reward the ability of the primary sector to produce common public goods (VV. AA. 2009).

sub-national institutions.

It seems quite clear that tools like modulation, originally designed as a temporary instrument to shift resources from the first to the second pillar, do not have any reason to be further developed and any adjustment among instruments needs to take place under a clear review of the whole CAP.

Looking at the role of old and new Member States, the whole process of CAP Reform has so far been the result of a political process started in the 1990s and led by the 15 old Member States. As a consequence, it has been mainly designed for their specific needs (with leading roles by France, Germany and, financially, the United Kingdom). The new Member States have 'paid' for access to the EU with a sort of silent acceptance of the rules, especially for the CAP, that has often been against their specific and legitimate interests (one such case for all is the limited access to direct payments and top ups). Of course now that the gradual process of enlargement is concluded, and in the current discussions on the future of the CAP and direct payments the new Member States are claiming a different and 'equal' position in the negotiations. Indeed, their position is already influencing the debate on the future of the CAP and on the budget review, since they are all net beneficiaries of the EU budget. Some of the larger New Member States (Poland, Hungary, Rumania) will play a major role in the future assets of the budget and the CAP. Not surprisingly, these countries are listed as 'gold diggers' in the Member States classification of Clapser and Thurston (2010) based on their position on EU budget reform: they are in favour of the status quo both in terms of the budget and CAP measures and adverse to any possible form of national co-financing of the CAP.

The final point has to do with the whole EU budget review and the weight that the CAP has in the ongoing debate. It is clear that the CAP, absorbing around 40 per cent of the EU, budget is the core around which any budget reform proposal spins. There is a wide literature about this issue that discusses different scenarios about EU financing. The hypotheses at stake are manifold. The most radical being a change in the Member State contributions, or a reduction of the resources devoted to agriculture and diverted towards other areas of the budget (competitiveness, cohesion, etc.). More conservatively, another option is the defence of the status quo for agriculture, with an internal switch from Pillar I to Pillar II, or even the creation of a new pillar for covering climate change and the environment. Key in this debate is the position of the individual Member States and their net positions in terms of partial balances (mainly agriculture vs. other policies), even though it is often underlined how the partial net balances are not able to catch the actual level of support received/contributed by each Member State.

Institutional and financial settings for the CAP post 2013

The previous section highlights how the CAP is still in a transitional phase today, despite the radical reform of 2003 and the modifications introduced by the Health Check. The almost integral decoupling of financial support from production requires the anchoring of EU expenditures for agriculture to the provision of public goods, which are coherent with the current requirements that European society expects from the agricultural sector.

It is not purely by chance that the European Commission felt the need, in the months preceding the publication of this book, to launch a public debate on the CAP post 2013, the results of which were debated by leading experts, stakeholders, NGOs, representatives of the food chain and institutions in a conference held in Brussels in July, 2010. Using the words of Commissioner Ciolos, the public consultation and the conference were supposed to outline objectives and strategies able to support the feeling that 'CAP is not only tailored for the farmers but for all European Citizens' (European Commission 2010).

The discussion showed widespread acknowledgement that maintaining food production capacity remains the core business of the CAP, despite the dramatic change in the global scenario experienced in the last 50 years. To confirm this statement, some participants observed that, even if very relevant changes have been introduced all around the EC Treaty, Article 33 (39 in the Treaty of Lisbon) setting out the internal objectives of the CAP has remained unchanged. Different opinions came out about the current meaning of food security and food production capacity, according to the international role of the European Union and the volatility of world food prices. Equally it has been recognized that European agriculture provides services and functions going well beyond those which the market recognizes and pays for. In other words, European agriculture provides its output while complying with a higher level of process and product standards, and providing more kinds of public goods such as environmental services and landscape protection. Thus, the key issue could be seen as structured around the twin themed approach that could be called Food and Environmental Security (Buckwell 2010).

The existence and nature of the CAP post 2013 will be greatly affected by these new conditions. In particular, we are talking about the new institutional setting introduced by the Treaty of Lisbon and the agreements that will be reached by the Member States regarding the new Multiannual Financial Framework (MFF). The two aspects are closely linked, especially taking into consideration the fact that the institutional changes introduced by the Lisbon Treaty create new relations between the main institutional actors even in their choices regarding the EU budget. The impact of such conditions on the configuration of a new CAP will be crucial because it is well known that the negotiations on the reform of the CAP look like 'a large group of diners at a restaurant, where the debate about what to order is matched only by disputes over how to split the bill' (Clasper and Thurston 2010).

According to CAP policy making, the Treaty of Lisbon introduces relevant changes concerning the legislative and EU budget procedures. First, the decision about the CAP will follow the rules of the 'co-decision' (now 'ordinary') procedure. Such rules establish a substantial balance of power between the Council and the Parliament in setting the EU Regulations. It will lead to a dramatic empowerment of the European Parliament (EP) in shaping the future CAP. It is indubitable that in the last 50 years the EP has played a marginal role in CAP decisions, providing only an obligatory (but not binding) judgement on the Commission proposals. Second, concerning the budget procedure, the expenditures on the CAP will no longer be 'obligatory expenditures' on which the Council keeps the right of having the last word. According to the Lisbon Treaty, EU expenditures on the CAP will also require an agreement between the Council and the Parliament. This change is obviously consistent with the new legislative procedure for the CAP.

The new institutional setting has implications for EU agricultural policy making. Over the last 20 years the EP has continuously tried, sometimes successfully, to enforce its influence over agricultural policy beyond its mere institutional mandate (Roeder Rynning 2003). In the eighties and nineties, new decision-making procedures concerning the EU budget, as well as consumer and environmental policy, gave the EP the means to affect CAP decisions externally despite its formal powerlessness in the agricultural domain. Furthermore, in the last few years, the emergence of food safety scares (e.g. BSE) provided a strong opportunity to make the EP voice more effective in shaping the recent CAP reforms. The EP participation in CAP decision-making, as stated in the Lisbon Treaty, will dramatically affect the traditional agricultural policy paradigm based on the sucess of the intergovernmental negotiations and the substantial institutional separation of CAP decision-making from broader issues of public policy (Bulmer 1998). Following the new legislative procedure and given that members of the EP represent at least 90 per cent of the population not involved in the farm sector, we would expect that the traditional intergovernmental profile of the agricultural negotiations will soften because of the overlapping with other prominent profiles (political parties, stakeholders, civil society, NGOs etc).

New CAP decision-making is also expected to strengthen the involvement of the so-called 'countervailing interests'. As noted by Swinnen (in this book) the enlargement in the number and the quality of political actors through the involvement of environmental organizations and consumer groups was a key element in the Fischler strategy to make its proposal for CAP Reform successful. With the assignation of formal policy-making prerogatives to the EP, a broader emergence of non-traditional political coalitions is to be expected. This should counterbalance the traditional prominence of the farm unions and offer political support for a more radical path to CAP reform.

The Treaty of Lisbon could also affect the institutional setting for the CAP through the change in budget decision-making. According to the previous budget procedure (in force before the Lisbon Treaty), the Council had the prerogative to freeze a high share of the budget for the CAP. Given the institutional feature of agricultural expenditure as 'compulsory expenditure', high spending on the CAP enabled the Council to maintain a large share of the budget under its absolute control. The Council's powers on the budget, together with the increasing national origin of 'own resources', made the intergovernmental feature prevail in budget negotiations and the 'net balances' among Member States become the key variable addressing policy choices. As in a vicious circle, the intergovernmental nature of the budget negotiations strengthened the net balance concern, as well as the net balance problem requiring a central role of the Council in the budget decision-making. In such a framework, given the worsening of the net balance problem originated by the EU regional and cohesion policies, the CAP remained the only way whereby Member States sought to obtain the most favourable net balance among contributions and appropriations from the EU Budget (IEEP 2009). Any attempt to reform the CAP had to face this severe impediment. The Fischler Reform was possible only after each Member State's appropriation from the first pillar budget of the CAP was defined by the national ceilings.

With the Lisbon Treaty, the distinction between 'compulsory' and 'non-compulsory' expenditures has been eliminated and the budget procedure requires agreement between the Council and the Parliament after two readings and a conciliation committee. In this new scenario, the CAP is no longer an effective device for preserving the absolute power of the Council with respect to the budget and the pressure for high CAP spending could be mitigated. Furthermore, the co-decision method on the entire budget spending should make the net balances conflict overlap with other arguments raised by societal demands that are widely represented in the EU Parliament. This could make it possible to shift resources from the CAP to other policies that address more relevant challenges such as economic growth, employment, climate change and foreign relations.

In such a framework, there could be some room for the several claims brought up in the public consultation on the budget review opened in 2008 by the Commission. Anticipating its concluding document, the Commission states that the principle of the 'European added value' should be at the heart of the budget and drive the future Multiannual Financial Framework (MFF) out of the 'net balances' and the *juste retour* arguments. A 'new, policy driven own resources' should allow the phasing out of all the net balance correction mechanisms, including the UK rebate. Regarding the agricultural policy, the Commission draft proposes shifting some of the money to other key policy challenges, assigning "larger responsibility" for current CAP spending to member states, or co-financing direct aid to farmers with national contributions (European Commission 2009). Of course, as long as the intergovernmental logic still prevails, despite the new institutional setting, such a severe change of the EU budget priorities and CAP spending will face a strong opposition in the Council. As already mentioned, the new Member States together with France, Spain and Greece are expected to oppose cutbacks in order to maintain the current CAP spending levels.

According to the institutional arrangements introduced by the Treaty of Lisbon, new political conditions could make a reshaping of the CAP possible. Potential significant reductions of the EU agricultural budget, along with the political strengthening of the new challenges in line with the societal claims, are expected to put pressure on further reform of the CAP. Such pressure would lead to substantial reform as long as the change of the institutional setting was capable of softening the intergovernmental feature of the CAP and EU budget decision making.

Conclusions

The Fischler Reform has undoubtedly introduced a breakdown element in the long and complex process of CAP reform. By dismissing the traditional model of coupled support once and for all, the 2003 reform has opened a transition period whose outcomes are still quite difficult to foresee, considering the crucial political and institutional courses of the EU building following the Treaty of Lisbon and the recent Union enlargements.

Now it is the right time to wonder what remains of the CAP, especially with reference to its first pillar, now that the old model has been abandoned. In other words, it is interesting to ask what is left of 'common', 'agricultural' and 'policy' in the current features of the CAP? Probably not much remains and this could be considered both positive and negative, but there are definitely only a few of the original principles of a 'common agricultural policy'. Let's briefly see why.

The first consideration is that the new CAP is diminishing the concept of a 'common' policy, to the extent that it progressively widens gaps for the national (and sometimes sub-regional) governments to filter into them and gain space at the expense of the EU institutions. This is particularly true for the some regulations and measures and in the way they are implemented (e.g., different hypotheses of regionalisation of the direct payments, different levels of implementation of conditionality and also possible differentiation in the competitiveness rules). It is not just a case of increasing reference to the equal treatment of farmers in the EU territory and to the competition rules on common markets. However, three elements still stand in favour of the commonness of the CAP: the redistributive function of the SPS (even though as underlined before, this is between individual subjects); the enhancement of the production of public goods at the European level (but this is a much debated aspect in the current discussion on the future of the CAP); finally, the resource allocation to both pillars which follow the solidarity principle.

The second consideration is whether the reformed CAP is really focused on agriculture and rural areas. Again, looking at the direct payments, it is hard to maintain, since decoupled payment has a fully subjective nature (it is a support to the single farmer who owns the land but might also not farm on it). Moreover, it can be capitalised on by the single farmer selling the land together with the entitlements. Also in this case, it is in the second pillar that at the moment one could find an agricultural feature of the intervention, especially considering measures for supporting and improving primary sector competitiveness and the quality of agricultural products.

Finally, is decoupled support a proper economic policy for agriculture and rural areas? This is hardly defensible, since its essential nature is that of an income support, leaving the market on its own in allocating resources. The only allocational function left is that of conditionality, but it is widely acknowledged as being rather weak, and moreover it is currently just an enforcement of an existing set of regulations. It is worth noting that direct payments play a potential role in the stabilisation of farmers' incomes, but in practice such an objective is not selective and targeted.

The agricultural policy has a more appropriate role and scope in Axis 1 of the second pillar. This is quite paradoxical if we consider that the fortune of the second pillar has been built on the idea of increasing a territorial and non-agricultural approach to support rural areas.

All in all, we can conclude that the 2003 Reform has widely-acknowledged credit, but can also imply a serious risk.

On one hand, clearing the way from the traditional support measures, it offers a chance to redesign an agricultural support consistent with the new demands expressed by society, even though it does not solve the financial problem of the 'juste retour'. The Fischler Reform has amplified the new philosophy launched by Agenda 2000, in its support of the primary sector and rural areas, focusing more on diversification and the production of public goods rather than on specific products, and at the same time enhancing synergies between pillars rather than encouraging overlapping.

On the other, losing the tight connections with the original goals of the Treaty of Rome (unchanged, probably not by chance), the Reform leads the generous EU budget for agriculture into a stormy ocean, in which the recent enlargements and the reform of the Treaty might prevent an easy journey towards a new CAP. In other words, the risk is that of a sudden and substantial reduction of financial support to the primary sector and to rural areas. This, in turn, might imply an unforeseeable reaction within some specific markets and, in specific areas of the EU, might even threaten food security.

References

Buckwell, A. 2010. *The public has spoken – over to you Commissioner,* Country Land & Business Association Magazine. Available at: http://www.cla.org.uk

Bulmer, S. 1998. New Institutionalism and the Governance of the Single European Market. *Journal of European Public Policy,* 5(3), 365-386.

Clasper, J. and Thurston, J. 2010. *Does the CAP fit? Budget reform, the common agricultural policy and the conflicting views of EU member states,* Farmsubsidy. Available at: http://www.farmsubsidy.org

Cooper, T., Hart K. and Baldock, D. 2009. *T*he Provision of Public Goods Through Agriculture in the European Union, in *Report for DG Agriculture and Rural*

Development, Contract No 30-CE-0233091/00-28, Institute for European Environmental Policy, London.

Copa-Cogeca. 2010. *The future of the CAP after 2013.* Brussels. Belgium.

Dwyer, J. and Bennet, H. 2001. Using Modulation to Support Rural Development. (Background Paper), The Countryside Agency – Institute for European Environmental Policy (IEEP), Brussels.

European Commission 2009. *The 2008/2009 Budget Review.* Draft 06-10-2009. Available at: http://www.euractiv.com/pdf/Draft%20document%20 reforming%20the%20budget%20oct%202009.pdf.

European Commission. 2010. *The Common Agricultural Policy after 2013 – Public debate.* Summary report. Available at: http://ec.europa.eu/agriculture/ cap-post-2013/debate/report/summary-report_en.pdf

European Court of Auditors. 2008. *Information note concerning Special Report N° 8/2008 'Is cross-compliance an effective policy?'* ECA/08/30. Luxembourg, 2008/12/09.

Henke, R. and Sardone, R. The fortune of modulation in the process of CAP reform. In this book.

Henning, H.C.A. 2008. EU enlargement: driver of or obstacle to future CAP reforms?, in *The perfect storm: the political economy of the Fischler Reforms of the Common Agricultural Policy* edited by J.F.M. Swinnen. Centre for European policy studies, Brussels.

Iagatti, M. and Sorrentino, A. 2007. La path dependency nel processo di riforma della PAC', *Agriregionieuropa,* 3(9). Available at: http://www.agriregionieuropa. univpm.it/dettart.php?id_articolo=219

IEEP. 2009. Beyond the Immediate Horizon – A CAP Fit for 2020', IEEP CAP2020 Policy Briefing N°6, October 2009. Available at: http://cap2020. ieep.eu/2009/10/16/beyond-the-immediate-horizon-a-cap-fit-for-2020

Kay, A. 2003. Path dependency and the CAP. *Journal of European Public Policy,* 10(3), 405-420.

Matthews, A. 2010. Understanding reform of the Common Agricultural Policy. *QA Rivista dell'Associazione Rossi-Doria,* 1.

Moehler, R. 2008. External influences on CAP reforms: an historical perspective, in *OECD (2000): A matrix approach to evaluating policy: preliminary findings from PEM pilot studies of the crop policy in the EU, the US, CANADA and MEXICO.* Directorate for Food, Agriculture and Fisheries of the Organisation for Economic Co-operation and Development. COM/AGR/CA/TD/TC(99)117/ FINAL. Paris, 2000.

Roederer Rynning, C. 2003. From Talking Shop to Working Parliament? The European Parliament and Agricultural Change. *Journal of Common Market Studies,* 41(1), 113-135.

Swinbank, A. And Zahrnt V. 2009. A CAP that better serves the public and the environment. Available at: http://www.europeanvoice.com/article/imported/a-cap-that-better-serves-the-public-and-the-environment-/66396.aspx

Swinnen, J.F.M. 2008. *The perfect storm: the political economy of the Fischler Reforms of the Common Agricultural Policy.* Centre for European policy studies, Brussels.

Swinnen, J.F.M. 2009. On the future of Direct Payments: paper presented at Workshop BEP, in *Reflections on the Common Agricultural Policy from a long-run perspective.* 26th of February 2009, Brussels.

Zahrnt, V. 2009. *Public money for public goods: winners and losers from CAP reform.* ECIPE working paper, n. 8.

PART 1
Future Scenarios for the CAP

Chapter 2

The CAP Reform Process in Perspective: Issues of the Post-2013 Debate[1]

Tassos Haniotis

This chapter briefly examines the main policy issues related to the prospects of the Common Agricultural Policy (CAP) post-2013. Still premature in terms of concrete proposals, the future direction of the CAP has nonetheless already generated a lively public debate, which in essence centres around two basic questions - do we need the 'P' and more importantly the 'C' in the CAP, and even if we do (most still agree to the need for an EU-wide policy) do we need its present two-pillar structure?

I will advance a positive answer to both questions, provided that the CAP continues to do what it so successfully has been doing since the mid-1990s - adjust and adapt to the challenges it faces. Today these challenges are not only related to CAP *per se*, in its post 'Health-Check' (HC) environment, but also extend beyond the wider institutional and economic setting within which the CAP evolves.

The institutional setting is characterised by:

- the new composition of both Commission and European Parliament;
- co-decision in the field of agriculture as a result of the Lisbon Treaty;
- continuing uncertainties about the prospects of a WTO agreement.

The economic setting has recently become more complex than before because:

- market prospects are influenced by the increased volatility of agricultural prices in the aftermath of the commodity boom and the (relative) bust of the past two years (Figures 2.1 and 2.2);

---

1 This chapter is based on a speech made on several occasions between November 2009 and May 2010, in preparation for the CAP post-2013 debate.

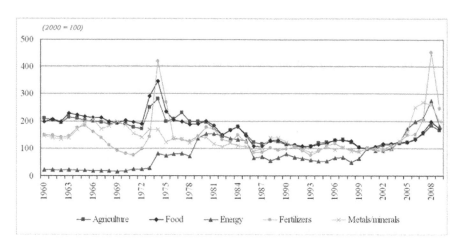

Figure 2.1   Evolution of commodity price indexes (1960-2009)

Source: World Bank

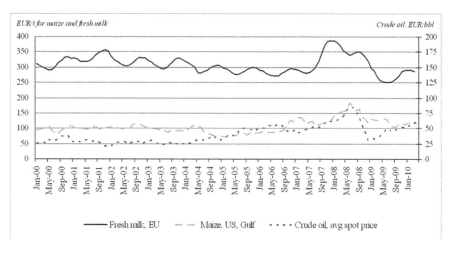

Figure 2.2   Recent trends of some commodity market prices

Source: European Commission – DG Agriculture and Rural Development, and World Bank

- the impact of the economic crisis on agriculture is more severe than initially thought because of the combined effect of lower output prices and higher input costs (Figures 2.3 and 2.4);

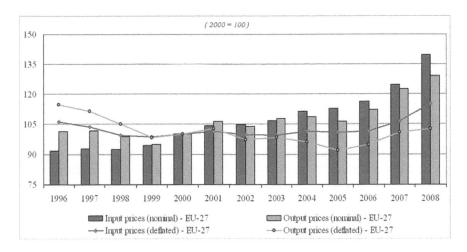

Figure 2.3   Recent evolution of agricultural input and output prices

Source: Eurostat

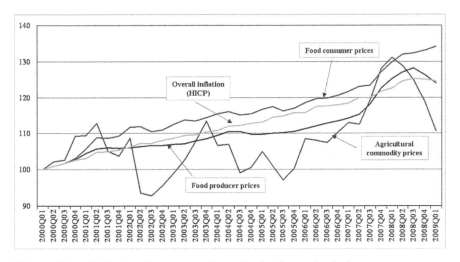

Figure 2.4   Price developments along the food supply chain

Source: European Commission – DG Economic and Financial Affairs, based on Eurostat and DG Agriculture and Rural Development data

- food security concerns are heightened, especially but not exclusively due to their link to climate change.

In my brief intervention, I will focus on the impact of the two key drivers of the debate about the future CAP - the impact of the economic crisis and of climate change - and the role of the CAP to enable EU agriculture to adapt to the combined effect of these challenges within the constraints of increasing limited budgetary resources. Yet a brief reminder of the present structure of CAP expenditure and the reform path that led to this is pertinent for both the debate on the future budget and the CAP.

Whatever the level or distribution of future CAP funds, the starting point of this debate will have to take account of the path of CAP reform and its results. A look at the three broad policy areas of the CAP today from a budgetary standpoint demonstrates the significance of direct payments (which represent more than 70 per cent of the total CAP budget, almost ten times the level of expenditure for the other Pillar I CAP component of market measures) and the new role of Pillar II Rural Development measures, recently further strengthened by the transfer of cuts from direct payment via modulation (Figure 2.5).

While the two-pillar structure of the CAP is characterised by some important differences (annual payments versus multi-annual programming co-financing, respectively), both pillars form an integral part of the same policy. Reform has been achieved by shifting away from the most trade distorting policy instruments to the least trade distorting ones, while in parallel decreasing the CAPs' share in the EU budget and the EU GDP.

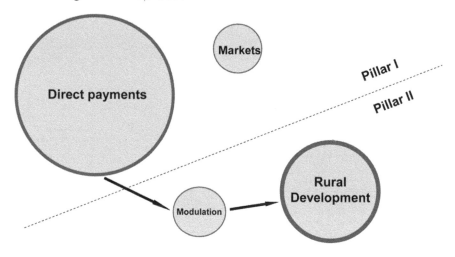

Figure 2.5   The CAP and the budget

Source: European Commission – DG Agriculture and Rural Development

Taking into account that the CAP is one of the very few policies exclusively financed by the EU budget, considering its overall share in EU-wide public expenditure puts CAP spending into context. Importantly, the trend of the CAP share in total EU public expenditure is already clearly on the decline, while the present role of modulation in strengthening Rural Development is worth mentioning.

With this background in mind, let's focus on the complex debate about the future of the CAP, and more specifically about its three broad areas of policy instruments - markets, direct payments and Rural Development - and the linkages between these policy areas.

- Market reforms set in motion both income pressures and structural adjustment; thus their link with direct payments and rural development.
- Direct payments focus the debate on their role, especially with respect to income and/or the delivery of public goods, and the implications for their future.
- Rural development is in turn linked to the 'baseline' stemming from cross-compliance, but also to the multiple needs for sustainable adjustment of our rural communities.

To get a better view of where we need to go with the CAP, a glance at where we are and where we are coming from could prove useful, with the aim of building upon the positive elements of the CAP reform process. When it comes to market instruments, the debate focuses on the pertinence of the remaining instruments; safety net intervention and quotas. Looking at CAP reforms in market intervention shows that we have come a long way in removing market distortions; support prices have declined in all sectors, and these cumulative declines are even more significant when measured in real terms (Figure 2.6). Yet this development did not

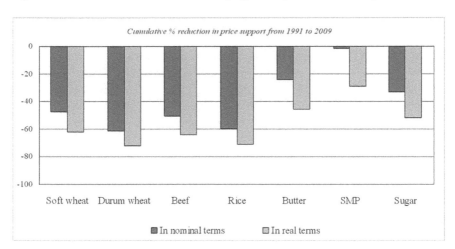

Figure 2.6   Cumulative reductions in EU price support

Source: European Commission – DG Agriculture and Rural Development

lead to a generalised decline in EU market prices, but gradually to their greater realignment with world market prices.

Taking the example of wheat (Figure 2.7), where reform started earlier, EU market prices follow world prices and the intervention price is clearly below normal price levels. Similar trends, albeit to different extent, are also evident in other reformed sectors. In beef, border protection plays a role in stabilising our domestic markets, but clearly intervention is a thing of the past. In dairy, where reform started later, and has a dual nature, first price cuts, now the phasing-out of quotas, intervention still has a role to play when prices drop, more so in butter than in SMP.

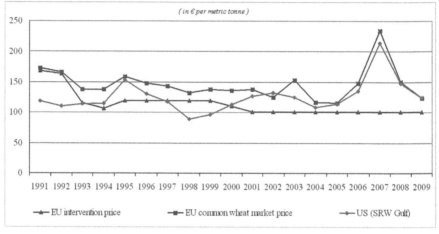

Figure 2.7   The evolving role of EU support prices – wheat

Source: European Commission – DG Agriculture and Rural Development

The cumulative impact of market reforms is more evident if we look at the net production surplus (the annual excess of production over consumption) in the EU when the CAP reform process started (the average of 1990-1994), and compare it to the corresponding average now. Although the EU is still a net exporter in most agricultural products, the previous negative impact of EU agriculture on world markets is long gone.

The above observations are pertinent to the debate of future CAP market policy instruments and indicate how less relevant guaranteed prices have become for support to farm income. Experience shows that the decline in farm support prices did not have a negative effect on farm income; the opposite happened because of the parallel reform of direct payments. On the contrary, the greatest reduction in the EU farm income in recent years occurred in 2009 and coincided with the aftermath of particularly high price levels.

This apparent paradox was influenced by what is called the 'farm price crisis' or 'food crisis' (both terms are used interchangeably to describe the dramatic increase, and then the dramatic decline of farm commodity prices). Clearly, it is not the presence of price variability as such that is of concern. This has been a permanent feature of agricultural markets, driven by the simple fact that the discontinuous, weather-dependent farm production interacts with the continuous, daily demand for food.

What has been of concern is the magnitude that farm price variations have taken in recent years and the fact that sometimes wide price swings seem to be linked more to parallel developments across all commodity markets than to the fundamentals of each individual market. As a result of this development, the signal that farmers receive about market direction has become unclear and often distorted and hampers their capacity to respond.

More importantly, from the agricultural producer's point of view of, is that while the increase in output prices followed (often higher) increases in input costs and were then passed through to consumer prices, the opposite has not occurred to the same extent when output prices declined. The result has been a double squeeze of farm income by what is a totally new development, the parallel boom and bust of all three commodity markets - agriculture, energy and metals/minerals. Though not necessarily generalised, this squeeze has clearly affected some sectors in a significant way, with the EU dairy sector being the most obvious example.

The 2007 food crisis and the current dairy sector situation have put emphasis on the need of maintaining some traditional market measures as a basic safety-net intervention mechanism (namely private and public storage) that does not influence the normal functioning of markets but reinforces a consistent food supply in Europe. Here, market measures played the role they were designed for when, as a result of the general shift of support away from product towards producer, such measures were retained as product safety nets in all major markets. The impact of using existing public intervention mechanisms has been positive in stabilising the price decline, and thus in reducing the extent of income losses producers could have faced as a result of the significant drop in the consumption of dairy products.

But these developments also brought the question of the role of a modern safety-net in targeting risks in agriculture into the forefront. This question is pertinent for the short-term, especially for those sectors more affected by the deterioration of market prospects due to the world economic recession; but it is even more pertinent for the long-term and the type of broad policy areas that we need to address.

Should we target revenue losses, or income losses (thus including costs in the equation)? How could such a scheme operate, especially since its administrative complexities rule out EU-wide implementation?

Needless to say both types of policy responses should be consistent with the same overall framework of our agricultural policy reform process, if the aim is to contribute to the solution of the problem and not to delay the consequences. Thus the link to the Single Payment Scheme (SPS) and especially decoupled direct support, which constitutes the bulk of our agricultural support providing an income cushion to our farmers.

These payments do not only provide a necessary income safety net to farming activity, but contribute in parallel, and jointly, to the provision of public goods from farming, such as scenic cultivated landscapes, grassland habitats, or carbon sequestration in soils. By doing so, they also contribute to climate change mitigation and adaptation.

Many critics of the CAP miss the real target of direct support, which is not to support farm household income, but to ensure a certain farm income stability which, in combination with cross-compliance, promotes sustainable farming activity. Member States have the exclusive competence to assess the fairness of household income distribution and tax its level. When it comes to EU farm income, the facts are very clear - it is well below EU non-farm income.

Direct payments aim to support farm income, in a non-distorting, non-product specific manner. Together with safety-net measures and targeted Rural Development measures, direct payments guarantee the continuation of what has been the overall contribution of the CAP, the presence of an environmentally and territorially balanced agriculture in the EU, especially compared to what it would be in the absence of support.

This evidence of this balance is shown by the results of economic model simulations, but also by a simple bird's eye view of EU agriculture. It is achieved because of the link between the basic income safety net role of direct payments, the market safety net provided by the remaining market measures and the increasingly important role of Rural Development measures in leading towards a sustainable structural adjustment of EU agriculture.

These positive effects have come as the result of the shift in support and its link with cross-compliance. And we consider that such support, albeit more balanced than it is today, is still required in the future to enable EU farming to continue to contribute to the delivery of those public goods that European citizens desire.

In the absence of some basic income support, many EU producers would opt for intensification of their production methods to retain their income while others would simply abandon production. The result would not be less production overall, but more intensive production, concentrated in the most competitive regions, which would result in fewer public goods overall and more challenges to meet climate change targets.

Secondly, the absence of clearly defined 'baseline' requirements, achieved through the link of direct payments with cross-compliance, would make the delivery of public goods through Rural Development less, rather than more effective - which is exactly the opposite of what we want to achieve with the future CAP.

Of course, the future balance between CAP measures will be one of the most crucial areas of the debate. The present system of direct payments responded successfully to the needs of adjusting the CAP to meet objectives related to 2003 challenges - market orientation, simplification and WTO compatibility of our support. But for it to be successful to meet 'post-2013' objectives, it needs further adjustments, for example in the direction of a more harmonised level of support.

We can no longer have a distribution amongst farmers based on historical production and it is also pertinent to look at the rebalancing of direct payments among Member States. But the difficulty in deciding the degree and the speed of such harmonisation should not be underestimated, as has been amply demonstrated by the Impact Assessment of the HC (Figure 2.8).

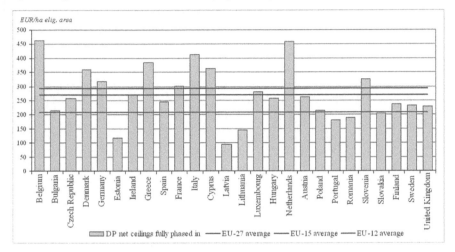

Figure 2.8   Direct payments per ha eligible area

Source: European Commission – DG Agriculture and Rural Development

To achieve this, we need to address many difficult questions - including the timing and transition towards convergence; whether or not to consider income and/or natural handicap criteria in the process; the extent and manner by which the gap between the status quo and a flat rate would be bridged.

Yet whatever forms these adjustments take, in our view they should still aim to address two fundamental needs - basic income support and support for the provision of basic public goods.

Finally, Rural Development, enhanced in terms of meeting climate change priorities, should continue to play a crucial role in complementing the process of CAP reform and the adjustment of EU agriculture by allowing MS to choose, from among a complete tool kit, the most appropriate combination of measures that enhance their competitiveness, improve their environment and sustain their rural communities.

To summarise, the broad range of complementary CAP instruments, including direct payments, will continue to be necessary and appropriate in the future to ensure the joint delivery of private and public goods through agriculture in order to:

- Firstly, meet food security concerns by retaining the capacity of EU agriculture to produce without influencing the product or production choice of farmers.
- Secondly, enhance all three aspects of a sustainable EU agriculture - economic, environmental and social.

Chapter 3

Reforming the CAP: an Agenda for Regional Growth?

Roberto Esposti

CAP, Lisbon and cohesion: a non-sectoral perspective

The main purpose of this chapter is to critically analyse the 2003 CAP reform from the perspective of the strategic objectives of the EU as defined by the Lisbon (and Gothenburg) Agenda. The attention is on the capacity of the CAP to generate economy-wide structural effects and stimulate regional growth.

The 2003-2005 reform (that is, the 2003 Fischler Reform of Pillar I and the 2005 Reform of Pillar II, the Rural Development Policy, or RDP) (Reg. 1782/2003 and Reg. 1698/2005, respectively), has been discussed, designed and approved in the context of a substantially redesigned long-term EU strategy, as defined at the Lisbon European Council in 2000 (the Lisbon Strategy or Agenda) and its successive reformulations (European Commission 2005a, p. 1, 2005b, 2007a, p. xiv). On request of the EU Commission itself, in 2002, a critical review of EU policies and instruments with respect to the Lisbon objectives (and, in particular, a faster knowledge-based growth) was carried out by a group of independent experts led by André Sapir. The consequent report (Sapir et al. 2003) inspired many other analyses on the appropriateness of the current major EU policies and on the need of substantial reforms.

Actually, the Sapir Report only marginally mentions the CAP and those critical issues that were under discussion in that period for the upcoming Fischler Reform. In one of the last pages, however, in making recommendations on the revision of the budget and funding mechanisms according to the alleged EU strategic objectives, the report concludes as follows: 'the CAP does not seem consistent with the Lisbon goals, in the sense that its value-for-money contribution to EU growth and convergence is lower than what is targeted for most other policies. Continuing to fund the CAP at present levels would amount to discounting its reduced contribution to the Lisbon goals compared with potentially much greater contributions from the other growth-enhancing policies' (Sapir et al. 2003, p. 166).

The EU Commission itself, in its cohesion reports (European Commission 2001a, 2004, 2007a), wonders whether the CAP has given any sort of positive contribution to improve cohesion across EU territories. In particular, the latter report (European Commission 2007a), with the reformed CAP (both pillars) already entered in force, still states that 'the negotiation on the budget of the Union for the period 2007-2013 has demonstrated the need for reinforced coherence and complementarity between the different elements of the Union intervention' (European Commission 2007a, p. 157).

The alleged small contribution of the CAP to regional growth and cohesion, therefore, remains a substantial argument against the CAP even after its 2003 reform. This may also explain why, when in 2005 the reform of Pillar II was approved and, almost contemporaneously, the debate on the EU budget for 2007-2013 programming period became particularly hot, DG Agriculture firmly emphasized that the CAP have been redesigned to make it work for Lisbon (European Commission 2005a, section 3). This ambition of the new CAP would not be only limited to the contribution to environmental sustainability according to the extension of the Lisbon Strategy made at 2001 EU Council in Gothenburg (the Lisbon-Gothenburg Strategy). Besides this key aspect, in fact, the alleged contribution of the new CAP would concern regional growth and cohesion itself: 'in the agricultural sector, and in rural areas, the EU is pursuing balanced economic growth and technological improvement and the creation of new jobs' (European Commission 2005a, p. 4). In Article 5 of Reg. 1698/2005, it is clearly stated the RDP has to be coherent with the objectives of other funds (thus, competitiveness and cohesion).

This attempt to give the reformed CAP a 'Lisbon justification', however, has not been fully successful, as evident in several documents of the EU Commission relaunching the Lisbon Strategy, that substantially neglect the role of the CAP. It is the case of the final Report of the Kok Group (European Communities 2004)[1] and the consequent communication from President Barroso for a new start for the Lisbon Strategy (European Commission 2005b).[2] In these documents, the main concern to reinforce the Lisbon justification of EU policies is on cohesion policy and the need to make it work in the same direction of competitiveness policy rather than interpreting it as of a purely distributional intervention (Sapir et al. 2003, pp. 146-147, European Commission 2007a, pp. 172-175). This latter vision is contrasted by those convinced that 'promoting cohesion is not only ethically correct, but economically efficient. Cohesion and competitiveness are mutually reinforcing goals' (Jouen and Rubio 2007, p. 12).

Such argument supporting the complementarity between competitiveness and cohesion policies, however, is never mentioned in favour of the CAP, even after the recent reform. This is also true for Pillar II for which, at least in principle, this leverage argument could also be valid and perhaps supported by major evidence (Mantino 2006). After all, this scepticism about the real integration of the CAP with

---

1   'The promotion of growth and employment in Europe is the next great European project' (p. 44). In this Report the word agriculture appears only once, while 'Common Agricultural Policy' never.

2   In this communication, neither 'agriculture' nor 'Common Agricultural Policy' are ever mentioned.

EU objectives and its (structural) policies finds a further sound confirmation in the debate on the HC (Health Check). The document released by the EC to present it says almost nothing on the contribution to cohesion and overall growth and competiveness; the Lisbon Strategy is never mentioned (European Commission 2007b).

As emphasized by Kuokkanen and Vihinen (2006, p. 11), it should be acknowledged that, besides official discourses and the large amount of financial resources, 'the CAP has been only a secondary element in the discussion about the Lisbon Strategy'. At the same time, at least in the perception of most scholars (Kuokkanen and Vihinen 2006, p. 16), also the Lisbon Strategy has also been a secondary element during the last CAP reform and remains such in the discussion about its future. In fact, we can conclude that if an exceptionalism holds for EU agriculture for its 'special role in relation to the state and the market when compared to other economic sectors' (Skogstad 1998), such exceptionalism is also valid for the CAP itself (Kuokkanen and Vihinen 2006),[3] a policy unrelated, if not incompatable, with other EU policies, that is not expected to contribute to sustainable growth across EU territories, where sustainability also implies, among other aspects, cohesion (growth convergence).

Nonetheless, the aim of the present chapter is to assess which kind of knowledge and empirical evidence we actually have on the real contribution of the CAP to EU growth and cohesion (that is, to what extent the CAP can be considered a Lisbon-related policy). A sort of exceptionalism also concerns research on the CAP in this respect: despite the political debate only a limited number of studies have been produced (Kuokkanen and Vihinen 2006, p. 11 e 16) on whether, and how, the CAP really contributes to regional growth and to achieving higher cohesion across EU territories and whether, in particular, the 2003-2005 CAP reform really represents a breakthrough in this respect.

---

3   This exceptionalism eventually acknowledges the historical bias of the CAP. After all, even the more radical reforms proposed over time with respect to the original characters of the CAP, as the Siena Memorandum for instance (Barbero et al. 1984), never mentioned a major contribution expected from the CAP in terms of overall EU growth potential and cohesion across regions and countries. Economic and social cohesion was not an explicit objective of the original CAP (although the fair standard of living for the agricultural population was mentioned in the Treaty of Rome) (Tarditi and Zanias 2001, Kuokkanen and Vihinen 2006, p. 7). Therefore, in CAP history the cohesion issues about the CAP have been mostly related to its redistributional effects between urban and rural areas and across social categories (farmers and non-farmers; large and small farms). At the same time, it should be also remembered that among the three general objectives of the CAP agreed at the 1958 Stresa Conference we also find '...to contribute to overall growth by allowing specialisation within the community and eliminating market distortions'.

Main conceptual issues

The lack of a well-established research tradition on the CAP contribution to regional growth and cohesion can not only be attributed to the above mentioned exceptionalism. Another major reason is that, as a sectoral policy, the assessment of CAP impacts outside the sectoral boundaries is conceptually and methodologically complex. In particular, the implementation is fully informative on who receives the support and how uses it, is often false. Analyses and empirical works mostly based on funds' allocation, and its alleged redistributional consequence, almost entirely miss the point. Funds allocation is just the first stage of the problem and these studies implicitly assume that allocation of funds across countries, regions, subjects and sectors, also informs on how this support is then redistributed across these units. This is unfortunately not true. Conceptually and methodologically, analysing the contribution of the CAP to growth and cohesion requires models with two basic characters. Firstly, they should admit different possible uses of the same support: a direct decoupled payment can be used by a farmer either to sustain household consumption (or saving) or to fund investments in his own agricultural production. Secondly, as a consequence, models have to be multisectoral and multiregional (i.e., open) to admit cross-sector and cross-region transfers.[4]

Once funds allocation across regions is appropriately computed, in order to analyse the impact of the CAP and its reform it is necessary to allocate funds within the regions, that is, among the different possible uses (Saktiṇa 2007). Unfortunately, we never know how moneies are really used once they are delivered to the first-level recipient, thus we need to assume some bi-univocal relation between the source of funds and their uses. This is, in fact, the very initial issue to be tackled when analysing the impact of CAP at the economy-wide level (Bonfiglio et al. 2006). Then, once allocation across uses is established, funds are transferred over the whole regional economy and to other regions according to its economic structure (presence and relevance of the different sectors and integration among them), the performance (factor productivity) of each sector and the integration with external economies. Eventually, the economy-wide effect (outcome)[5] of the policy under study depends on: the (observable) amount of funds allocated to the

---

4   'Their [of conventional policy evaluation methodologies] major limitations are in the assessment of only direct effects on agriculture, excluding indirect and induced effects that, via the circular flow of the regional economy, the supported program could induce. To assess these higher order effects a regional multi-sector model is needed' (Felici et al. 2008, p. 3).

5   Among possible outcomes targeted by agricultural and rural policies, we may include also environmental and consumer concerns, or market stabilisation. These are evidently aspects of major relevance in evaluating the CAP and its reforms. However, they are not considered here as the attention is specifically put on regional growth and economy-wide implications.

region,[6] the (unobservable) allocation of funds within the region across different uses and the (modelled) transfer to other sectors and regions. Dynamically and mainly through the tax system, the outcome itself (growth and cohesion, in the present case) can, in turn, affect funds allocation across and within regions. Such evaluation framework can be thus summarized as a sequence of analytical phases, that is, Source, Use, Transfers and Outcome (Figure 3.1).

Figure 3.1   Sequence of analytical phases in evaluating the economy-wide effects of the CAP

Though in a stylized way, we may firstly try to represent the connection between policies, according to their sources (funds and measures), and uses. Figure 3.2 provides this representation for three major EU policies, that is, structural policy (Structural Funds and Cohesion Fund), Pillar I (EAGF) and Pillar II (EAFRD) of the CAP. We can identify seven different uses of funds, each with a different implication in terms of transmission over the economy. Both pillars may be intended as contributions either to income of agriculture households or to agricultural investments. Which of them actually prevails depends on how the policy is designed and delivered. In addition, we may distinguish between conventional investments (physical capital) and investments in human capital, knowledge or R&D, whose aim is to induce technological progress and, therefore, increase factors' productivity. For simplicity, and according to discussion above, we call these latter Lisbon-related investments.

CAP payments, however, can be delivered to non-agricultural uses, i.e. households non involved in agricultural production as well as investments in sectors other than agriculture. This is evidently possible for Pillar II funds, as some measures are explicitly dedicated to non-agricultural sectors within the rural economy (for instance, Axes 3 measures of current RDP). Though marginally, payments directly flowing to non-agricultural subjects can also be possible for Pillar I, either as administrative/bureaucratic costs or as coupled support in favour of agents operating in the downstream sectors of the supply chains (for instance, aids to product transformation).

---

6   This very first allocation across regions is the real focus of large part of the literature and often encounters serious problems in terms of policy data availability at the regional level; this will be discussed in section 4.1.

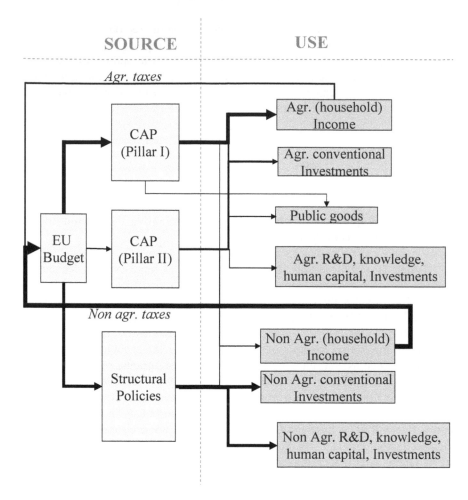

Figure 3.2   Flow of funds from policies (source) to uses (and vice versa through taxes); price support is excluded

A final use of CAP funds is the payment of public goods. These measures cover costs, usually beard by farmers, aimed at improving or reconstructing some public good (mainly, but not exclusively, environmental goods). Therefore, they can be considered neither as additional income (as they cover additional costs) nor as investments, since they do not necessarily generate a demand of capital goods and their eventual positive impact on factors' productivity, if any, is not necessarily limited to the single farm or the agricultural sector.

Similarly, we can allocate structural funds to different possible uses.[7] They are mainly intended as investments in non-agricultural sectors (Esposti and Bussoletti 2008) and can be distinguished, as well, in conventional and Lisbon-related investments. However, a smaller part of these funds may directly sustain household income (Sapir et al. 2003). These uses may evidently interest agriculture and agricultural households, though in a small amount, and probably less than proportionally, due to the at least partial substitutability between CAP and structural policies (for instance, investments in human capital for farmers are already included in Pillar II of the CAP, thus partially rule out analogous investments founded by the ESF).

Figure 3.3 exemplifies the reallocation of CAP funds according to most relevant changes occurred in CAP design since 1992. From the introduction of direct payments in 1992, all CAP reforms brought a relevant reallocation across the different uses of Figure 3.3 (European Commission 2001b, ch. 6). This is evidently the case of the 2003 (or Fischler) Reform: decoupling entails a substantial (albeit to a variable and unknown extent) reallocation form agricultural investments to agricultural household income; mandatory cross-compliance, as additional costs, a transfer from agricultural income and investments to public goods and, possibly, to non-agricultural income or investments in form of bureaucratic and administrative costs; modulation an explicit transfer from Pillar I to Pillar II which implicitly means a transfer from agricultural income to either agricultural or non-agricultural investments and to public goods.[8]

The 2005 reform of Pillar II, with the consequent attribution of funds in the context of the 2007-2013 financial perspectives, should have resulted in a more strategic approach, namely a more specific attention to Lisbon-related investments. At the same time, for the period 2007-2013, a reallocation of funds (at least in relative terms) from the CAP to structural policies should redirect money towards non-agricultural uses (European Commission 2007a). Finally, within the HC measures, beside further reallocation due to full decoupling and larger modulation, the gradual extension of the regional payments (regionalization) to all countries might imply a reallocation of funds across regions and, consequently, a reallocation across uses given the different structures of regions.

Providing a quantification of such funds reallocation across uses as effect of CAP reforms is, however, particularly difficult. An attempt can be made by comparing the last year before the implementation of the 2003 Reform (therefore 2004) with the first year of full application of both reformed Pillar I and II (2007). Such comparison is tentatively displayed in Figure 3.4. Allocation is made by assuming that decoupled payments support agricultural income while coupled

---

7 The same can be done for Cohesion Fund, in principle, though this is not a regional policy strictu sensu.

8 In Pouliakas et al. (2007), for instance, modulation is modelled as a transfer of support from agricultural household income mainly to investments in the construction sector (but also education, business services and public administration).

direct payments support conventional agricultural investments. However, we do not actually know how much of these direct coupled payments, are reallocated to income or to Lisbon-related investments (for instance, education).

For Pillar II funds, allocation among uses is achieved by re-classifying the respective measures (for 2004 and 2007) according to their content and objective as detailed in Table 3.1.[9] This reclassification has been then applied to regional RD Programmes to obtain the allocation in terms of real funds. This cannot be done at the EU scale as detailed enough information for many regions or countries is lacking. Values are thus obtained by applying to the whole Pillar II budget the distribution across measures observed in the Italian case. Table 2.2 reports such distribution over this reclassifying scheme also for two very different Italian regions (the richest and the poorest, one in the Competitiveness Objective, the other in the Convergence Objective) and for Latvia, who is expected to represent a very different situation in terms of policy implementation not only for its geographical position but also as a NMS. As may be appreciated, the differences across these different cases are not too large; therefore, the use of the Italian averages instead of the real EU proportions should not bias too much the attribution of Pillar II funds across uses.

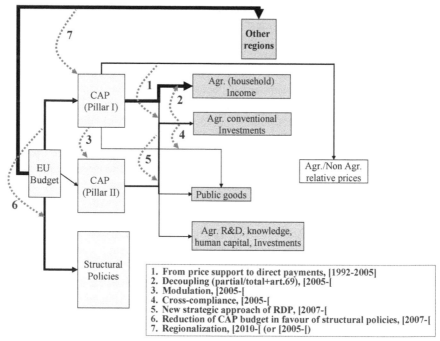

Figure 3.3   Major changes in CAP budget allocation over the period 1992–2010

---

9   An analogous interpretation of 2007-2013 RDP measures according to the Lisbon Strategy's aims can be found in EC (2005a, p. 3).

Though several effects are still to be observed after 2007 (for instance, modulation from Pillar I to II), we may appreciate that over these three years the main reallocation concerns a reduction of agricultural investments towards agricultural household income, though we are not really in the condition to observe to what extent this occurs. In fact, the effect of decoupling is, at least partially, compensated by the change in the distribution of Pillar II funds across measures and, thus, uses. A reallocation towards public goods can be observed as well (though the contribution of cross-compliance in this respect can be hardly measured). A further relevant effect, induced by larger funding and different design of Pillar II, is the reallocation towards non-agricultural uses, prevalently investments. Among these latter (both in agriculture and in other sectors) a larger share of Lisbon-related investments can be appreciated (about EUR1 billion more in 2007 with respect to 2004), though in absolute this remains a minor use of total CAP funds.

Under the extreme assumption that all decoupled payments go directly to agricultural income and all coupled payments to agricultural investments (here including the purchase of all agricultural inputs),[10] the reforms would have implied a net decrease of these latter for more than 20 billions euros. Evidently, this change in the use of funds may have caused major impact within the regional economy, in particular on agricultural production itself (in terms of employment, production, productivity, etc.) and on those sectors that are more vertically integrated (either upstream or downstream) with agriculture. Furthermore, the real impact in terms of investments can only be detected once all the transmission effects generated by this initial change in funds allocation are expressed.

Once funds are allocated across units and uses, the hardest task in evaluating the economy-wide effects of policies is the analysis of their transmission within and between regional economies. Figure 3.5, though in a very stylized form, represents the set of linkages transmitting funds from the initial allocation to other sectors and regions, eventually producing the outcome in terms of sectoral and regional growth.[11] This transmission occurs through savings and demand induced by either consumption or investments. Demand also depends on relative prices which, in turn, may be affected by Pillar I measures (price intervention) and, together with investments and technical progress (i.e., Lisbon-related investments), may condition factors' productivity in each sector. Differentials in factor productivity

---

10   For a more detailed motivation of this assumption see Bonfiglio et al. (2006, pp. 126-127). Balamou et al (2008, pp. 6-7) present a similar assumption in this respect.

11   For simplicity in Figure 7.2, and in relative discussion, we do not consider the possible transmission of support to lower prices of production inputs. This aspect may actually be relevant in analysing the effect of decoupling in particular on land market and price also because it may, in turn, affect agricultural structure (for instance, farm size). For more details on these aspects see also Courleux et al. (2008). Some CGE models, however, do take into account the effects on the land market (Roberts 2008, Finizia et al. 2005).

across sectors and regions then provoke reallocation of factors and, finally, differentials in sectoral and regional output (and income) growth.

Reconstructing this whole set of relations in a coherent methodological framework also allowing for empirical policy analysis is challenging. Nonetheless in the last few years several fruitful efforts and steps forward in this direction has been made and now this kind of evaluation, at least partially, can be practically achieved. Within regional economics literature there is, in fact, a long tradition of models aimed at analysing regional policies but not specifically designed and applied to the CAP. Recently, such models have also been designed to study the impact of EU structural policies within regional economies and on regional growth. As an example, we can mention the HERMIN model (Bradley et al. 2003).[12] These approaches have been progressively improved to include both General Equilibrium (GE) and New Economic Geography (NEG) features (van Bork and Treyz 2005). As models of multisector and open regional economies, even in their current state their extension to agricultural and rural policy evaluation could be of interest,[13] though it is relatively recent and should still be improved. Some research projects have recently tried to put together the best practitioners in the field trying to apply these kinds of approaches to CAP issues.[14] Among others, we can mention the TERA and the ADVANCED-EVAL research projects, both funded under the 6th EU Framework Programme. Several proposals, solutions and approaches emerged in these projects may deserve further attention and improvements in the future.[15]

---

12   It is worth noticing that the HERMIN model (Bradley et al. 2003) is based on endogenous growth literature to capture the long-run supply-side (that is, growth) impact of policies (EU structural funds, in particular). Therefore, it can be considered as a relevant step in the direction of growth models suitable for empirical policy analysis. The extension of this model to incorporate the CAP and RDP, however, has not been developed so far.

13   See Finizia et al (2005), Roberts (2007) and Balamou et al. (2007) for more details and in depth review of this CGE literature. Other approaches also use CGE, or SAM, models to analyse the distributional effects of the CAP across different types of farms or on the productive performance of agriculture itself after the 2003 reform (Rocchi et al. 2005).

14   For bi-regional (rural-urban) SAM models see Roberts (1998) and Psaltopoulos et al. (2006). For attempts to include the typical issues of the NEG literature (agglomeration economices, imperfect competition, etc.) see Felici et al. (2008), van Bork and Treyz (2005) and Thissen (2005) as examples.

15   More details can be found in the respective research sites: www.tera.it and www. adavanced-eval.eu. Not discussed here for the lack of published empirical applications so far, but of potential interest for the future, is the development, within the ADVANCED-EVAL project, of an Agent-Based Model combined with a General Political Economy Equilibrium model (CGPE-ABM) to evaluate RD policies.

Table 3.1 **Reclassification of RDP measures according to the different uses**

	RDP 2004: measure codes	RDP 2007: measure codes
Agricultural Household Income	d, e.1, e.2, m, x.1, x.2, z, ab	1.1.3, 1.3.1, 1.3.2, 2.1.1, 2.1.2, 2.1.3, 2.2.4
Agricultural Investments (non-Lisbon)	a, j, k, u, v, ac, aa	1.2.1, 1.2.2, 1.2.3, 1.2.5, 1.2.6, 1.3.3
Agricultural Investments (Lisbon-related)	b, c, g, l, y	1.1.1, 1.1.2, 1.1.4, 1.1.5, 1.2.4, 4.1.1
Public goods	f, h&i, i.2, q, t	2.1.4, 2.1.5, 2.1.6, 2.2.1, 2.2.2, 2.2.3, 2.2.5, 2.2.6, 2.2.7, 4.1.2
Other sectors - Household Income	-	-
Other sectors - Investments (non Lisbon)	n, o, p, r, s, w	3.1.1, 3.1.2, 3.1.3, 3.2.1, 3.2.2, 3.2.3, 4.1.3, 4.2.1
Other sectors - Investments (Lisbon-related)	-	3.3.1, 3.4.1, 4.3.1

Source: Elaboration on European Commission (2007c)

Table 3.2 **Reclassification and allocation of Pillar II funds (2007) in Italian regions and Latvia (see Table 3.1)**

	Italy	Italian richest region - Lombardy (Competitiveness)	Italian poorest region - Calabria (Convergence)	Latvia
Ag. Income	8.7%	8.4%	5.2%	11.6%
Ag. Investments (non-Lisbon)	29.6%	28.3%	35.0%	38.4%
Ag. Investments (Lisbon-related)	9.6%	6.0%	7.1%	4.8%
Public goods	37.2%	45.5%	37.3%	24.9%
Other sectors - Income	0.0%	0.0%	0.0%	0.0%
Other sectors - Investments (non Lisbon)	14.5%	10.8%	14.2%	20.4%
Other sectors - Investments (Lisbon-related)	0.4%	1.0%	1.1%	0.0%
Total	100%	100%	100%	100%

Source: Elaboration on data from Saktiņa (2007), Sotte and Ripanti (2008), Sotte and Camaioni (2008)

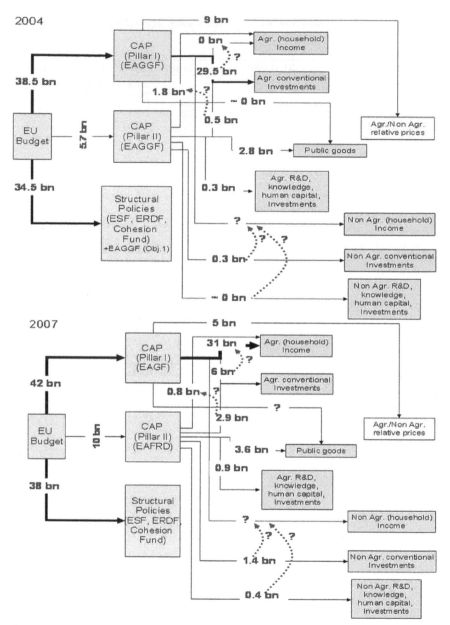

Figure 3.4   Tentative comparison of EU funds' allocation across uses in 2004 and 2007 – billions (bn) euros

Source: European Commission

Note: Data on RDP refer to 2005 and 2007, for 2005 also include the LEADER initiative; pre-accession funds and co-financing not included.

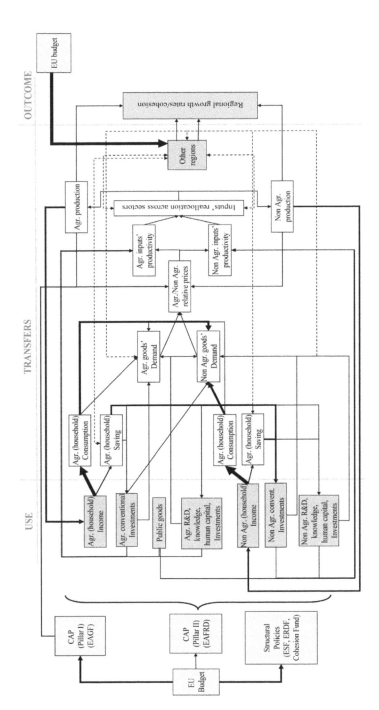

Figure 3.5   Transmission of support over the regional economy and across regions

Evidence

We may finally ask whether empirical research has been really able, so far, to answer the main questions raised in the previous section: Does an increase in CAP support (in whatever form) favour regional growth? Did the 2003-2005 CAP reform contribute in this direction?

The first question is somehow a general and primitive matter on which the other question also depends. A longer tradition of empirical studies thus exists on this issue, though it sometime presumes a positive answer. We thus start this review of the empirical evidence from two extreme perspectives on the first question and then we move towards evidence emerging from multisectoral and multiregional models.

The distributional argument

The empirical analysis on the territorial or regional impact of CAP has become a major research concern in the last 15 years (Sotte 1995, Laurent and Bowler 1997). This can be explained by the fact that the construction of disaggregated regional data on CAP support was a major problem, in particular when this support was primarily delivered in the form of market intervention. In these early studies how CAP expenditure is distributed across EU regions represents the critical question. Though this distributional concern was one of the major criticism raised about the CAP already in early eighties (see the Siena Memorandum, for instance; Barbero et al. 1984), rigorous empirical studies on this aspect only started in the mid-nineties and few pioneering works provided a major impulse in this direction.[16] In particular we can mention the studies made by Tarditi and Zanias in the late nineties (Tarditi and Zanias 2001) and by the EC in preparation of the Second Report on Socio-economic Cohesion (European Commission 2001a).

These two studies firstly made clear that the computation of CAP support distribution across regions would not, by itself, exhaust the issue. A more rigorous calculation should also take into account how regions contribute to the EU (thus, the CAP) budget. When price support represented the largest part of CAP support, analysing this side of the distributional issue was particularly difficult. Not only did regions contribute directly to the CAP budget through taxes; they also contributed indirectly through higher food prices paid by their consumers. But, at the same time, their farmers indirectly received a further support due to higher prices. Consequently, computing the net support each region received from the CAP meant disaggregating the gross contribution of the CAP region-by-region and, then, calculating how much each region paid and received directly and indirectly. This calculation was not (and it is still not) easy and, therefore, most of

---

16   See also Kuokkanen and Vihinen (2006, p. 6-10) for a detailed survey of the literature.

the research effort was just spent in such direction in order to better understand the distributional implication of the CAP across EU territories (regions).

Empirical evidence emerging from these early studies is, in fact, controversial. Tarditi and Zanias (2001) concentrate on the impact of price support. On the one hand, they detect positive effects (that is, favouring poorer areas) in terms of territorial distribution both among countries and within each member country at regional level.[17] On the other hand, despite this positive distributional outcome, price support reduces overall regional efficiency, thus hampering competitiveness and economic development. This latter conclusion anticipates the discussion in part 4.

If we limit our attention to the purely distributional issue, the EC preparatory study for the Second Cohesion Report (European Commission 2001a) provides a quite different conclusion. The difference may be explained by the fact that this latter study focuses more on direct payments (and the impact of 1992 reform) rather than price support. Nonetheless, it eventually concludes that 'if we take into account both direct payments and price support, it is evident that the distribution of the support has changed not significantly' with the 1992 reform (European Commission 2001b, chapter 6, p. 7). As a consequence, 'the regions and farms, producing more, [continue to] receive also the bulk of the price support. This implies an uneven distribution of support at territorial level and between farms' (European Commission 2001b, chapter 5, p. 5).

This latter interpretation on the distributional implication of the CAP became progressively prevalent even when support increasingly moved from price support to coupled direct payments and, finally, to decoupled payments. In 2003 the ESPON research network (founded by the EC) started a research project aimed at analysing the distribution of CAP support over the EU space (NUTS III regions) with specific emphasis on the possible effect of 2003 reform and on the differences emerging between Pillar I and Pillar II. The underlying hypothesis is that decoupling could imply a reduction in the uneven distribution of support across territories and that Pillar II may play a major role in this respect, its funding being mostly and allegedly directed towards less developed EU regions.

The conclusions of this ambitious research work were published in Shucksmith et al. (2005) with the title 'The CAP and the regions'. The message, at least with respect to Pillar I, is a confirmation of the 2001 EC study: Pillar I of the CAP works against cohesion as more money (at least if measured as support per ha) goes to richer regions; that is, it behaves as a regressive policy.[18] The 2003 reform (actually started in 2005) is not expected to significantly change this outcome

---

17   Tarschyis (2007) somehow generalizes this conclusion by stating that 'if the gross disparities before taxes and public expenditures are compared to net disparities after these public interventions [...] we will find that [...] regional disparities are much smaller than normally assumed'.

18   'Pillar I support acts in such a way that it does not contribute towards the economic and social cohesion objectives of the EU [...] Pillar I support per hectare goes

(Shucksmith et al. 2005, p.138).[19] The picture emerging for 2000-2006 Pillar II is not much different. Though large differences emerge across measures and countries (INEA 2007), the eventual outcome remains the same: no clear negative relation occurs between Pillar II support and regional income.[20] Albeit poorer regions tend to be more rural, such areas of richer regions show a better capacity to attract EU resources, and this makes Pillar II at best neutral in distributional terms (Shucksmith et al. 2005, p. 66).

Partially supported by other studies (Anders et al. 2004), the idea of the CAP working against cohesion on the basis of such purely distributional argument, soon became prevalent[21] (Roberts 2008), and it also gained space in official documents. The Sapir Report itself (Sapir et al. 2003, p. 58) states that 'adding CAP [both pillars] and the other internal spending programme funds to cohesion policy disbursement, the simple correlation between total Community fund disbursement and GDP levels per head across the 17 macro-regions drops to -0.4 (in 1991) and to -0.2 (in 1995 and in 2000)'. Quoting the above mentioned ESPON study, in its Forth Cohesion Report the EC states that 'CAP [...] market policy support tends to benefit the more developed rural areas [...] concentrated more in core regions in northern and western Europe and less in the peripheral regions in the east and south'. These official documents acknowledge the possibility that Pillar I of the CAP may work contrary to cohesion objectives across the EU, and that, though the empirical evidence is often based on the CAP before the 2003 reform (that is, on data up to 2005), this evidence also suggests that such reform is not doing very much to remove this inconsistency.

These studies on the distribution of the CAP support across EU territories, and its eventual change after the recent reform, have the major merit of focusing their attention on the practical issues underlying the correct calculation of such distribution. At the same time, however, the conclusions of this stream of literature

---

unambiguously to richer regions, support per worker is distributed more ambiguously' (Shucksmith et al. 2005, p. 58).

19   'The results suggest that the MTR CAP reform proposals would have increased CAP direct payments more in [...] prosperous areas' thus 'suggesting that the overall impact of the MTR proposals on farm incomes would be territorially neutral' (Shucksmith et al. 2005, p.138). The same conclusion, as mentioned before, was also reached by the EC in its Second Cohesion Report with respect to the impact of 1992 reform (European Commission 2001b, chapter 6, p. 7).

20   '...Contrary to expectations, Pillar II support is inconsistent with cohesion objectives, favouring the more economically viable and growing areas of the EU' (Shucksmith et al. 2005, p. 66).

21   '...Despite the rural development policies, it is questionable if the CAP as a whole promotes cohesion. The contradictory effect of the CAP on cohesion is mentioned in some textbooks on the EU' (Kuokkanen and Vihinen 2006, p. 7). '...The CAP has also been associated with negative [...] effects. The distribution of income effects has been found to be contrary to the principles of cohesion with the richest areas and farmers benefiting most' (Balamou et al. 2008).

may be questionable for two main reasons. Firstly, the alleged inconsistency of CAP with respect to cohesion objectives largely depends on how CAP contribution is computed.[22] It can not be considered in absolute values (due to the large heterogeneity in regional size) and, while it may be true that support per ha is larger for richer region, this is not necessarily the case in terms of support per AWU (Agricultural Working Unit) or of 'CAP intensity', that is per unit of regional GDP or per head. The second and more important problem of this distributional argument is that it implicitly assumes than one additional euro spent in the i-th region compared to the j-th region gives the former an higher growth impulse. As a consequence, if richer regions receive a larger support the growth gap with poorer regions is expected to increase. In general terms, this 'distributional' literature over-emphasizes the political issue of funds' allocation across territories, while underestimates the economic relevance of how these funds are spent (Pillar I or II, which kind of measures, etc.) within the regional economies, thus activating a complex transmission across subjects, sectors and regions.

The counter-treatment hypothesis

Tarditi and Zanias (2001) and, more generally, Hall et al. (2001), already realized that the allocation of CAP funds across regions does not exhaust the issue about the contribution of CAP to growth and cohesion. Even if the net contribution of CAP to EU's poorer regions is positive, this does not necessarily imply that these regions grow faster. On the contrary, CAP support may prevent regional economies from achieving a more productive sectoral structure and more efficient production processes, i.e., from being more competitive.

In Tarditi and Zanias (2001), this negative effect of CAP on regional competitiveness may eventually offset the positive net distribution towards poorer regions they observe. In their analysis, CAP's negative feedback mostly operates through price support that makes agricultural production, agricultural mix and allocation of productive resources across sectors inefficient.[23] It would follow that, as market intervention and price support was progressively replaced by

---

22   It may be noticed that, differently from Tarditi and Zanias (2001) and EC (2001), the analysis carried out by the ESPON study does not really compute net transfers but only considers gross transfers. This may in fact generate some misleading evidence. Moreover, it does not consider the whole amount of EU funds received by any region, in particular the lagging ones, and thus may cause a further misspecification problem, as will be underlined in the next section.

23   In this regard, Kuokkanen and Vihinen (2006, pp. 7 and 11) comment on Tarditi and Zanias's work as follows: 'According to the authors, the impact of the agricultural price policy of the EU is a result of different and contrasting effects both in terms of equity (income distribution) and efficiency (competitiveness and economic development) [...]. Farm price support generates large distortions in the domestic market at inter- and intra-sectoral level, reducing EU competitiveness [...]. In the long term, price support hinders structural adjustment in rural areas'.

direct support in 1992 and 2000 reforms, this negative effect on regional growth is expected to vanish. This should become even more evident after the introduction, in 2003, of decoupled payments whose declared objective, among others, was to favour re-orientation of regional agricultures to market according to their specific specializations and comparative advantages.

In terms of inter-sectoral factor allocation, however, direct payments, both coupled and decoupled, may maintain their distorting effects. As far as they remain somehow linked to agriculture, their major effect is to increase remuneration of agricultural labour, capital and land which will eventually reduce the productivity gap with other sectors and, therefore, the progressive loss of resources from agriculture to other uses. If this may definitely occur under coupled payments, it is still not clear as to what extent it remains valid under decoupled support. This depends on how this support is really independent on maintenance of production factors within agriculture.

Beside the results of Tarditi and Zanias (2001), whose applicability to the current CAP is quite limited, as mentioned, we may find only few studies that have tried to empirically assess this negative effect of CAP on regional growth and, eventually, on growth cohesion. Here, we may mention two recent works by Bivand and Brunstad (2006) and Esposti (2007) who have analysed regional convergence across EU and the role of CAP in this respect within a conventional neoclassical aggregate growth model.

In Bivand and Brunstad (2006) we find both the basic argument and the empirical evidence in favour of this interpretation. In their results, 'regions with lower net transfer to agriculture experience slightly faster growth than regions with larger net transfers' (Bivand and Brunstad 2006, p. 288). This evidence is fully consistent with the authors' expectations: 'we expect the level of agricultural policy support to be negatively related to regional growth, because higher levels of support are likely to slow the reallocation of labour and capital to non-agricultural sectors' (Bivand and Brunstad 2006, p. 287).[24]

Starting from the same intuition of Bivand and Brunstad (2006), Esposti (2007) develops a model where CAP support received by each region is additional to other EU funds. In particular, as the attention is on the impact of CAP on convergence, CAP support is investigated together with Objective 1 funds to detect whether the effects of these funds are reciprocally reinforced or offset. Within an aggregate growth model, CAP affects regional growth by compensating the lower labour

---

24   This concept of CAP as a policy that mainly operates within regional economy by artificially maintaining high employment levels in agriculture is also shared by Urwin (1991) who states that the CAP has been since the beginning of the eighties a social rather than economic policy, that aims at maintaining the agricultural employment, as the industrial and service sectors could not absorb the surplus labour supply (Kuokkanen and Vihinen 2006, p. 12).

productivity of agriculture.[25] As in Tarditi and Zanias (2001) and in Bivand and Brunstad (2006), this could slow down regional growth and thus, if structural funds do operate in favour of convergence, CAP eventually generates a counter-treatment effect. However, estimates obtained do not support this hypothesis of counter-treatment. The conclusion turns out to be that CAP has a substantially neutral effect in terms of regional growth and cohesion.

It has to be noticed that both quoted works actually refer to years before the 2003 (2005) reform (1996 in Bivand and Brunstad; 1999-2000 in Esposti). Therefore, their results could be hardly extended to the new CAP. Nonetheless, while Bivand and Brunstad, as Tarditi and Zanias, concentrate on CAP support deriving from market intervention, which makes the extension of results to the current CAP much less reliable, Esposti only considers regional support from direct payments.[26] As CAP 'artificially' improves agricultural labour productivity thus retaining labour in the primary sector, the results observed in Esposti could also be valid for decoupled payments and, as such, for the CAP reformed in 2003.

Despite their specific results and shortcomings, these studies are of major interest here as they may eventually share the same conclusion with the literature of the purely distributional argument: CAP works against cohesion. This conclusion, however, is reached from opposite directions. According to the distributional argument, the CAP acts against cohesion because poorer regions receive less support, the assumption being that the CAP does favour regional growth. Therefore, the major problem becomes how to allocate it across EU territories. According to the counter-treatment hypothesis, on the contrary, the CAP may act against cohesion just because it reduces the regional growth performance or, at least, reduces the growth enhancing effect of other EU funds. Consequently, redistributing the support in favour of poorer regions would not solve the problem. From the point of view of the mere distribution of funds across regions, the 1992, 2000 and 2003 reforms had a minor impact (European Commission 2001b, Shucksmith et al. 2005). Nonetheless, they did have, as discussed extensively, major implications on how these funds are delivered within the regional economies, who and how uses them and, eventually, how they affect regional development processes.

Some evidence from multisectoral and multiregional models: Does the CAP reform eventually work for regional growth?

These contrasting approaches on the impact of CAP and its recent reforms on regional growth performances and disparities provide interesting but incomplete evidence on the basic questions raised earlier. Though still in its initial stages, the stream of research analysing such impact within multisectoral and multiregional

---

25   The Second Cohesion Report itself (European Commission 2001a) agrees on the fact that, due to lower productivity in agriculture, the relative poorer performance of some regions is often linked to the higher degree of employment in agriculture.

26   See Esposti (2007, p. 124) for a more detailed explanation.

models is able to offer a more complete picture (Pouliakas et al. 2007, Esposti 2009). First of all, it confirms that CAP may be not neutral in terms of regional growth and, above all, its impact is not necessarily univocal. An increase in overall support may even reduce regional growth implying that allocation of funds across regions is not by itself informative on the consequence in terms of cohesion (or convergence). More than allocation across regions, what seems critical in terms of growth implications is the allocation of assigned funds within the region. In particular, the change in Pillar I support from price support to coupled direct payments and, finally, to decoupled payments, implies an increasingly negative impact on agricultural production and employment that, if not compensated otherwise, may transmit to vertically integrated (either upstream or downstream) sectors. At the same time, due to factor reallocation outside agriculture, it may favour growth of other sectors which would eventually compensate the first negative impact.

These results are of major interest for the discussion on the effect of 2003-2005 reform and, above all, on the direction to be taken with the HC and post-2013 CAP reform. At the same time, they should be commented carefully as they evidently do not fully consider all the possible implications of the reform itself. In particular, the adopted methodologies can usually not adequately represent the impact that some measures (of both pillars) may have on farm efficiency and productivity. Decoupling itself, as well as several Lisbon-related RDP measures, may improve these performances. The same argument also holds with respect to public (mostly environmental) goods. Even considering the contribution of the CAP only from a strictly growth perspective, it should be acknowledged that supporting the creation of public goods within agricultural production may have a positive growth impact both on the primary sector and on other economic activities.

Research in this field should aim at improving the currently adopted methodologies in these directions. At the same time, efforts should continue on more practical aspects. The systematic collection of data and the construction of a complete and consistent regional dataset reporting the allocation of CAP gross and net support is an unfinished work, and it is expected to improve (European Commission 2001b). It must be acknowledged, in fact, that the different results reported in the literature may still depend on the different computation of regional CAP support. More and better statistical information (obtainable, for instance, by improving the set of information collected through the FADN) is also needed on how CAP support is used by first-level recipients (mostly farmers) among different alternatives (saving, consumption, different kinds of investment in agricultural activity) depending on how this support is delivered (through prices, coupled or decoupled direct payments, Pillar II measures, etc.).

Though having in mind the need to improve the adopted research tools and the partial and incomplete evidence they provide, we may still try to draw at least two policy implications from these results. The first one is that the enlarged EU is extremely heterogeneous across its regions and this makes the answer to questions above strongly region-specific. Empirical results show that this answer may even be opposite moving from one region to another due to their large intrinsic and structural differences. The second policy implication has mostly to do with the inter-regional and urban-rural distribution of effects. Evidence shows that, besides the initial allocation of funds, what really matters in assessing the impact of CAP across territories is how funds are then transmitted over space. This spatial redistribution inevitably affects the enduring debate on the regional allocation of CAP funds and its coherence with cohesion objective (the 'distributional argument'). This effect is strongly region-specific also, and it depends on the degree of openness of the regional economy under study and on the integration with the bordering regions. Therefore, a viable solution in this respect can hardly be generalized. But one possible policy development could be to allow strongly integrated regions (especially within urban-rural or core-periphery patterns) to also have an integrated management of CAP resources, at least of those admitting regional programming.

References

Anders, S., Harsche, J., Herrmann, R. and Salhofer, K.. 2004. Regional Income Effects of Producer Support under the CAP. *Cahiers d'Economie et Sociologie Rurales*, 73, 103-121.

Balamou, E., Pouliakas, K., Roberts, D. and Psaltopoulos, D. 2008. *Modeling the rural-urban effects of changes in agricultural policies: a bi-regional CGE analysis of two case study region*s, Paper prepared for presentation at the 107th EAAE Seminar: Modelling of Agricultural and Rural Development Policies, Sevilla, Spain, 29 January-1 February.

Balamou, E., Psaltopoulos, D. and Thomson, K.J. 2007. *Modelling agriculture, tourism and policy in a CGE environment*, Deliverable n. 4/a, TERA Specific Targeted Research Project, FP6-SSP-2005-006469 (http://ww2.dse.unibo.it/tera).

Barbero, G., Bergmann, D., Bublot, G., Koester, U., Larsen, A., Mahé, L-P, Marsh, J., Ritson C., Sarris, A., Tangermann, S. and Tarditi, S. 1984. *The Siena Memorandum on the reform of the Common Agricultural Policy*, Conclusions of the Workshop on 'The reform of Common Agricultural Policy', 17-18 Freburary, Certosa di Pontignano (Italy).

Bivand, R. and Brunstad, R. J. 2006. Regional growth in Western Europe: detecting spatial misspecification using the R environment. *Papers in Regional Science*, 85, 277-297.

Bonfiglio, A., Esposti, R. and Sotte, F. (eds.) 2006. *Rural Balkans and EU integration. An I-O Approach*. Milano: Franco Angeli.

Bradley, J., Morgenroth, E. and Untiedt, G. 2003. *Macro-regional Evaluation of the Structural Funds Using the HERMIN Modelling Framework*, Economic and Social Research Institute (ESRI), Working Paper 152, Dublin.

Esposti, R. 2007. Regional Growth and Policies in the European Union: Does the Common Agricultural Policy Have a Countertreatment Effect? *American Journal of Agricultural Economics*, 89, 116-134.

Esposti, R. 2009. *Reforming the CAP: Something to do with Territorial Cohesion?* Presentation made for the Invited Seminar at the European Commission, DG AGRI, 30th September, Brussels.

Esposti, R. and Bussoletti, S. 2008. The Impact of Objective 1 Funds on Regional Growth Convergence in the EU. A Panel-data Approach. *Regional Studies*, 42, 159-173.

European Commission. 2001a. *Unity, Solidarity, Diversity for Europe, Its People and Its Territory*, Second Report on Economic and Social Cohesion, Brussels: European Commission.

European Commission. 2001b. *Study on the impact of community agricultural policies on economic and social cohesion*, Directorate-General for Regional Policy, Brussels.

European Commission. 2004. *A New Partnership for Cohesion. Convergence, Competitiveness, Cooperation*, Third Report on Economic and Social Cohesion, Brussels: European Commission.

European Commission. 2005a. *Putting rural development to work for jobs and growth*, Special Edition Newsletter, Directorate-General for Agriculture and Rural Development, Brussels.

European Commission. 2005b. *Working together for growth and jobs. A new start for the Lisbon Strategy*, Communication from President Barroso in agreement with Vice-President Verheugen to the Spring European Council, Brussels.

European Commission. 2007a. *Growing Regions, Growing Europe*, Fourth Report on Economic and Social Cohesion, Brussels: European Commission.

European Commission. 2007b. *Preparing for the 'Health Check' of the CAP reform*, Communication from the Commission to the Council and the European Parliament, COM(2007) 722, Brussels.

European Commission. 2007c. *Rural development in the EU, Statistical and Economic Information*, Report 2007, Directorate-General for Agriculture and Rural Development, Brussels.

European Communities. 2004. *Facing the challenge. The Lisbon strategy for growth and employment*, Report from the High Level Group chaired by Wim Kok, Luxembourg.

Felici, F., Paniccià, R. and Rocchi, B. 2008. *Economic Impact of Rural Development Plan 2007-2013 in Tuscany*, Paper presented at the XII EAAE Congress - Ghent: 'People, Food and Environments: Global Trends and European Strategies', 26-29 August.

Finizia, A., Magnani, R., Perali, F., Polinori, P. and Salvioni, C. 2005, *Construction and simulation of the General Economic Equilibrium Model Meg-Ismea for the Italian Economy*, Quaderni del Dipartimento n. 17, Dipartimento di Economia, Finanza e Statistica, Università di Perugia

Hall, R., Smith, A. and Tsoukalis, L. (eds.) 2001. *Competitiveness and Cohesion in EU Policies*. Oxford: Oxford University Press.

INEA. 2007. *Territorio e imprese nell'articolazione del sostegno comunitario all'agricoltura italiana*, Osservatorio sulle Politiche Agricole della UE - INEA, Roma.

Jouen, M. and Rubio, E. 2007. *The EU budget: What for?*, Synthesis Paper of the Seminar held on 19th April 2007 in Brussels, Paris: Notre Europe (htpp://www.notre-europe.eu).

Kuokkanen, K. and Vihinen, H. 2006. *Contribution of the CAP to the general objectives of the EU*, Background Note 4 of the Specific Support Action (6th FP), SASSPO-SSP4-022698.

Laurent, C. and Bowler, I. 1997. *CAP and the Regions: Building a Multidisciplinary Framework for the Analysis of EU Agricultural Space*, INRA Editions. Paris.

Mantino, F. 2006. Evaluating Structural, Territorial and Institutional Impacts of Rural Development Policies in Italy: Some Lessons for the Future Programming Period (2007-2013), in *Coherence of Agricultural and Rural Development Policies*, edited by D. Diakosavvas. Paris. OECD, 281-290.

Pouliakas, K., Phimister, E., Roberts, D., Thomson, K.J., Balamou, E., Psaltopoulos, D., Hyytiä, N., Kola, J. and Galassi, F. 2007. *Application and results of individual CGE analysis*, Deliverable n. 8, TERA Specific Targeted Research Project, FP6-SSP-2005-006469 (http://ww2.dse.unibo.it/tera).

Psaltopoulos, D., Balamou, E. and Thompson, K.J. 2006, Rural/Urban Impacts of CAP Measures in Greece: An Inter-regional Social Accounting Matrix (SAM) Approach, in *Coherence of Agricultural and Rural Development Policies*, edited by D. Diakosavvas. Paris: OECD, 153-169.

Roberts, D. 1998. Rural-Urban Interdependencies: Analysis Using an Interregional SAM Model. *European Review of Agricultural Economics*, 25, 506-527.

Roberts, D. 2007. *Design of CGE model*, Deliverable n. 4, TERA Specific Targeted Research Project, FP6-SSP-2005-006469 (http://ww2.dse.unibo.it/tera).

Roberts, D. 2008. *The territorial impacts of rural and regional policies: what do we know?* Presentation at the XII EAAE Congress - Ghent: 'People, Food and Environments: Global Trends and European Strategies', 26-29 August.

Rocchi, B., Stefani, G. and Romano, D. 2005. Distributive impacts of alternative agricultural policies: a SAM-based analysis for Italy. *Cahiers d'Economie et Sociologie Rurales*, 77, 85-112.

Saktiņa, D. 2007. *Relevance of structural policies and territorial factors (comparative analysis)*, Deliverable n. 13, TERA Specific Targeted Research Project, FP6-SSP-2005-006469 (http://ww2.dse.unibo.it/tera).

Sapir, A., Aghion, P., Bertola, G., Hellwig, M., Pisani-Ferry, J., Rosati, D., Viñals, J. and Wallace, H. 2003. *An Agenda for a Growing Europe. Making the EU Economic System Deliver*, Report of an Independent High-Level Study Group established on the initiative of the President of the European Commission, Brussels.

Shucksmith, M., Thomson, K.J., and Roberts D. (eds.) 2005. *CAP and the Regions: Territorial Impact of Common Agricultural Policy*. Wallingford: CAB International.

Skogstad, G. 1998. Ideas, Paradigms and Institutions: Agricultural Exceptionalism in the European Union and the United States. *Governance: An International Journal of Policy and Administration*, 11, 463-490.

Sotte, F. and Camaioni, B. 2008. *The rural development programs 2007-13 of the Italian regions. A quali-quantitative analysis*, Paper prepared for the 109th EAAE Seminar '*The CAP after the Fischler Reform: national implementations, impact assessment and the agenda for future reforms*', Viterbo (Italy) 20-21 November 20-21st.

Sotte, F. and Ripanti, R. 2008. *I Psr 2007-2013 delle regioni italiane. Una lettura qualiquantitativa*, Working paper n.6, Gruppo 2013 - Forum Internazionale dell'Agricoltura e dell'Alimentazione, Roma.

Sotte, F. 1995. *The Regional Dimension in Agricultural Economics and Policies: Proceedings of the 40th EAAE Seminar*. Ancona, CNR-RAISA.

Tarditi, S. and Zanias, G. 2001. Common Agricultural Policy, in *Competitiveness and Cohesion in EU Policies*, edited by R. Hall, A. Smith and L. Tsoukalis. Oxford. Oxford University Press, 179-216.

Tarschyis, D. 2007. *What should be the redistributive function of the EU budget?* Paper presented at the Seminar 'The EU budget: What for?' held on 19th April 2007 in Brussels, Paris: Notre Europe (htpp://www.notre-europe.eu).

Thissen, M. 2005. RAEM: Regional Applied general Equilibrium Model for the Netherlands, in *A survey of spatial economic planning models in the Netherlands. Theory, application and evaluation*, edited by F. van Oort, M. Thissen and L. van Wissen. Rotterdam: NAi Publishers, 63-84.

Urwin, D. 1991. *The Community of Europe: A History of European Integration since 1945*. London and New York: Longman.

van Bork, G. and Treyz, F. 2005. *The REMI model for the Netherlands*, in *A survey of spatial economic planning models in the Netherlands. Theory, application and evaluation*, edited by F. van Oort, M. Thissen and L. van Wissen. Rotterdam: NAi Publishers, 25-44.

PART 2
Political Economy of CAP Reform

Chapter 4

A Political Economy Perspective on the Fischler Reform of the CAP[1]

Johan F.M Swinnen

In 1995, Franz Fischler, a then largely unknown Austrian politician, became EU Commissioner in charge of the CAP. There was surprise that a new member state had been given the powerful Agricultural Commission chair, but no major expectations surrounded his arrival in Brussels. However, a decade and two tenures later, Fischler was recognised by friend and foe alike as the architect of the most radical reforms of the CAP.

There were several reforms implemented over the two terms of Commissioner Franz Fischler (1996–2004). Some reforms, such as Agenda 2000, were very important, but his name is most closely associated with the reform of 2003. At the time the 2003 reform was generally referred to as the 'mid-term review' (MTR), a term that in hindsight does not do justice to the extent and substance of the reform package. The key innovation of the 2003 reforms[2] was the introduction of the single farm payment (SFP), decoupling a large share of CAP support from production. These reforms are assessed by many experts as the most radical reforms of the CAP since its creation.

One of the most radical features of the reforms was the timing and the audacity of the European Commission in proposing them.[3] After the decisions on the Agenda 2000 reforms commentators suggested that the reforms were insufficient (Brenton and Núñez Ferrer 2000). Yet, for many involved, the MTR was considered exactly what the term said: a review to check halfway through the Agenda 2000 implementation period (2000–06) whether any (minor) adjustments were needed.

---

1   This paper draws some elements on the main conclusions of my book "The Perfect Storm". I refer to this book (Swinnen, 2008) for many details and more extensive analyses.

2   Two new instruments, "cross-compliance" and "modulation", were also introduced. The reforms also changed market organisations in dairy and rice. In addition, there was a shift from quantity and the public regulation of markets and prices to a policy focused on quality, marked-based initiatives and rural development.

3   Other elements of the 2003 reforms were less radical. Total farm support remains essentially the same as before and there is little change in the distribution of CAP benefits across countries and farms. The reforms had no effect on EU border protection, except for the rice sector (Olper 2008): despite all of Fischler's emphasis on rural development, the budget for rural development was lower at the end of his tenure than it was before the 2003 MTR. There was also no ceiling on payments to farms.

When Fischler announced his plans in the summer of 2002, most opponents and member states expressed shock and dismay – Pirzio-Biroli (2008) refers to farmers' organisations considering the proposal a "sort of Molotov cocktail". The reform proposals went much beyond what they considered a 'review' or 'minor adjustments'. The Commission was accused of going beyond its mandate.

Conceptual framework

What made these radical reforms possible? To explain this, I make use of a conceptual framework drawing on the theoretical research on EU decision-making (e.g. Crombez (1996), Steunenberg (1994)) which has been applied to CAP decision-making by Pokrivcak, Crombez and Swinnen (2006). They develop a theory of CAP reform and identify the conditions which create what they refer to as "the optimal reform context". According to this theory, CAP reform is more likely when (a) an external change moves policy preferences in a pro-reform direction, (b) this external change is large, (c) the policy preferences of the EU Commission are pro-reform, and (d) the EU voting rules require a lower majority. External changes alter the political preferences of member states but effective policy adjustments will only occur if the change is large enough, because of the decision-making procedures in the EU that induce a 'status quo bias'. The final outcome will also depend on the preferences of the Commission – which sets the agenda. The Commission can make a proposal within a certain policy range. Hence, if the Commission has pro-reform preferences, it can pick the strongest reform option that is possible within the policy range that can be approved.

The external change that is required for internal change, and thus the likelihood of reform, is directly related to the voting rules. Under the unanimity rule, those most opposed to reform hold an as-if veto over the reform decision. Under simple majority rule, any change that affects the preference of the median (country) voter will lead to a reform. Under a qualified majority rule, an external change needs to be sufficiently large for a minimum coalition of countries to be better off with a policy change than with current policies.

Differences in quality of the political agents, for example due to experience or to political capital of various agents, may obviously also play a role. In addition, consecutive enlargements of the EU have affected the number of member states and with it the heterogeneity of preferences of member state governments (Henning and Latacz-Lohmann 2004).

In the rest of the chapter I use this framework to argue that radical CAP reform was possible in 2003 because several factors contributed to "an optimal reform context".

External changes and pressures for CAP reform

Several studies show how changes in external factors, such as world market prices and exchange rates have earlier induced CAP reforms (e.g. Olper 1998, 2008). Josling (2008) argues that in the early years of the CAP the main external pressures on the CAP were macroeconomic factors and that budgetary and trade relations became important later.

The Agenda 2000 reforms were intended to address the CAP constraints related to the enlargement process, but the reforms were diluted in the political negotiations. However by 2001, even with the diluted Agenda 2000 decisions, enlargement was less likely to create a conflict with WTO commitments than predicted five years earlier (Swinnen 2002). The likelihood of a WTO conflict depended more on the outcome of the new negotiations. If, as a result of a new WTO agreement, the EU needed to significantly reduce export subsidies or change the implementation of the direct payments, the CAP would have to be reformed, irrespective of enlargement.

The WTO was not the only trade arrangement affecting the CAP. Opposition to several multilateral and regional trade initiatives, such as the Everything but Arms (EBA) initiative further raised awareness of the international effects of the CAP and contributed to the sense of a need to reform the CAP.

Increased consumer demands for food quality and safety were reinforced by several food safety crises in the years before the MTR. The food crises had considerable political impacts. In Germany, Belgium and the U.K, ministers resigned or governments lost power in the wake of the scandals. Extensive media coverage provoked strong reactions of consumers and the general public, which contrasted sharply with their rather passive attitude vis-à-vis traditional agricultural policy issues. These crises put food safety, animal welfare and environmental concerns on top of the agricultural policy agenda. There were calls for the overhaul of the CAP: government subsidies for farming practices that did not adhere to appropriate food safety, environmental and animal welfare standards were unacceptable.

These various elements, some traditional and others new, combined to increase pressure for CAP reform. Franz Fischler himself summarised these developments as "the CAP had lost its legitimacy among the EU public". The CAP was seen as hurting EU trade interests, having negative effects on the environment and unable to address food safety concerns of EU consumers. These elements compounded to a call for radical changes in the CAP at a time when ministers of finance and other members of the European Commission were demanding significant CAP budget cuts. Hence, at the start of the Prodi Commission there was a view among many that the CAP budget should be drastically reduced.

Institutional reforms, changes in voting rules and in the political actors

The voting rules had been altered by several institutional reforms of the EU prior to 2003, including the Single European Act (SEA) and the Treaties of Maastricht and Nice. Most relevant for our analysis are the changes in voting rules initiated by the SEA. In the 1990s, qualified majority voting was increasingly used for minor CAP decisions but major CAP decisions were often still decided by unanimity. In this respect, the 1999 CAP reforms (Agenda 2000) were a watershed: for the first time a major country (France) was outvoted in relation to a major CAP reform. For the MTR decision-making, this change in EU decision-making rule was very important: Fischler and his team spent a lot of effort trying to put together a winning coalition and breaking a blocking minority coalition.

There were also important changes in who was involved in the decision-making and in the reform preparations. First, the enlargement of the EU in the previous decades had affected preferences and the distribution of votes in the EU. The accession of Sweden, Finland and Austria to the EU in 1995 reduced the share of votes of the established players, such as France and Germany. In addition, none of these three countries were 'natural allies' of France, in opposing CAP reforms. The pro-reform camp was reinforced with Sweden, which had gone through a process of radical liberalisation of its agricultural policy in the early 1990s. Finland and Austria, however, supported farm subsidies as their small farms in disadvantaged areas depend on subsidies. However, as high-income countries with small-scale farmers, many based in mountainous or Arctic regions, Finland and Austria were more sympathetic to supporting rural development and agri-environmental policies than subsidies for quantity – which mainly benefited larger producers.

Second, the anticipation of eastern enlargement played a role as well. The Commission realised that reform would not become easier after enlargement.[4] This gave a sense of urgency to the reforms.

Third, a large share of the EU Commissioners wanted CAP reform. President Romano Prodi and several Commissioners wanted the share of the CAP in the EU budget to be substantially reduced. Commissioner for Trade Pascal Lamy wanted the CAP to be reformed in order to allow the EU to take the initiative in the Doha round. In addition, the Commissioners were reflecting consumer and environmental concerns.

Fourth, Commissioner Fischler himself was in his second term and had gained experience. The Agenda 2000 negotiations had also made him better prepared for new reforms. He was keen to leave a legacy of having put the CAP on a course that he considered sustainable and consistent with his view of European agriculture. A CAP that is more in line with rural development, the environment, the production of quality and safe food, different from the "old CAP", which was focused on quantity, output and prices. In a rather unexpected way, the food safety and environmental crises of 1999–2001 had reinforced this agenda.

---

4   See Henning and Latacz-Lohmann (2004) for a formal analysis of this issue.

Fifth, the 2003 MTR discussions also transformed the politics-as-usual in the CAP. Traditionally, farm unions put pressure on their agriculture ministers and on the Commission and tried to obtain as much as the other ministers would allow. But the MTR brought consumer groups and environmental groups to the political negotiations more forcefully than before. In fact, Fischler reached out to these groups to establish public support for a reformed CAP. He deliberately designed a media strategy and a series of presentations to win their support. His view was that even in countries that were not in favour of the CAP per se the public was still very much in favour of policies that improved the rural environment, enhanced animal welfare, ensured food safety and food security, etc.

Finally, an important potential source of opposition against the reforms was within the Commission itself: the traditional thinking of the DG AGRI administrators. The preparation of the reforms was kept within a tight circle of six top officials. In addition, a small group of policy analysts within DG AGRI were asked to assess the effects of some of the proposals, but without being fully informed.

Earlier institutional changes played a role. The administrative reforms had removed the hold of France on the top job in DG AGRI and many of the old-style DG AGRI officials formed in the early years of the CAP, had left and younger persons had joined in the past decade. Thinking within the DG was much more open to, for example, environmental and economic arguments (Moehler 2008).

A complex reform puzzle

Although there were strong pressures to reform and institutional changes had enhanced the opportunity for reform, the success of the reforms was far from certain at the outset. Most thought that the chances were slim *ex ante*.

The timing was complex.[5] The interaction of elections in the Member States, eastern enlargement, WTO negotiations, the EU budget negotiations and CAP reform was complex. But because the second term of Commissioner Fischler ran out in January 2005, 2002–2003 must have looked like the best – or the only possible – timing for CAP reform.

The reform process was complex also because of the interaction, and sometimes contradiction, of the various elements. The various demands for reform in some sense appeared to weaken rather than reinforce one another. For example, the food safety crisis significantly contributed to the demand for reform of the CAP, and probably more than anything else put CAP reform on the political agenda and raised public awareness about agricultural policy. Yet, the reform ideas coming out of this agenda tended to go in the direction of more regulation rather than less, and of more subsidies (albeit redirected) rather than less. For example, few of the environmental groups who pressured for a radical rethinking of agricultural

---

5   For a detailed review, see Swinnen (2001).

policy in the EU considered trade liberalisation and WTO negotiations a positive development. Their stance contrasted with the more traditional pressure for CAP reform, mostly from economists and some politicians, arguing for less regulation and lower subsidies, and favouring more liberalised trade and markets.

There was reinforcement in the sense that '*something* needs to be done', but there was much less common ground on *what*. Nonetheless, in Fischler's mind the various pressures were crystallizing into a consistent reform strategy.

Franz Fischler: strategist and tactician

Fischler not only had a clear strategic view on where he wanted the reforms to go, he also used masterful political tactics in getting there. Experience mattered as well. Experience with the Agenda 2000 reforms made Fischler better prepared for the reform battle. France had opposed the main reform proposals and voted against them in 1999. Yet, there was a qualified majority of votes in favour of the Commission's reform package. To the surprise of many, Chirac managed not only to bring up the issue during a meeting of the European Council in Berlin in March 1999, but also to re-open the compromise decision. In the final negotiations, and despite Fischler's opposition, he succeeded in convincing the other heads of state to approve a weakened version of the reforms.[6] Fischler had learned from this that he had to anticipate any potential political obstacle and strategy in order to avoid the fate of the Agenda 2000 reforms.

Preparation of the reforms

Fischler had also learned that if specific proposals come out early, vested interest groups will mobilise quickly and reform efforts may be undermined. Therefore, Fischler put together a small inner circle of six officials to prepare the entire reform package. A small group of senior Commission officials, drawing on analyses by experts within the Commission administration, prepared the details of the proposals. Everybody else was kept in the dark or on a need-to-know basis. The extensive in-house discussions and analyses made DG AGRI well prepared when the discussions came out into the open. Since nobody outside a small Commission circle had expected Fischler to propose full decoupling, the opponents had little preparation and little analysis. In contrast, the DG AGRI team was ready to address any critique and comment on their proposals with careful analysis and counterarguments, all of which had been prepared in the previous years and months.

While hints about the reforms were given to the public in 2001, there was a complete communication stop in the spring of 2002, during the period leading up to French elections, which took place in May and June of 2002. Commission officials were forbidden to speak in public on CAP reforms. Fischler announced his plans in the summer of 2002. The opponents of the reforms publicly expressed shock and dismay. Opponents had underestimated Fishler's determination.

A key element of Fischler's strategy was to build support for the reforms from non-traditional political coalitions. Instead of focusing on the farm unions, Fischler gave presentations, interviews to the media and participated in conferences to secure support from environmental organisations and consumer groups. When the reform proposals were announced, the traditional negative reactions emerged from farm unions. In the past they had dominated the political discussions on the CAP. Now other organisations joined the debate and presented a different view. As a result, the discussions were more balanced than in previous reform efforts.

Counting the votes: the role of the Iraq war

Qualified majority voting had become the rule for CAP reform decisions. After some time three groups emerged: the 'pro-reform group', which included the UK, Sweden, the Netherlands and Denmark; a 'middle group' that included Greece, Belgium, Luxembourg, Finland, Austria and Italy; and the 'anti-reform group' with France, Spain, Germany, Portugal and Ireland. The middle group was not considered a major problem by The Commission because it included Greece, which held the presidency and wanted a successful summit, Belgium, which did not have a unified voice with three ministers of agriculture, and others which either had few votes or who were not completely opposed but had specific problems with the reforms.

The anti-reform group was strong with three large countries (France, Spain and Germany) and it easily controlled a blocking minority. The anti-reform group was not a natural coalition, however. Renate Künast, the politician from the Green Party who was German minister responsible for the CAP, had been a vocal advocate of a more environmentally friendly CAP – and should have been a natural ally of Fischler. This was indeed the case initially, but a Franco-German political alliance between Gerhard Schröder and Jacques Chirac Prevented this.

During the Iraq war the governments of Spain under Prime Minister Aznar, Italy under Prime Minister Berlusconi and the UK under Prime Minister Blair joined US President Bush in "the coalition of the willing". France and Germany strongly opposed the Iraq war.

Furthermore, on the eastern enlargement of the EU, France and Germany had opposing views but managed to come to an agreement. Germany was a strong proponent of eastern enlargement while France was not enthusiastic. A deal was made. As one Commission official summarised it, "the French agreed with the enlargement if the Germans agreed to pay the bill" – which included future CAP financing for the EU-15. This was cemented in the 2002 Brussels Council meeting where, among other things, a decision was made with far-reaching implications for the CAP: CAP Pillar I payments were fixed until 2013 to ensure French farmers and politicians that CAP benefits would continue to come long after enlargement.

All this mattered for the CAP reform proposals of Fischler. France was opposed. For Schröder, who was in some international isolation with this strong anti-Iraq war stance, maintaining the general Franco–German international political coalition because of the Iraq war and enlargement was more important than the preferences of his Green Party coalition partner on CAP reforms.

Facing this strong anti-CAP reform coalition, Fischler then decided to use the Iraq war for his own purposes. As Chirac was using the war coalition to keep Schröder on his side in the CAP debate, Fischler went out to seek support for his reforms from the opposite side, more specifically from the Blair–Aznar camp. Fischler went to Blair to ask him to approach Aznar and convince him to switch sides and support the CAP reforms. Blair, who had supported the CAP reforms all along, agreed on one condition: the Commission had to drop the capping of support to large farms – which would hurt the large UK farms and landowners– from the package of reform proposals. Fischler agreed and Blair convinced Aznar to switch sides in the CAP reform debate. This left the opposing coalition severely weakened.

After this, Künast managed to re-take the initiative in Germany on CAP reforms and to change the German stance. But she asked for adjustments of the reform proposals in exchange for her support. Künast liked the idea of cross-compliance, but wanted to implement the decoupled payments not based on what farms had received in the past (which became known as the 'historical model') but instead pay farms in the same region the same payment, independent of what they had received in the past (later termed the 'regional model').

Fischler opposed this idea, not so much for the economic effects, but rather for the political ones. He feared that the redistribution of subsidies among farms, which was implicit in the regional model, could increase the opposition to the CAP reforms on the part of farm interests. Olper (2008) emphasizes the lack of redistribution of support by the Fischler reforms as a key element of the reform proposals. By limiting the redistribution and emphasising the more efficient instrument of decoupling for enhancing farm incomes, Fischler had avoided a lot of potential opposition.

Nonetheless, Künast was adamant on this issue because she wanted to use the new subsidy system to support more extensive and organic farming systems, which traditionally had not been receiving as much support as the intensive, conventional production systems.[6] Fischler ultimately gave in to Künast's demands in order to obtain the German votes. In the final proposal, Germany was allowed to introduce the regional model.

In the end, France found itself isolated in its opposition. And even within France, the opposition was no longer unanimous. Facing a loss in qualified majority

---

6   Instead of 30 per cent as proposed by the Commision, the Council decided to cut the beef support price only by 20 per cent; instead of a proposed cut of 20 per cent for cereal support prices, it was decided at 15 per cent; and finally, the dairy reforms were postponed to 2005, instead of starting in 2000 as proposed by The Commission.

voting, France then tried to join the winning camp and to extract compensations and adjustments from the Commission reform proposals. But they were in a weak negotiating position and finally ended up with little.

Franz Fischler: Killer or saviour of the CAP?

The previous discussion depicts the Fischler team at the Commission as one to introduce major changes to the CAP. The acrimonious reactions following the launch of the proposals involved accusations of the Commission siding with opponents who wanted to 'scrap the CAP'. Yet, Fischler and his team had a very different view. They saw their proposals not as an instrument to kill the EU's agricultural policy, but instead as a way of saving it. Pirzio-Biroli (2008, p. 124) puts it thus:

> "Scrapping the CAP [was] not an option. ...The Fischler reform was aimed at helping the CAP and its farmers reconcile the needs of modernisation and restructuring with the acknowledgement of their community function, and the recognition of the positive externalities generated by agriculture, and rural activities and spaces....Fischler acted in the conviction that the EU needed to keep a strong agricultural policy, but periodically update it in order to adapt it to new realities."

From this viewpoint, it is interesting to note that according to Fischler himself, the concept of decoupling was not chosen for the reasons most often mentioned by economists, i.e. to reduce distortions, but because it was the best thing to do to save the CAP.

Economists, based on their focus on efficiency and reducing distortions, had long preferred non-distortionary (lump sum) transfers, of which the type of decoupled payments that Fischler proposed were a welcome improvement. These economic arguments were never convincing for Fischler. He came to conclusions in favour of decoupling from a more political reasoning. Fischler looked at the reforms from the standpoint of how he could save the support for European agriculture in the twenty-first century with new demands and new constraints being imposed on European farmers and with new opportunities. Decoupling in this way was an attractive choice for several reasons. It was an improvement in terms of the efficient use of EU funds, given taxpayer pressures on the budget. Decoupled support improved net income gains for farmers compared to payments per hectare or per animals – when much of the support is dissipated through an induced increase in the prices of land and other inputs. Fischler saw the salience of the argument not from an economic perspective, but from a political one. If the CAP was using half the EU budget, he considered it essential to convince the EU taxpayer that this money was well spent and was effectively used. Decoupling also reduced trade distortions and improved the environment as it reduced incentives to use land intensively and the introduction of cross-compliance further enhanced the environmental benefits. Fischler and his team emphasised in interviews that while the WTO negotiations were not an initial motive for decoupling, obviously, once they started thinking about the option, they saw it could be very useful for the Doha negotiations as well.

Finally, while discussions on the importance of the reforms focus mostly on the 2003 MTR (and to some extent on the Agenda 2000 reforms) Fischler himself sees the achievements of the 2003 reforms much more in tandem with the 2002 budgetary agreement. In his view, the 2003 reform (proposals) allowed him to convince those most opposed to the CAP within the European Commission to agree to a much smaller budget cut than they had asked for. By proposing a series of bold reforms that reduced the negative effects of the CAP on the environment, on market distortions and on the WTO negotiations, and that enabled the CAP to fit within a concept of sustainable rural development, Fischler and his team had reduced the ammunition of those demanding large budget cuts and created a new support base for the CAP. In this way he was able to convince the Commission to limit cuts for the financial period to 2013.

The European Council summits in 2002 cemented Fischler's achievements. It was decided that the CAP would continue to receive generous funding from the EU budget. For 2007–13 the total budget for market interventions and direct payments was fixed at the 2006 level in real terms. Fischler and his colleagues saw this as a major achievement: the CAP budget was saved up to 2013 – and the cuts were much less than demanded at the start of the Prodi cabinet in 2000. From this perspective his reforms had 'saved the CAP' instead of scrapping it.

Concluding comments

The 2003 MTR under Commissioner Franz Fischler was the most radical reform in the history of the CAP, albeit that not all aspects were substantive reforms. The reform puzzle was complex, as was the timing.

Three (sets of) factors came together in the period around 2002. They created "the perfect storm" for radical CAP reforms. The three factors were (1) the effect of institutional reforms; (2) changes in the number and quality of the political actors, and, (3) strong pressure to reform from external factors. The main pressures came from the WTO and other trade negotiations, the budget, food safety and environmental concerns and to a lesser extent enlargement. These elements, some traditional and others new, combined to increase pressure for change. Institutional change (the Single European Act) introduced qualified majority voting for CAP decision-making. The enlargement of 1995 reduced the share of the votes of the established players in the EU and entered Sweden as a strong voice for CAP reforms. The politics-as-usual of the CAP was transformed as consumer groups and environmental groups played a more prominent role in the debate. Finally, many of the old-style DG AGRI officials had left and younger persons had joined in the past decade, such that thinking within the DG AGRI was much more open to environmental and economic arguments.

The combination of Fischler's experience, his strategic vision, his political tactics and the Commission officials' effort and preparation played a vital part. The reforms were prepared in relative secrecy by a small inner circle of officials while experts within the Commission administration were calculating the effects of the reforms.

The proposals initially faced a strong anti-reform group, with three large countries (France, Spain and Germany) controlling a blocking minority. But Fischler used shrewd political tactics to manoeuvre Spain out of the anti-reform group which later also brought other countries to support the reforms.

Farm unions were taken by surprise by the Fischler proposals and were unprepared. They also faced a new political environment in which environmental and consumer groups were taken seriously by political leaders. However, decoupled payments are more effective in transferring income to farmers, and farmers may ultimately have realised this, which could contribute to explain their limited opposition.

Finally, Fischler and his team saw their reforms not as an instrument to reduce the importance of the CAP, but as a way of saving it. Bold reforms to reduce its negative effects on the environment, on market distortions and on the WTO negotiations reduced the pressure for large budget cuts and created a new support base for the CAP. In this way major budget cuts for the next financial period were avoided. From this perspective, the Fischler reforms contributed to the survival of the CAP, rather than to its demise.

References

Brenton, P. and Núñez Ferrer, J. 2000. EU Agriculture, the WTO and Enlargement, in *Negotiating the Future of Agricultural Policies: Agricultural Trade and the Millennium* edited by S. Bilal and P. Pezaros. The Hague: Kluwer Law International, 113–31.

Crombez, C. 1996 Legislative Procedures in the European Community. *British Journal of Political Science*, 26, 199-228.

Henning, C.H.C.A. and Latacz-Lohmann, U. 2004. Will enlargement gridlock CAP reforms? A political economy perspective. *EuroChoices*, 3 (1), 38–43.

Josling, T. 2008. External Influences on CAP Reforms: An Historical Perspective, in *The Perfect Storm: The Political Economy of the Fischler Reforms of the Common Agricultural Policy* edited by J. Swinnen. Centre for European Policy Studies Publications, Brussels.

Möhler, J. A. 2008. The Internal and External Forces Driving CAP Reforms, in *The Perfect Storm: The Political Economy of the Fischler Reforms of the Common Agricultural Policy* edited by J. Swinnen. Centre for European Policy Studies Publications, Brussels.

Moyer, H.W. and Josling, T. 2002. *Agricultural Policy Reform—Politics and Process in the EU and US in the 1990.* London, Ashgate Publications.

Olper, A. 1998. Political Economy Determinants of Agricultural Protection Levels in EU Member States: An Empirical Investigation. *European Review of Agricultural Economics*, 25(4), 463–87.

Olper, A. 2008. Constraints and Causes of the 2003 CAP Reform, in *The Perfect Storm: The Political Economy of the Fischler Reforms of the Common Agricultural Policy* edited by J. Swinnen. Centre for European Policy Studies Publications, Brussels.

Pirzio-Biroli, C. 2008. An Inside Perspective on the Fischler Reforms, in *The Perfect Storm: The Political Economy of the Fischler Reforms of the Common Agricultural Policy* edited by J. Swinnen. Centre for European Policy Studies Publications, Brussels.

Pokrivcak, J., Crombez C. and Swinnen J. 2006. The Status Quo Bias and Reform of the Common Agricultural Policy: Impact of Voting Rules, the European Commission, and External Change. *European Review of Agricultural Economics*, 33(4), 562-90.

Steunenberg, B. 1994. Decision Making Under Different Institutional Arrangements: Legislation by the European Community. *Journal of Institutional and Theoretical Economics,* 150, 642-69.

Swinnen, J. 2001. Will Enlargement Cause a Flood of Eastern Food Imports, Bankrupt the EU Budget, and Create WTO Conflicts?. *EuroChoices*, Spring.

Swinnen, J. 2002. Transition and Integration in Europe: Implications for agricultural and food markets, policy and trade agreements. *The World Economy*, 25(4), 481-501.

Swinnen, J. 2008. *The Perfect Storm: The Political Economy of the Fischler Reforms of the Common Agricultural Policy.* Centre for European Policy Studies Publications, Brussels.

Syrrakos, D. 2008. An Uncommon Policy: Theoretical and Empirical Notes on Elite Decision-Making During the 2003 CAP Reforms, in *The Perfect Storm: The Political Economy of the Fischler Reforms of the Common Agricultural Policy* edited by J. Swinnen. Centre for European Policy Studies Publications, Brussels.

Tangermann, S. 2001. Agricultural Policy Making in Crisis Times: BSE and the Consequences in Germany. *EuroChoices*, Spring.

Chapter 5

The Budget and the WTO: Driving Forces behind the 'Health Check'?[1]

Carsten Daugbjerg and Alan Swinbank

Introduction

Until the late 1980s there was little doubt among CAP analysts that budgetary concern was the major driving force capable of generating CAP reform. The MacSharry reform of 1992 triggered a debate on the driving forces of CAP reform in a new era in which farm trade had become more fully integrated in the WTO trade regime with its distinct *Agreement on Agriculture*. During the debate two camps crystallised – one emphasising the WTO, and the other the budget, as the root cause of the MacSharry reform. The debate re-emerged after the Agenda 2000 CAP reform but with less strength, presumably because Agenda 2000, in general, is seen as a limited reform, or perhaps rather as a simple policy adjustment. The 2003 Fischler reform raised the same questions once again (see Swinbank and Daugbjerg 2006 for an overview of these debates).

In 2007 the EU launched a new, mini, reform of the CAP, which it called the 'Health Check', and debate on the driving forces behind CAP reform re-emerged once again. In this chapter we revisit 'the usual suspects': the budget and the WTO. The Chapter proceeds as follows. First we introduce these potential drivers of reform. Second we set the scene by outlining policy developments from 2003 to 2008. Then we introduce, briefly, the Health Check package, and suggest that it was rather more ambitious than is usually assumed. The discussion then moves on to consider the potential drivers that shaped and conditioned the Health Check debate, before moving to a brief conclusion.

Two potential drivers of reform

Since the launch of the Uruguay Round in 1986, CAP reform and GATT/WTO developments have tended to progress in unison, although the sequencing of events has been problematic, and establishing cause and effect is difficult. Nonetheless, there is some evidence to suggest that the MacSharry, Fischler, and 2005 sugar

---

1 An abridged version of a paper presented at the EAAE Seminar, which also resulted in Daugbjerg and Swinbank (2010): a paper that also discusses paradigm change.

reforms, were prompted by the EU's perception of the need to make progress in the GATT/WTO. These CAP reforms in turn facilitated progress in GATT/WTO: the MacSharry reforms enabled the EU to accept the Uruguay Round *Agreement on Agriculture* (URAA), the Fischler reforms allowed the EU to adopt a more offensive negotiating stance in the Doha Round (on domestic support in particular), although it is still uncertain whether the Doha Round can eventually be brought to a successful conclusion, and the sugar reform enabled the EU to curb its exports of subsidised sugar and comply with a Dispute Settlement ruling (Daugbjerg and Swinbank 2009). Can similar WTO pressures be observed in the run-up to the Health Check? Since the setback at the Cancún Ministerial Conference in September 2003, the Doha Round negotiations had suffered one crisis after another and the round was in danger of *de facto* collapse after the failed, but intensive, negotiations in Geneva in 2008. However, each intensive negotiating round had moved negotiators closer to a compromise. Can the Health Check be seen as a concession of the EU designed to pave the way for an agriculture agreement?

Since world market prices remained high in 2007 and well into 2008, EU expenditures on export subsidies and intervention purchases were correspondingly low. At first sight this may rule out budgetary pressure as a likely candidate explaining the Health Check. However, with a 2009 debate on the Financial Perspective post-2013 in sight, high world market prices prompted some ministers to question whether the EU still needed to support its farmers at the present level. So, one could argue that high world market prices had opened a window of opportunity for a CAP reform that would release funding for other purposes in the EU.

The historical context

The 2003 CAP reform was agreed prior to the ill-fated WTO Ministerial meeting in Cancún that had been seen as an important milestone in wrapping-up the Doha Round by the official deadline of December 2004. In adopting the package Farm Ministers said:

> This reform is ... a message to our trading partners ... . It signifies a major departure from trade-distorting agricultural support, a progressive further reduction of export subsidies, a reasonable balance between domestic production and preferential market access, and a new balance between internal production and market opening. ... The CAP reform is Europe's important contribution to the Doha Development Agenda (DDA), and constitutes the limits for the Commission's negotiating brief in the WTO Round. Its substance and timing are aimed at avoiding that reform will be designed and imposed in Cancun and/or Geneva –which could happen if we went there empty handed (Council of the European Union 2003: 2).

In adopting the Fischler package EU Farm Ministers mandated a review of certain aspects of the reformed CAP in about 2007/08, long after the scheduled completion of the Doha Round. For example Article 64(3) of Regulation 1782/2003 provided for a review of the partial decoupling option in the Single Payment Scheme (SPS).[2] The 2003 package was quickly followed in 2004 with an extension of the SPS mechanisms to Mediterranean crops; the new Commissioner, Mariann Fischer Boel, spent much of 2005 pushing for sugar reform; and Member States were busy implementing the SPS.

But the UK was agitating for more, and attempted to trigger a new CAP reform debate in 2005 as Member States tried to decide on the size, and scope, of the EU's budget (its *Financial Perspective*) for 2007-2013. In particular, at the European Council meeting in June 2005, the British Prime Minister, Tony Blair, appeared to be saying that the UK would be willing to give up its budget rebate, first negotiated in 1984, in exchange for a sizeable cut in the CAP budget (*Agra Europe*, 24 June 2005: EP/5-EP/7). Speaking to journalists, Commission President José Manuel Barroso noted the commitment for a mid-term review of the SPS in 2008 (*ibid.*, EP/7). As the debate intensified, French President Jacques Chirac was characteristically staunch in his defence of the CAP in refusing the British 'offer' (*Agra Europe*, 15 July 2005). The need for further CAP reform was widely debated, not least in the *Financial Times*. Bertie Ahern (2005), the Irish Prime Minster, in September 2005 argued that, as well a being 'unwise and unfair to ask farmers to accept another radical reform now', tactically it would 'handicap the EU' in the forthcoming WTO negotiations in Hong Kong, and 'remove the motive for other big food producers to move towards reform of their agricultural sectors.'

The 2007-2013 *Financial Perspective* was eventually agreed at the European Council in December 2005, with: i) the British rebate more-or-less intact; ii) no CAP reform, but a much reduced budget for Pillar II expenditure; and iii) the Commission being asked 'to undertake a full, wide ranging review covering all aspects of EU spending, including the CAP, and of resources, including the UK rebate, to report in 2008/9' (Council of the European Union, 2005, paragraph 80). But just before this decision had been reached, the UK Government published a document setting out its *Vision* for the CAP (HM Treasury & Defra 2005); which prompted a caustic response from Mariann Fischer Boel, and her comment that: 'There will be full debate on the future direction of the CAP, before the end of the current budget period in 2013. And we have already programmed *a health check* for the reforms in 2008 to 2009' (as quoted in *Agra Europe*, 9 December 2005: EP/7; emphasis added). However it was not until June 2006 that *Agra Europe* began to make regular use of the term 'Health Check' (9 June 2006: EP/1). Later the Commissioner was to talk about the Commission proceeding on the basis of 'one vision, two steps'. First there would be the Health Check; and second 'a look ahead to the CAP after 2013, within a general review of the European Union budget' (Fischer Boel 2007).

---

2   Official Journal of the European Union, L270, 21 October 2003.

The health check: unambitious?

In the run-up to the November 2007 launch of the Health Check the Commission had been keen to suggest that the forthcoming review of the CAP would be technical in character, focusing on simplification. No-doubt it was keen to dispel fears in the farming community that this 'mid-term review', like its predecessor under Franz Fischler, would turn into a major CAP reform. Major changes to the policy, it was hinted, would only arise in the context of the budget debate, to be launched in 2009, on the *Financial Perspective* post 2013.

In launching the debate on the Health Check in November 2007 the Commission said:

> The 2003 Reform was the first step to make the CAP fit for the 21st century. Consensus on all the elements of the 2003 Reform could not be reached in one go. Indeed, this is why a number of review clauses were ... foreseen in the final agreement ... . These review clauses, without implying a fundamental reform of the existing policies, allow the possibility of further adjustments in line with market and other developments (Commission of the European Communities 2007: 3).

One way in which circumstances had changed since 2003 was that world market prices for oil, metals, and agricultural commodities were much higher than they had been. Export subsidies on dairy products had not been used since June 2007, and intervention stocks were depleted. Thus in its formal Health Check proposals in May 2008 the Commission concluded that 'that any remaining supply controls of the CAP (namely, dairy quotas and set-aside) should be removed' (Commission of the European Communities 2008).

Despite the Commission's desire to postpone discussion of more substantial reform until 2009, it could be argued that the Health Check proposals were rather more ambitious than might have been expected. As widely leaked beforehand, the Commission used this occasion to announce that it will not propose an extension of the milk quota regime beyond 2015, and it suggested a gradual increase in quota in the interim to prepare the sector for a 'soft landing'. It proposed a further shift of funds away from the SPS to Pillar II; but the system of 'modulation' which it favoured taxed more heavily larger farm businesses, making this a politically sensitive issue in some Member States. It wanted a simplification of the SPS, including abolition of the set-aside requirement (popular amongst farmers because of soaring world market prices, but problematic to environmentalists), a shift to fully decoupled SPS payments for some specific crop payment schemes, and a substantial reduction in the partial decoupling option, particularly for arable crops, that had been favoured by France and other southern Member States. Intervention mechanisms would also be weakened.

In promoting the package to the European Association of Agricultural Economists, the Commissioner's Deputy Head of Cabinet suggested that whilst 'better regulation and simplification' was an important political priority, so too was a move 'towards more market orientation' and competitiveness. For this a key aspect was

> decoupling of farmers' support from production decisions. Decoupling allows EU farmers to make their choices in response to market signals … . It brings EU agriculture much closer to world market [sic] without distorting them because of their Green Box compatibility … … keeping some arable payments partially coupled has brought no market benefits but has added red tape. In this sector and several others, it's time to phase in full decoupling (Borchardt 2008: 2, 4).

The package that the Agriculture Council agreed in November 2008, after the French Presidency had tabled two compromise papers, approved a number of the Commission's moderately ambitious proposals, but backtracked on others. *Agra Europe* (21 November 2008: EP/1) suggested it was a 'watered-down Health Check deal'. Partial decoupling of the SPS for the main arable crops would end in 2010, and for some specialised crops and most livestock payments by 2012. Set-aside has been abolished. Milk quota will be increased by 1 per cent per annum, with the expectation that the quota regime will end in 2015, but two further reviews are scheduled to take place before then; and, as a sweetener to Italy, Italian producers will had their 5 per cent increase immediately. The level of modulation agreed was rather less than the Commission had proposed, and it is only farm businesses in receipt of SPS payments over EUR300k per year that will be taxed more heavily. The extra funds for Pillar II (Rural Development), from 2010, generated by modulation will, in part, be used to fund four new priorities (climate change, renewable energy, water management and biodiversity), as proposed by the Commission, and one old challenge: the dairy sector which had suffered from a sharp collapse in milk product prices. Member States would be allowed to channel up to 10 per cent of SPS funds into particular commodity sectors (under the so-called Article 68 provisions), with the expectation that some Member States would use this to aid dairy farmers. Intervention arrangements were maintained more-or-less unchanged.

Budget constraints?

The CAP reform decisions of June 2003 were taken in the context of a budget framework agreed in October 2002, and lasting beyond the *Financial Perspective 2000 – 2006* through to 2013 (Swinbank and Daugbjerg 2006). Although this did not create immediate financial problems for the CAP, with increased expenditures on direct payments in the new Member States there was some concern that CAP budget constraints would tighten by the end of the decade.[3] According, Article 11 of the SPS regulation (1782/2003) included a *Financial Discipline* under which, on an annual basis from 2007, the Council was mandated to reduce the direct payments if there was likely to be any breech of the annual budget ceilings set in October 2002. Enlargement proceeded as planned, with increasing levels of expenditures on direct payments in the new Member States, and new compensation payments had been brought into the SPS, but there had been no need to invoke the Financial Discipline.

Indeed, as a result of high world market prices, expenditure on export subsidies and intervention purchases had fallen away sharply, and there had been competing claims put forward for appropriating the 'unspent' CAP money: on extra support to farmers in the new Member States for example, or for agricultural development in the Third World (*Agra Europe*, 20 June 2008: EP/2). Thus the Health Check was not discussed at a time of CAP budget crisis.[4] It would not reduce CAP expenditure; although a result of 'modulation' and the redirection of funds from Pillar I to Pillar II is to *increase* overall EU taxpayer expenditure on agricultural support because of the requirement on Member States to co-finance Pillar II.

In March 2007 the European Council had agreed that 20 per cent of EU energy supplies should come from renewable sources by 2020 (excluding nuclear power), and that a mandatory minimum 10 per cent blend of biofuels would be used by all Member States in 'transport petrol and diesel' by 2020 (Council of the European Union 2007). With a world 'food crisis' apparently looming, these and other biofuel policies worldwide were increasingly seen as problematic. Responding to these concerns in an open letter to his colleagues on the ECOFIN Council, just before the Commission tabled its Health Check proposals, Britain's Chancellor of the Exchequer proposed, *inter alia*, 'a fundamental reform of Europe's agricultural sector' including 'phasing out of all elements of the CAP that are designed to keep EU agricultural prices above world market levels' and 'an end to direct payments to EU farmers'; and 'a close examination of the direct and indirect effects of EU biofuels policy, including a full assessment of its effect on food prices …' (Darling 2008). From the other extreme the French response to high world food prices seemed to be to defend the CAP. For example the French Agriculture Minister Michel Barnier, interviewed by the *Financial Times* (28 April 2008: 1), had

---

3   Corrado Pirzio-Biroli (2008: 106), chief of staff in Fischler's cabinet, implies that the Fischler reforms were, in part, designed to outflank the Commission President's preference for a 30 per cent cut in the CAP budget in the 2007-2013 Financial Perspective.

4   With much weaker world market prices, however, the 'old' CAP support mechanisms subsequently kicked back into life. Three-hundred thousand tonnes of grain (nearly all maize) were offered to intervention between 29 December 2008 and 4 January 2009, bringing the EU close to its self-imposed ceiling on maize intervention; and export subsidies on milk products were reintroduced in January 2009 (*Agra Europe*, 9 January 2009: EP/4; and 23 January 2009: EP/2)

suggested that the CAP was a good model for others to follow: 'It is a policy that allows us to produce to feed ourselves'.

The Health Check was of course not unrelated to the budget, but it was certainly not driven by a budgetary constraint. However, the budget may have played an unexpected role in the Health Check. The 2005 debate over the 2007-2013 *Financial Perspective*, with the UK's persistent calls for CAP reform, might explain why the Commission upgraded the level of ambition in its Health Check proposals, whilst insisting that any substantive discussion on the CAP after 2013 should form part of the 2009 budget review.

WTO pressures?

Whether or not motivated by the desire to make progress in the Doha Round, the June 2003 reform package, with the associated commitment to extend its 'objectives and approach' to 'the so-called Mediterranean products, such as olive oil, tobacco or cotton' (Council of the European Union 2003: 2), was an important step in the EU's approach to the Doha Round. Most importantly the SPS (when 'fully' decoupled) shifted the bulk of blue box, and some amber box, support to the green box; and it gave the EU some scope to negotiate tariff and export subsidy reductions on dairy products. Nonetheless Swinbank and Tranter (2005) questioned whether the SPS met all the requirements of the green box.

One concern was addressed by the 2007 fruit and vegetable reform, which removed (from 2010) the planting restrictions on SPS land that might otherwise have been challenged in a manner analogous to the problems the US faced over *Upland Cotton*.[5] The fruit and vegetable reform also removed the processing aids previously paid on various processed products, channelling the budget funds into increased SPS payments for the growers concerned (European Commission 2007). Had the processing aids remained in place, the EU could have been vulnerable to a WTO challenge that they fell foul of the *Agreement on Subsidies and Countervailing Measures*, being Prohibited Subsidies 'contingent ... upon the use of domestic over imported goods'.

In 2005 the WTO's Dispute Settlement Body ruled that the EU had subsidised its export of sugar to a greater extent than was allowed, and the EU agreed to respect the ruling and bring its exports into line with its export subsidy commitment. This involved eliminating the export of so-called C sugar, and severely cutting back on the volume of sugar exported with payment of export subsidies. It did so by reducing EU production, as a result of a quota buy-out scheme that was facilitated, in part, by a phased 36 per cent reduction in the support price. Political agreement on the sugar reform was achieved in November 2005, just ahead of the WTO

---

5 However potential problems remain because the SPS is an annual payment, to farmers, who have to have land at their disposal in 'agricultural' production, and cross compliance applies.

Ministerial in Hong Kong, allowing Mariann Fischer Boel to suggest that the deal would strengthen the EU's negotiating position in Hong Kong as it 'demonstrated that the EU could tackle one of the bastions of its common agricultural policy as well as respect WTO rulings' (as quoted in the *Financial Times*, 25 November 2005: 10). Although some progress was made in Hong Kong –it was agreed, for example, that export subsidies would be eliminated by 2013 if an overall agreement could be achieved– it was not sufficient to conclude the Doha Round, and the negotiations limped on.

By the time the Doha talks entered a deep freeze in December 2008 the EU had come a long way from its position in January 2003 when it had indicated only a very limited willingness to engage in reductions in domestic support, export subsidies and import protection. As well as the elimination of export subsidies by 2013 (which the EU had first accepted in May 2004), Crawford Falconer's draft *Modalities* envisaged tariff cuts of about 70 per cent in the highest tariff band (although the detail of Special Safeguard mechanisms, sensitive product status, and safeguards on preference erosion, significantly complicate the picture). It also envisaged a 70 per cent reduction in the EU's Aggregate Measurement of Support (AMS), and individual AMS product caps; a 75 to 80 per cent reduction in a new aggregate measure of *Overall Trade-Distorting Support* (OTDS); and a capping of blue box expenditure at 2.5 per cent of the average total value of agricultural production in a 1995-2000 base period (WTO 2008).

Whether this package would have proved acceptable to its trading partners had the negotiations progressed is unclear; but what is evident is that the green box status of the SPS is crucial for the EU's acceptance of the proposed AMS, OTDS, and blue box reductions. Is the Health Check in any way associated with this; or is it unrelated to the WTO?

It must be conceded that the Commission's November 2007 communication contains neither the place-name 'Doha' nor the acronym 'WTO', although in the introduction to its legislative proposals of May 2008 the Commission does refer to the WTO twice: once explaining the 2003 reform ('The main objective was providing a direct payment system that allows farmers to be market oriented, as simple as possible from an administrative point of view and compatible with WTO'), and second the need to respect green box provisions if a relaxation of the Article 69 rules were to allow Member States more spending discretion (Commission of the European Communities 2008: 5, 6). However, the Health Check proposals, and the emerging Doha package, were both moving in the same direction.

Klaus-Dieter Borchardt (2008) drew attention to the Commission's proposal for 'further decoupling', moving from a situation in which some arable payments were 'partially coupled' by phasing in 'full decoupling'; and this would be achieved in 2010 as a result of the package agreed. It seemed fairly certain that the EU would declare the bulk of SPS expenditure in the green box; but it was less clear how partially coupled payments would be declared. Presumably the partially coupled part (for example 25 per cent of the old arable payment that had been retained in France) would remain in the blue box until fully decoupled in 2010. But can partially coupled SPS payments be split between the green and blue boxes; or did partial coupling imply that the whole of the partially coupled SPS payment should remain in the blue box (all the old arable payment in France, prior to 2010, for example)? And might concerns of this sort have prompted the Commission's quest for full decoupling in the Health Check?

Borchardt (2008: 5) also noted the ambition of the Health Check to weaken the remaining elements of market price support: 'Market orientation further needs the conversion of our traditional market instruments, like intervention, private storage, export refunds and quotas, into a genuine safety net. These instruments should not be applied any more as 'price setters' on the markets but as instruments that keep farmers in business in case of dramatic market disruptions'. At the time, the recent world market price spikes, and the expectation that prices would continue at higher levels than prevailed in the early 2000s, meant that CAP market price support was not important for most commodities. Whilst this situation prevailed, the EU could agree to the elimination of export subsidies and to substantial reductions in import protection, without serious erosion of EU market prices. But, if world market prices were to revert to previous levels, the Falconer package would limit the EU's ability to maintain its traditional levels of market price support. Thus yet again the Health Check proposals to weaken market price support mechanisms could be seen as an important element in the Commission's Doha strategy; but this was thwarted by the Council decision.

If the Doha package was to be agreed it would 'lock-in' the CAP reforms the EU has agreed to-date. If the Doha Round is not concluded the WTO system, with its existing URAA, will continue to apply, and aspects of the present CAP could be challenged. Paradoxically, the green box status of the SPS would not, in these circumstances, be particularly problematic, as the blue box and the EU's AMS allowance are sufficiently commodious. But without leverage from the WTO, it is unclear whether budget or other 'internal' concerns would be sufficiently powerful to drive further CAP reform, or even prevent backsliding towards the 'old' CAP.

Concluding comments

Although the bitter debate in 2005 over the *2006-2013 Financial Perspective* possibly influenced the Commission's approach to the policy reviews built into the 2003 Fischler reform package, the Health Check was not about the budget. The budget debate demanded by the UK had been deferred to 2009, or later; there was no immediate budget crisis; and the Health Check proposals would simply reallocate the CAP budget between Pillars I and II.

Instead, the primary goal of the Health Check was to move European agriculture onto a more competitive footing, more compatible with any likely Doha agreement; and, with buoyant world market prices, to release the productive potential of European agriculture through the abolition of set-aside and milk quotas.

The fact that the CAP is to be re-examined in the context of a budget review raises the possibility of a real budget constraint post-2013, with *finance ministers* determining the size of a CAP budget within which agriculture ministers must operate, rather than the past practice of finance ministers being called upon to finance a CAP largely out of their control. The senior Commission official in charge of the Directorate General for Agriculture has made clear his personal view that, post-2013, the CAP should continue with its two-pillar structure, with decoupled support payments and rural development. Whilst it would be important for farmers 'to be as market-oriented as possible', some sort of safety net should still be retained, but not necessarily based upon price support. He suggested that such a system 'should be more targeted at 'special cases' than the current system of publicly available intervention purchasing'. While the Health Check did not question the basis structure of CAP support, he expected that the debate on the post-2013 CAP would be lively, and conceded that 'many will be thinking 'outside the box'' (*Agra Europe*, 9 January 2009:EP/1). So the debate *within* the Commission is likely to be as lively as that without.

With the Doha Round still not concluded, a number of questions remain unanswered. Will the Health Check's mini-reform facilitate completion of the Doha Round, or –in the case of a Doha failure, and with no Peace Clause in place– will it better equip the CAP to withstand further attack in Dispute Settlement proceedings? And what role will the WTO play in the forthcoming budget review, and the shape of the CAP post-2013? Without a final Doha agreement in place there is no external constraint to lock-in the decoupling of EU farm support achieved between 2003 and 2008. How this will play out on the CAP in the future is too soon to say, but there is a slight risk that the CAP will backtrack to a pre-2003 version.

References

Ahern, B. 2005. We must stand by the Common Agricultural Policy. *Financial Times*. 26 September: 23

Borchardt, K.D. 2008. *Health Check*, text of a speech prepared for the Brussels Session of the EAAE Congress, 28 August 2008.

Commission of the European Communities (2007) *Communication from the Commission to the Council and the European Parliament: Preparing for the 'Health Check' of the CAP reform*. COM (2007)722 (Brussels: CEC).

Commission of the European Communities (2008) *Proposal for a Council Regulation establishing common rules for direct support schemes for farmers under the common agricultural policy and establishing certain support schemes for farmers ...* COM(2008)306/4 (Brussels: CEC)

Council of the European Union. 2003. *CAP Reform – Presidency Compromise (in agreement with the Commission)*. 10961/03. Brussels: Council of the European Union

Council of the European Union. 2005. *Submission from the Presidency to the European Council. Financial Perspective 2007-2013. Provisional Version. 16 December*. Official website of the UK EU Presidency. http://www.eu2005.gov.uk, accessed 19 July 2007

Council of the European Union. 2007. *Brussels European Council 8/9 March 2007 Presidency Conclusions*. 7224/1/07 REV 1. Brussels: Council of the European Union

Darling, A. (2008) letter to Dr Andrej Bajuk, Minister of Finance, Slovenia. 13 May. Available at http://www.hm-treasury.gov.uk/media/4/1/chx_letter130508.pdf, accessed 10 September 2008

Daugbjerg, C. and Swinbank, A. 2009. *Ideas, Institutions, and Trade: The WTO and the Curious Role of EU Farm Policy in Trade Liberalization*. (Oxford: Oxford University Press)

Daugbjerg, C. and Swinbank, A. 2010. Explaining the 'Health Check' of the CAP: Budgetary Politics, Globalisation and Paradigm Change Revisited, *Policy Studies*. 31 forthcoming

European Commission. 2007. *CAP reform: fruit and vegetable reform will raise competitiveness, promote consumption, ease market crises and improve environmental protection*. Press Release IP/07/810 (Brussels: CEC)

Fischer Boel, M. 2007. European Agricultural Policy Facing up to New Scenarios. To the Confederazione Generale dell'Agricoltura Italiana,Taormina, Italy, 23 March. SPEECH/07/182. Available at http://ec.europa.eu/commission_barroso/fischer-boel/speeches/index_en.htm.

HM Treasury and Department for Environment Food and Rural Affairs (Defra). 2005. *A Vision for the Common Agricultural Policy*. London: HMSO

Pirzio-Biroli, C. 2008. An Inside Perspective on the Political Economy of the Fischler Reforms in *The Perfect Storm: The Political Economy of the Fischler Reforms of the Common Agricultural Policy* edited by J.F.M. Swinnen. Brussels: Centre for European Policy Studies.

Swinbank, A. and Daugbjerg, C. 2006. The 2003 CAP Reform: Accommodating WTO Pressures. *Comparative European Politics,* 4(1), 47–64

Swinbank, A. and Tranter, R. 2005. Decoupling EU farm support: does the new single payment scheme fit within the green box?. *The Estey Centre Journal of International Law and Trade Policy*, 6(1), 47-61

World Trade Organization. 2008. *Revised draft modalities for agriculture*. TN/AG/W/4/Rev.3 (Geneva: WTO)

Chapter 6

Options of Financing the CAP – Consequences for the Distribution of Farm Payments

Franz Sinabell, Erwin Schmid and Hans Pitlik

Introduction

Until 1992, trade restrictions, market price support and supply management policies were the major tools of the CAP. To mitigate the well known weaknesses of this policy conception a process of 'decoupling' was initiated with the Mac-Sharry reform in 1992. Direct payments (DP) were granted to producers of arable crops, beef and veal, sheep meat and goat meat as compensation for lower administrative prices. In the Agenda 2000 reform, this process continued by including the milk sector and by establishing the program for rural development (the "second pillar" of the CAP). After the 2003 CAP reform fully or at least partially decoupled "single farm payments" (SFP) were implemented to mitigate the negative effects of price policy as well as payments based on historical areas and heads of livestock after 1992 (OECD 2006a, 2006b).

During the last 15 years direct payments have become the most important fiscal policy tool of the EU. In 2006, direct payments amounted to EUR33.1 billion, which was equivalent to 31 per cent of the EU's total operating expenditure (EUR106.6 billion). According to the Economic Accounts of Agriculture, the share of direct payments in the factor income of agriculture amounted to 26.5 per cent in 2006.

Direct payments can be divided in two major classes:

a. decoupled direct payments (DPP) which are subdivide into Single Farm Payments (SFP, EUR14.2 billion – mainly in EU-15 Member States), and Single Area Payments (SAP, EUR1.7 billion – mainly in New Member States); and
b. output linked direct payments are granted for plants (EUR12.0 billion) and livestock products (EUR5.7 billion).

Within DP the share of decoupled payments has increased recently, because the milk quota premiums had been fully decoupled by 2007, and due to the phasing in of area payments for the new Member States that entered the EU in 2004. The share of DDP will likely further increase because the Commission has pledged to further reduce trade distorting internal support measures (see EU offer at the G4-summit in Potsdam 2007).

In the 2003 CAP reform, the Commission fell short of establishing a single scheme that could prevail in all Member States. During the decision making process several concessions were made: at the Member States discretion not all direct payments need to be decoupled (e.g. those for suckler cows) and premiums per hectare agricultural land may be different according to the payments a farm has received during the reference period 2000-2002 ("historic model"). Single Area Payments which were introduced in the New Member States have the characteristics of decoupled payments: they are not linked to outputs and involve payments of uniform amounts per hectare of agricultural land in the Member State, up to a national ceiling resulting from the accession agreements.

Fully decoupled payments are considered to have no allocative effects. This reasoning allows the EU to classify them as "green box" payments which are deemed to be minimal trade distorting in the WTO. An implication of such characteristics is that DPs can be considered to be elements of a distributive policy.

Usually, the political rationale of distributive policies is to improve the income distribution by transferring money from richer to poorer households in order to correct market outcomes according to politically determined equity objectives. For a long time, agricultural economists (e.g. Koester and Tangermann 1976) have advocated the introduction of decoupled direct payments as an important step to mitigate the negative effects of market price support, including the mitigation of the regressive distribution effects of output linked support.

According to Article 33 of the Treaty, one goal of the CAP is "to ensure a fair standard of living for the agricultural community, in particular by increasing the individual earnings of people engaged in agriculture" while simultaneously guaranteeing adequate consumer prices. This vaguely formulated goal seems to point in the direction of a fair distribution of support across farming units.

In this article the overall distributive effect of direct payments is addressed. In a first step we discuss when and under which circumstances direct payments for re-distributive purposes should be a responsibility at the European Union level. The following section provides a brief description of the relevant literature regarding the distributive consequences of the CAP on farm household incomes and their spatial implications as well as the relevant data sources. Then we compare the distribution of direct payments for farm holdings across EU member states in the period 2000 to 2006. Using concentration ratios (which are similar to Gini coefficients), we also look at the distribution within EU member states. A short summary as well as an outlook, embedded in the overall framework of the CAP, concludes this contribution.

Direct payments from the view of Fiscal Federalism

The theory of Fiscal Federalism aims to identify policy agendas that should be addressed at the EU level, national levels, and sub-national levels. From a purely economic point of view, the assignment of responsibilities for certain policy fields on different government levels depends on the costs and benefits from (de) centralisation (e.g. Olson 1969, Oates 1999, Breuss and Eller 2004). While the existence of significant economies of scale and intergovernmental spillovers from certain policies point to more centralisation, decentralisation is better suited for policies with high preference heterogeneity among regional jurisdictions and for competition among political entities. Taking the special nature of a supra-national body like the European Union into account, Pelkmans (2006) additionally argues that a co-ordination of national policies is often superior to a complete centralisation.

With respect to agricultural policies the current division of responsibilities between the EU-level and the national levels is far from being in line with most normative requirements (e.g. Tabellini 2003, Grethe 2006). To be clear, market policies including all regulations that aim to keep the common agricultural market open, should be the primary responsibility of the EU level. This is a logical consequence from enormous economies of scale that can be realised by a common market. Policy liberalisation in the aftermath of the MacSharry and Agenda 2000 reforms contributed to a better realisation of the advantages of a common market and is therefore welcomed by many economists. However, for most remaining agricultural policy fields, a common policy at the European level does not appear to be the best solution from the view of Fiscal Federalism. This holds for most policies of the Second Pillar i.e. the program for rural development, but especially for the institution of direct payments.

Historically, direct payments had a clear connection to agricultural production. By more and more de-coupling since 2005, the character of direct payments clearly changed: it cannot be classified as a production subsidy any more, but instead it turns out to be an instrument of both personal as well as sectoral income policies. Against the background of Fiscal Federalism theory the central question emerge, whether these income support policies for the agricultural sector should be in the primary responsibility of the European Union. In general, such a centralisation of re-distributional competences at a supra-national level is seen very sceptical. First, there appears to exist a substantial preference heterogeneity with respect to redistribution among the member states. This is particularly the case for sectoral redistributive policies, as the importance of the agricultural sector differs significantly across the EU. Secondly, there are no sizeable international spillovers and no economies of scale from agricultural income support policies at the national level. Cross-compliance i.e. the link between direct payments and environmental and animal welfare issues might serve as a rationale for centralisation. However, benefits of these policies are more of a local nature and usually do not spill over national borders. Thirdly, international mobility of recipients of direct payments appears to be rather low, and fiscally induced migration is not important. Therefore, fiscal externalities of national income redistribution policies in the agricultural sector can be neglected. All these arguments are clearly in favour of a national responsibility for agricultural income support policies. Consequently, the changing character of decoupled direct payments also requires a re-nationalization of this part of agricultural policy from the viewpoint of Fiscal Federalism.

Literature and Data Sources

Selected studies on distributional aspects in agriculture

Over recent years, OECD has repeatedly looked at the various dimensions of the distribution of agricultural incomes. OECD (1999) analyses the distributional effects of agricultural policies in the mid-90s by comparing the distribution of support in relation to output and income in OECD countries. The report concludes that the distribution of market price support is very similar to the one of output and that differences across regions are less than those across farm types or size classes. Moreover, the distributional patterns have shown little change over the last ten years. Kurashige and Hwan Cho (2001) examine the incidence of low income as well as the impact of social security policies of OECD countries in agriculture. Based on various indicators they find out that "low income" is higher among farm households than among non-farm households and despite generous support in many OECD countries the income distribution of farm households shows a higher degree of inequality than of non-farm households.

Allanson (2003) explores the redistributive impact of Common Agricultural Policy reform 2000 with reference to the distribution of farming incomes in Scotland. The main result is that direct payments have exacerbated the inequality of farm incomes in Scotland in 1999/00. Allanson (2007, 2008) analyses the redistributive effect of "horizontal inequity", being the differences between the level of support received by farms of a given type and the level of pre-support income: again, the provision of support increased the average size of farm income differentials throughout the period 2000/01 to 2004/05. Similarly, in a recent study on Tuscany (IT), Allanson and Rocchi (2007) find that the provision of support increased absolute income inequality within the agricultural community because the distribution of transfers was both vertically and horizontally inequitable.

There are only a small number of studies which lead to other conclusions. One example is Keeney (2000), a study of Irish agriculture based on individual farm records. Keeney demonstrates that the direct payments of the MacSharry reform induced a more equal distribution of family farm incomes in Ireland.

The territorial dimension of CAP expenditures has been analysed by Shucksmith et al. (2005). Looking at the regional distribution of CAP payments and their contribution to cohesion objectives, the authors found that CAP payments do not support territorial cohesion, because more prosperous regions get higher levels of

CAP transfers. This holds not only for market based support, but also – although somewhat less pronounced – for support through rural development programs. At a similar result with respect to the distribution of farm support between continental and Mediterranean agriculture arrive Mora and San Juan (2004).

With hardly any exceptions, studies looking at distributional effects of the CAP reveal that the current instruments of the CAP do not prevent a substantial part of farmers from being among the poorest citizens of EU Member States. At the same time, direct payments to high-income farm units and regions contribute to pronounced income inequalities in this sector. This survey also shows that a cross country comparison of direct payments before and after the 2003 CAP reform has not yet been made.

Data Sources and Methods

Established information systems measuring the effects of CAP on farm incomes are hardly adequate for analyzing distributional outcomes (Court of Auditors 2004):

- The income indicator of the farm accountancy data network (FADN) – 'farm family income' – is tricky to interpret, because many agricultural holdings are organized as companies. In addition, the sample of farms providing the information is considered to be not representative.
- The economic accounts for agriculture (EAA) is a satellite account of the national accounts. Its main indicators are 'factor income' and 'net entrepreneurial income'. Besides the fact that the quality of data supplied by some Member States seems to be poor, these indicators are only provided at sector level. Distributional comparisons can therefore only be made across countries or with other sectors, but not among farm holdings within the farming sector of a country.
- The same is true for the statistics on the income of the agricultural households sectors (IAHS; see Eurostat 2002). The methodologies of the underlying concept are not harmonized which 'cast[s] doubt on the possibility of comparing data supplied by member states' (Court of Auditors 2004). In general, IAHS allows comparing non-farm household incomes with farm-household incomes, yet not in all member states.

Aggregated data on the distribution of direct payments across EU Member States have been published regularly since they were introduced and can therefore be set in relation to other variables of interest like the number of farms or persons engaged in farming. The most up-to-date figures on the distribution of direct payments across farm holdings were published by Eurostat in 2008. In 2006, EU expenditures for the Common Agricultural Policy amounted to EUR49.9 billion (47 per cent of the total budget). Direct payments (EUR34 billion) had the largest share, followed by market related expenditures (EUR8 billion) and payments for the rural development program (EUR7.7 billion). Both, the volume and share of direct payments have increased since the CAP reform in 1992. In the year 2000, direct payments amounted to EUR24.1 billion and increased to EUR32.5 billion in 2005. Given that farm payments have been increasing and that structural change has taken place at an average annual rate close to 2 per cent, payments per annual working unit (AWU) have been increasing until the entry of ten new Member States in 2004.

In the year 2000, the average payments per recipient were below EUR2,000 in Portugal and Italy and were highest in Denmark (EUR10,585) and the UK (EUR19,272). The EU-15 average was EUR6,331 (ranging between 1,747 in Greece and 21,429 in the United Kingdom) five years later. Direct payments per holding were considerably lower in the new Member States that entered the EU in 2004 (on average EUR723 – from 232 in Cyprus to 11,397 in Czech Republic). Therefore the mean of direct payments per holding in the EU dropped from EUR5,017 per holding to EUR4,682 between 2000 and 2006.

In preparing the 2003 CAP reform, EU Commissioner Franz Fischler released for the first time fairly detailed data about the distribution of direct payments to foster a political climate to curb the size of high-end CAP payments. EUROSTAT publishes the number of recipients and the volume of transfers aggregated in 12 classes. Comparing the holdings getting less than EUR5,000 with those getting more can be used to show that a small number of recipients got a relatively large share of all direct payments in 2000: 953,000 holdings received more than EUR5,000 amounting to EUR15.5 billion. About 21 per cent of holdings getting such support received 82 per cent of all direct payments. Until 2006 the distribution has become more unequal: 1.3 million farms (18 per cent of the 7.3 million recipients) got EUR27.9 billion (84 per cent of direct payments).

Methods and Results

The distribution of direct payments is quite different in the EU member states. Figure 6.1 shows a comparison of selected countries of EU15 in 2002 as well as the change in distribution for EU15 between 2002 and 2006.

In order to depict the distribution of DP within EU Member States and thus being able to compare between countries we use mean, median and concentration ratios (CR). Using the method described in Bleymüller et al. (1991: 15ff), the median payments per Member State were calculated. A high ratio between mean and median of payments is a simple indicator of an unequal distribution.

A more sophisticated measure of (in) equality is the concentration ratio (CR). It has the same interpretation as the Gini-Coefficient, but it is calculated in a slightly different way (Rasche et al. 1980, see the appendix). High levels of CR indicate that a small number of recipients get a large amount of payments while a low CR indicates a more equal distribution.

Figure 6.2 shows the three measures: The horizontal axis indicates the mean (indicated by x) and median (|) payment per holding in the EU 25 Member States

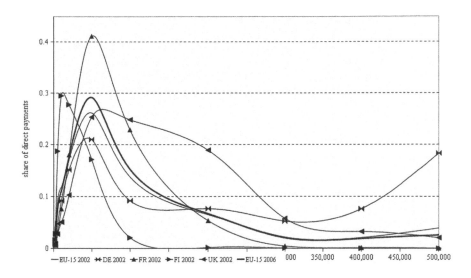

Figure 6.1   Distributions of direct payments in EU-15 and selected member states in 2002 and 2006

Source: Commission of the EU, own calculations.

Note: Figures are truncated at EUR500,000, the presented volume of payments is for the open class EUR500,000 and above. The graph is based on classified data with varying class sizes, therefore the real, but unknown distribution may look slightly different.

in 2006. The vertical axis showing the CR is used to rank them according to the concentration of payments within the countries. The overview shows that even if the difference between median and mean is very large in absolute terms (like in the United Kingdom or in Germany) the CR may be relatively moderate compared to other countries (like Malta, Slovakia or Portugal). Given that the CR is relatively high in the Member States that have entered the EU in 2004, it is evident that the CR in the EU has increased between 2000 and 2006.

Summary and Discussion

The comparison of CRs between the years 2000 and 2006 shows, that (1) the CAP reform 2003 has not evened the distribution of decoupled direct payments, and (2) that there is no uniform pattern of change. The concentration of direct payments towards a small number of recipients in EU-15 Member States did not change between the two years (the CR was 78 in both years). This is the result of two antagonistic developments: in some countries like France, Ireland, Austria the measure of inequality was lower in 2006 compared to 2000 while the opposite was true in countries like The Netherlands, Denmark, Sweden, and Italy. Given that

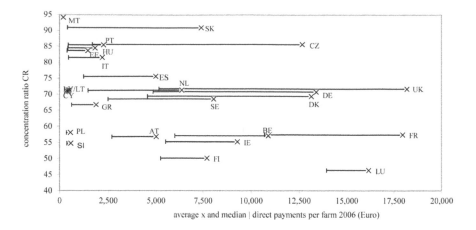

Figure 6.2   Concentration Ratios (CR), medians (|) and means (x) of direct payments in 2006

Source: own estimates

there are only few observations since the Single Farm Payment was introduced it is too early to draw conclusions on the distributive effects of the historical versus the area based scheme.

Admittedly, CAP payments, among them direct payments, are hardly motivated by distributive considerations alone. Currently they are justified to ease the process of integration for the agricultural community of Member States that have recently entered the EU. Another purpose is to facilitate structural adjustment of farms that are exposed to freer market conditions after decades of CAP interventions. Moreover, as direct payments are only granted if standards of good agricultural and environmental condition i.e. cross compliance are complied, such payments have an environmental and animal welfare facet as well.

The current debate about abandoning the historical model provides possibilities to improve the distribution of DP. If this process takes places, eventually the distribution of DP will reflect the distribution of land which is highly concentrated in many countries, as well. To shift more resources to the program for rural development could therefore change the distribution of farm payments more substantially. At the same time, taking into account the principle of "fiscal equivalence" (Olson 1969) could give guidance for the question which of the issues currently addressed by direct payments should be addressed at the EU or Member States levels.

References

Allanson, P. 2003. The Redistributive Effects of Agricultural Policy on Scottish Farm Incomes. *Journal of Agricultural Economics*, 57 (1), 117-28.

Allanson, P. 2007. Classical Horizontal Inequities in the Provision of Agricultural Income Support. *Review of Agricultural Economics*, 29 (4), 656-71.

Allanson, P. 2008. On the Characterisation and Measurement of the Redistributive Effect of Agricultural Policy. *Journal of Agricultural Economics*, 59 (1), 169–187.

Allanson, P. and Rocchi, B. 2007. An Analysis of the Redistributive Effects of Agricultural Policy in Tuscany with Comparative Results for Scotland. *Dundee Discussion Papers in Economics*, No. 193, Sept. 2006, revised Jan. 2007. Dundee : University of Dundee.

Bleymüller, J., Gehlert, G. and Gülicher, H. 1991. *Statistik für Wirtschaftswissenschaftler*. München: Franz Vahlen.

Breuss, F. and Eller, M. 2004. The Optimal Decentralisation of Government Activity: Normative Recommendations for the European Constitution. *Constitutional Political Economy*, 15, 27-76.

Court of Auditors 2004. Special Report No 14/2003. *Official Journal of the European Union*, C 45/1, 20.2.2004

Eurostat 2002. *Income of the agricultural households sector, Report 2001.* Luxembourg: Eurostat.

Grethe, H. 2006. *Environmental and Agricultural Policy: What Roles for the EU and the Member States?* Keynote paper for the conference "Subsidiarity and Economic Reform in Europe" Brussels, November 8-9 2006.

Keeney, M. 2000. The distributional impact of direct payments on Irish farm incomes. *Journal of Agricultural Economics,* 51, 252-265.

Koester, U. and Tangermann, S. 1976. Alternativen der Agrarpolitik. Eine Kosten-Nutzen-Analyse im Auftrag des Bundesministeriums für Ernährung, Landwirtschaft und Forsten, *Landwirtschaft – Angewandte Wissenschaft*, Heft 182, Münster-Hiltrup.

Kurashige, Y. and Hwan Cho, B. 2001. *Low Incomes in Agriculture in OECD countries*. Working Party on Agricultural Policies and Markets, AGR/CA/APM(2001)19/FINAL, OECD, Paris.

Mora, R. and San Juan, C. 2004. *Farmers Income Distribution and Subsidies: Product Discrimination in Direct Payment Policies for Continental and Mediterranean Agriculture*. Dpto. Economía, Universidad Carlos III de Madrid. Available at http://www.uc3m.es/uc3m/dpto/CJM/farmersincome.pdf [accessed June 2010].

Oates, W.E. 1999. An Essay on Fiscal Federalism. *Journal of Economic Literature*, 157, 1120–1149.

OECD 1999. *Distributional Effects of Agricultural Support in Selected Countries*, AGR/CA(99)8/FINAL, OECD, Paris.

OECD 2006a. *Decoupling: Policy Implications*. Paris: OECD. Available at http://www.oecd.org/dataoecd/34/10/39283467.pdf [accessed 4 June 2010].

OECD 2006b. Special Issue on Decoupling Agricultural Support, *OECD Papers Vol 5*, No.11. Paris: OECD.

Olson, M. 1969. The Principle of Fiscal Equivalence: The Division of Responsibilities among Different Levels of Government. *American Economic Review, Papers and Proceedings*, 59(2), 479-487.

Pelkmans, J. 2006. Testing for Subsidiarity. *Bruges European Economic Policy Briefings*, 13, College of Europe, Bruges.

Rasche, R.H., Gaffney, J., Koo, A.Y.C. and Obst, N. 1980. Functional forms for estimating the Lorenz Curve. *Econometrica*, 48(4), 1061-1062.

Shucksmith, M., Thomson, K. J. and Roberts, D. (eds.) 2005. *The CAP and the Regions: the Territorial Impact of the Common Agricultural Policy*. Oxfordshire, Cambridge: CAB International.

Tabellini, G. 2003. Principles of Policymaking in the European Union: An Economic Perspective. *CESifo Economic Studies*, 49, 75-102.

APPENDIX – Lorenz Curve Estimation and Concentration Ratio Computation

The data of Eurostat provides ten classes of farms (x) and direct payments received (y), of which cumulative proportions are calculated. We use the functional form proposed by Rasche et al. (1980) to estimate Lorenz curves. The explicit functional form is as follows:

(1) $$y = \left[1 - (1-x)^{\alpha} \right]^{1/\beta} \quad \text{where } 0 < \alpha \le 1, \; 0 < \beta \le 1;$$

The computation of the Concentration Ratio (CR) is based on the functional form specified in equation (1). It is defined:

(2) $$CR = 1.0 - 2.0 \int_0^1 \left[1 - (1-x)^{\alpha} \right]^{1/\beta} dx,$$

substituting variables

(3) $$u = 1 - (1-x)^{\alpha},$$

this is equal to:

$$CR = 1.0 - 2.0\left(\frac{1}{\alpha}\right)\int_0^1 (1-u)^{1/\beta} u^{1/\alpha - 1}\, du$$

(4)

$$= 1.0 - \frac{2.0}{\alpha} B(1/\alpha, 1/\beta + 1)$$

,

where B represents the beta distribution. It ranges between zero (absolute equality) and one absolute inequality).

Chapter 7

The European Commission in the CAP Decision Making: a Case Study on the Sugar Reform

Matteo Iagatti and Alessandro Sorrentino

Introduction

The last ten years have represented a period of major changes for the Common Agricultural Policy (CAP). The first elements for substantial change were introduced with the so-called MacSharry Reform in 1992. More specifically, a market orientation process of the CAP started with the establishment of partially decoupled aid and a significant reduction of guaranteed prices for farmers. The core elements of the 1992 intervention found their last expression in the guidelines and instruments shaping the Fischler Reform in 2003. On this occasion the decoupling was extended to a wider range of agricultural products and it was deepened, with the complete decoupling between financial support and farmers' choices. Simultaneously, this aid has been linked to the compliance of a set of measures concerning the environmental sustainability of the whole sector.

Along with the evolution of CAP policy instruments, the role of the European Commission (COM) within the decision-making process has been changing, too. In order to understand such a change, a comparison can be made between the so-called 'price marathons', which characterised the negotiations during the first phase of CAP evolution, and the negotiations that led to the MacSharry Reform, and particularly to the Mid Term Review (MTR). In the first case, the Commission acted as a simple mediator between Member States' preferences on the level of institutional prices. On the other hand, during the 1992 Reform, and in the definition of the MTR, the COM has been able to propose new schemes of intervention in agriculture safeguarding the core nature of the reform, notwithstanding the difference in the Member States' positions and the consequent pressures towards a 'status quo' solution. From this point of view, it is likely that the COM has developed several instruments to close the gap between its position and the ones expressed by the Member States, building a complex system of side-payments to steer the negotiation to a positive political solution.

The aim of this chapter is to draw a clear picture of how the COM manage the CAP definition process and specifically how it acts in order to close the gap between Member States' positions and its preferred political solution.

This chapter focuses on the sugar CMO Reform of 2006, which will be used as case study to point out the main elements of the CAP decision-making process and the role of the COM. Firstly, we define the theoretical framework in which we have inserted the elements which emerge from the case study. Secondly, we point out the institutional channels provided for the COM by the consultation procedure to exert its active role. In order to meet the objectives of this chapter, we propose an analysis of the documents produced in the different steps of the legislative procedure by the various institutions taking part in the CAP definition process. For each document, we report the issue in question, the correspondent phase of the procedure and the bodies involved in the discussion. Lastly, we gather all the information in order to point out the winners and losers of the 2006 Sugar CMO Reform and the role the COM played. Some final remarks will conclude the chapter.

Theoretical background: Putnam's conjectures and the CAP

The contribution by Putnam (1988) greatly helped to clarify many aspects of international relations. He conceived the entanglements between international negotiation and domestic politics as a game of ratification of international treaties, defining level I as the international *«bargaining between the negotiators, leading to a tentative agreement»*; and level II as the *«separate discussions within each group of constituents about whether to ratify the agreement»* at a domestic level. In this framework he defined *«the win set for a given level II constituency, as the set of all possible level I agreements that would win when simply voted up or down»*. In such a framework, the final agreement is possible only if the actors' positions overlap a determined issue. The size of the win set, and hence the bargaining position, are affected by issues operating at the so-called level II. The negotiators operating at level I are constrained by the domestic implications of the agreement reached inside the international arena, hence, the more the domestic positions are *'flexible'* over the considered issue, the more the negotiator will increase his/her bargaining power.

We have focused this analysis on the level I determinants and more specifically on how a *'preferential'* dialogue between some Member States and the COM can affect the distribution of the bargaining power among Member States.

Cavallo et al. (2007) have pointed out that the dialogue with the COM was one of the main determinants of the Member State's bargaining power in the CAP reform process. The contribution we propose can be considered as an extension of those results toward a better specification of the instruments used by the COM and hence of the effects that the COM's active role has on shaping the CAP reform process.

Bargaining scenario for the analysis

The bargaining scenario in which we perform our analysis is quite relevant for the characterization of the actors' respective roles and the impact of their activity on the final outcome. More specifically, Member States positions and COM policy preferences have to figure out a bargaining context in which the COM active role is decisive for a positive solution of the legislative process. Cavallo et al. (2007) have described the institutional settings and the bargaining scenarios in which the COM can perform an active role along the negotiations. In such a scenario the Member States positions tend to be spread and rather distant over the possible policy outcome and the COM has to promote a specific solution for the policy in question. A specific policy outcome might be imposed by strong pressures insisting over the COM coming from various sources such as: stakeholders and interests groups, international commitments (i.e. WTO and surroundings), EU budget constraints, consistency with other EU policies, etc.

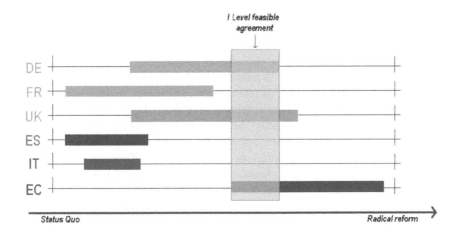

Figure 7.1   CAP bargaining scenario with an active role of the COM and a wide range of position for the Member States

The outlined scenario is represented in Figure 7.1, where the horizontal stripes represent the actors' (Member States and COM) win-sets at level I and the vertical band indicates the sample of the feasible potential agreements. From this figure, it clearly emerges that some Member State states win sets, namely Italy Spain and France, are out of the range of the final outcomes the COM can accept, given its win set. In such a contex, it is likely that the COM acts in order to move the positions of the most reluctant Member State towards an agreement that encompasses its preferences. This is a very simplified but not just an '*imaginary*' scenario. As stated in the introduction, the process of definition of the MTR could provide quite a good example.

Rules of the game: the Consultation procedure

Although the political scenario discussed above gives some general explanations for the active role of the COM, the aims of our analysis impose a more detailed description of the institutional opportunities granted to the COM by the legislative procedure involved in the CAP first pillar definition. In this paragraph we will briefly describe the relevant steps of the consultation procedure, highlighting in each step in which directions the COM plays an active role. The procedure formally starts with the submission to the Council of an official proposal for the regulation. The proposal can be amended by the Member States and approved by the Council under the qualified majority rule or refused. Once the Council reaches a political agreement, the COM together with the Committees is appointed by the Council to manage the drafting phase of the applicative regulation and the implementation phase of the new measures.

The proposal, which the COM submits to the Council, is the result of a drafting process which begins long before the official submission. Usually the proposal drafting starts with a document in which the COM communicates to the other legislative partners the purpose of an intervention, the current situation of the sector involved and the possible solution for the new legislation. Based on this document, a discussion among the actors starts. The Member State and the stakeholders, by various means, communicate their opinions on the perspective proposal to the COM, and based on impact studies, the possible effect of the reform. Despite its informal nature, the drafting phases of the proposal represent a decisive step in which the COM is the centre of a complex net of communications. From its privileged position the COM takes stock of the Member State and stakeholder positions, and evaluates the relative practicability of different solutions together with the strength of the oppositions. During this phase, the proposal is reshaped on the base of its feasibility and accordingly to intermediate compromises between the COM and the Member State. Once the COM has submitted the proposal to the Council, opinions from the European Parliament (EP) and the Economic and Social Committee and Committee of Regions are required. In our analysis, the role of the EP and the Committees is disregarded given the non-binding nature of the amendments and the opinions proposed, although their advice over the direction of the reform could be used by some of the actors to strengthen their positions. The Commission evaluates the amendments which may be proposed by the EP and consequently modify the proposal which is submitted to the Council in this new form. At this point, the controversial aspects of the proposal are taken into consideration; the Member State can propose amendments voted under the unanimity rule. Before the proposal is discussed and amended by the Council, the important work of mediation is undertaken inside the Special Committee for

Agriculture (SCA). The SCA is the equivalent of the Coreper, which accomplishes the same function for all the other EU intervention areas. The formal vote inside the Council concludes the political process but not the legislative procedure. The regulation drafting and the implementation phase are, by appointment, directly under the control of the COM. This means that in those phases new compromises are possible and also that implicit, and in some cases bilateral compromises, which could be settled in the previous phases, are made concrete. Of course the nature of such compromises must be technical and administrative but they may have facilitated the achievement of a political solution.

The framework we have developed so far tries to give an overall picture of the work of the COM during the legislative procedure. The nature of the policy, the distribution of preferences and bargaining powers between Member States and European Institutions can affect the channels in which the COM can operate. What we want to highlight here is that, despite the rigid framework imposed by the procedure in terms of competences and power, there is a political level of the decision making process that the COM can explore in order to influence the outcome towards its preferred solution.

Case Study: The Sugar CMO reform

The case study we propose examines the definition of the sugar CMO Reform started in September 2003 and concluded in June 2006. This preliminary explorative analysis is based on the documents produced from the Council and the Commission during each phase of the procedure. We used the status quo, the Commission proposals and the Presidency Compromise as *bench marks* of the sugar CMO reform in which the views and the positions expressed by al the partners are synthesized in a single outcome. We have also explored the discussion within the Committees and Working Groups involved in the CAP definition where Member States and Commission representatives express their positions over each political and technical issues.

The sugar reform: an evolutionary overview In this subsection we present the evolution of the issues involved in the reform in each phase. The sources of information are the Commission introductory document (COM (2004) 499), the official proposal (COM (2005) 263) for the sugar CMO reform, and the agreement reached inside the Council (Doc. n°14982/05). We report the status quo situation in each aspect as well as the proposed modification and the final agreed outcome. The objective is to highlight the main components of the negotiation and answer basic questions such as: how much does each step of the procedure differ from the previous one? Which elements have the Commission been able to defend during the procedure? How far does the final agreement go if compared to the status quo? Which aspects can be defined as crucial for the Commission?

In Figure 7.2 we have synthesized each issue in its fundamental components highlighting what has been changed in the Commission proposals (in light grey) and what has been changed or added by the compromise in the Council (in grey).

On a general level firstly we bring attention to the fact that the Commission considered it necessary to produce two distinct documents for the proposal concerning the sugar CMO reform. In the first one – (COM (2004) 499) – the Commission outlined the general guidelines for the reform, briefly indicating some of the instruments for a sustainable policy. In June 2005, the Commission submitted a new version of the document which represents the official proposal for the regulation. Between the issuing of these documents almost a year of intense discussion took place. The comparison between the above mentioned documents shows important modifications to the instruments proposed in the direction of a more 'acceptable' base for further negotiations. The Commission, thanks to the time spent evaluating the proposals, had the chance to 'weigh up' the Member States positions and evaluated the most suitable solution in order to satisfy its own vision for the future of the sugar sector and the constraints imposed by Member States preferences and by the international arena. This procedure helped the Commission to conduct a wider discussion proposing the best structure for a widely accepted and effective reform. In greater detail, the value of the proposals as negotiation tools emerges clearly, taking into considering the options for quotas and prices. In the first communication (COM (2004) 499) a combined reduction of quotas and prices was conceived, mixing two of the initially separated reform scenarios described in the introducing document of 2003.[1] Such a solution encompasses two instruments which will result in a double oriented policy. On one hand, the price reduction stands as a market orientation toll in the direction of liberalization. On the other, the reduction in quotas imposes stronger constraints to producers, especially for those who operate in the most productive areas.

The second information we derive from the table is the structure of the reform that the Commission wanted to propose. The core of the intervention was a mix of instruments which aimed to accomplish two objectives. The first was the reduction of the market surplus in order to decrease the budget expenditures and to fulfil the international trade commitments, the second was the mitigation of the negative effects of such reductions on beet growers and sugar producers by means of a set of old *ad hoc* instruments.

Institutional price reform was based on the abolition of the intervention price and the creation of a reference price for the whole sector. The reference price served for the definition of the minimum price for beet growers, and represents the level at which private storage mechanism and the price for border protection are triggered. Reference price reduction was conceived in three years with a 37 per cent price cut. As regards quotas, the reduction was planned in four years for a total quota cut of 16 per cent.

---

1   COM(2003) 554

As market balance tools and restructuring scheme the Commission proposed the development of a private storage mechanism, together with the possibility to transfer quotas among Member States and the institution of an aid for sugar producers which withdraws their quotas.

The restructuring fund was planned to be co-financed by the Commission and the Member States. The income reduction for beet growers due to price reduction was mitigated by a direct payment which covers 60 per cent of the loss. No coupled aid, even as a transitory measure, was conceived in the Commission proposal.

		Status Quo	Elaboration of the COM Proposal		Compromise/Regulations
			COM (2004) 499	COM (2005) 263	
Prices	Sugar Reference (€/t)	631,9	506 → 421	631,9 → 385,5	631,9 → 404,4
	Reduction/Years	-	25 → 37% / 3	0 → 39% / 4	0 → 36% / 4
	Sugar Beet Min (€/t)	43,63	32,8 → 27,4	32,8 → 25,05	32,9 → 26,3
	Reduction/Years	-	25 → 37% / 3	24,8 → 42,5% / 4	24,5 → 39,7% / 4
Quotas	Level (mln t)	17,4	16,1 → 14,6	17,4	17,4
	Reduction/Years	-	7,5 → 16% / 4	0	0 / -
	Supplementary Quotas (mln t)	-	0	1,1	1,2
	Merging A/B	-	yes	yes	yes
	Quota Declass.	yes	no	no	no
	Quota Transf.	no	yes	no	no
	Duration Regime	2006	2009	2014	2014
Market Balance Tools	Sugar Intervention	yes	no	no	yes, transitory, 60% ref. price
	Private Storage	no	yes	yes	yes
	Carry-over	yes	yes	yes	yes
Restructuring Scheme	Aid Sugar Producers (€/quota)	-	350	730 → 420	730 → 520
	Duration	-	5	4	4
	Contribution to the Fund (€/quota)	-	175	126,4 → 64,50	126,4 → 113,3
	Duration	-	5	3	3
	Others Measures	-			Regional Diversification Aid increase if severe quota reduction Transitory aid for full time refineries Compensation for machinery suppliers
Payment Beet Growers	Overall EU Envelope SFP (60% income loss)	no	1340 mln €	1542 mln €	1542 mln €
	Coupled Aid (30% income loss)	-	no	no	Ms with quota red. + 50% FEAOG guar. Financing 2006 → 2014 Temporary National Aid Authorized financing 2006 → 2014

Figure 7.2   **Resume of the evolution of the issues involved in the sugar CMO reform**

Starting from the structure described so far, the Commission has been able to safeguard only part of the initial proposed reform. More specifically, price restructuring/reduction has been the object of minor modifications and many of them have been introduced by the Commission itself. The agreement by the Council slightly adjusted the price cut and introduced a temporary maintenance of the intervention price at a level of 60 per cent of the reference. Substantial changes with respect to the 2004 proposal have been introduced by the Commission on: quotas, the restructuring fund and the amount of payments for beet growers. The Commission cancelled the compulsory quota cut and the possibility for quota transfers among Member States, extending the quota system to 2014 instead of 2009. In its official proposal, the Commission also established larger and longer subsidies for sugar producers together with a smaller financial contribution by the industry. Lastly, the Commission increased the envelope for the direct payment of the beet growers (+13 per cent).

The Presidency Compromise amended all those aspects. More specifically, it has introduced supplementary quotas for some Member States, slightly increased the aid to the sugar industries and introduced a temporary coupled aid for those Member States which voluntarily reduced their quota. Major modifications have been introduced by the Council in the restructuring scheme. The Council added several important aspects as regional diversification of the aid, with aid increasing in cases of severe quota cuts and a set of compensations for full time refineries and machinery suppliers.

The steps of the sugar reform process provide a clear picture of the different roles played by the main EU institutions. The core of the reform has been substantially preserved, even if only through price reduction, with no compulsory quota cut. The Commission itself substantially adjusted the initial proposal by making significant changes in the key and permanent pillars of the reform: prices, quotas, and the key parameters of the restructuring fund. The Council introduced minor changes on prices and transitional measures concerning the restructuring fund and the payments to the beet growers targeted to soften the impact of the reform and accompany the structural adjustment of the whole sugar industry. More specifically, the maintenance of a transitional coupled aid and the extension of the restructuring fund to refineries and machinery suppliers have reduced the impact of the reform on the industry profitability and retained a certain digressive level of protection for some years. Such additional measures are all temporary and have not prevented the overall objectives of the reform which consist in extending the decoupled scheme to new CMOs and reducing the gap between the domestic and the world market. From this point of view, the final agreement did not vary too much from the Commission's proposal in either a conservative or a reforming direction.

Such a picture suggests that most of the main negotiations among Member States occurred between the first and the second proposal with the strategic management of the Commission. The multidimensional nature of the reform made several technical and technical-political arguments available to the Commission to guide Member States with radically different positions towards a sustainable compromise, which is not so far, at least in the key issues, from the original ideas of the Commission on the sugar reform.

The Sugar Reform: winners, losers and the role of the European Commission   Our analysis starts from the division of the negotiation in three fundamental phases of discussion. First of all we analysed the pre-proposal discussions in which, from the options presented by the Commission in the document COM (2003) 544, Member States raised questions and comments revealing their preferences. Secondly, we examined the discussion between the proposals evaluating Member States Positions with respect to the instruments raised by the Commission to study the sugar reform. Lastly, we evaluated the regulations agreed by the Council, pointing out which Member States can be considered winners and losers. We have summarized the results obtained from the documental research, pointing out which aspects of the agreed reform can be considered as side payments for the losers.

Pre-Proposal phase: Revealing Member States Preferences Before the submission of the first proposal from the Commission, an important discussion on the basis of the document COM (2003) 544 took place. In the cited document, the Commission proposed three options for a new asset of the sugar CMO, together with a detailed impact assessment. In this phase, Me,ber States raised general questions and preliminary evaluations of the instruments proposed by the Commission. The discussion can be divided in two main parts. On one hand were the preliminary responses to the options proposed, on the other, the concerns raised by the Commission about the timing of the reform.

As stated above, the questions raised and the opinions on the options remain on a general level. The provisional nature of the position expressed by the Member States does not prevent a preliminary classification of the countries into groups supporting each option. In Table 7.1 we have summarized the three options, the Member States supporting each option and the specific remarks justifying their support. A more detailed situation can be outlined for the timing of the reform. From the data collected it a situation emerged in which the most competitive countries support an intervention, even if by different means.

Table 7.1 Member States' support of the options proposed for sugar reform

Commission Options	Member States' Remark	Member States
Quota retention Small price reduction	Maintenance of the current production Opening EU market (EBA, ACP)	Finland, Romania, UK
Price reduction quota abolition	Considered the most politically realistic	France, Netherlands, Germany
Liberalization	Other options are not sufficiently far- reaching	Denmark

This is shown in Table 7.1, where Romania and UK are in favour of a quota reduction. The Netherlands and Germany are more 'attracted' by a price reduction with quota abolition. In this context we found no southern European country in favour of any of the reform options. This emerges clearly if we read the two tables in parallel. In Table 7.2 Greece, Spain and Italy conceive the reform as coming too early if compared with the situation of the sugar market. On the contrary, the front which agrees with the Commission timing is composed by all the Member States which support at least one option for the reform.

We can now summarize the information obtained using the theoretical framework developed above. We can thus use the win-set concept to depict the scenario of the pre-proposal stage. We divided the Member States in four groups, and derived their position starting from the Commission's options which was used as a *'proxy'* of its position in the early stage of the negotiations.

In more depth, the Member States gathered in group 1 are all against the proposed intervention and their main concerns are related to the continuation of the production of sugar in certain areas. Group 2, which prefers option 1, is in favour of a 'soft' reform, especially in terms of price and quota cuts, which have to be the as minimal as possible in order to accomplish the strict requirements of consistency with the MTR.

The Commission position ranges from the 'minimum for the MTR accomplishment' and the 'most radical reform options," liberalization. In this range the Commission will structure its proposal encompassing the highest range of possible positions under the constraint of a minimum level of reform. Indicatively the suitable proposal is represented by the vertical band.

Between the proposals

The discussion in the period between the Commission proposals reveals important elements. Firstly, the concerns and preferences of the Commission are clarified and secondly the Member States expressed their position more specifically. Thirdly, the value of the first proposal as a negotiation tool emerges and is specifically related to the Member States Position.

As far as the guidelines are concerned, the Commission clearly stated which factors are consider crucial in sugar reform in document n°15445/04. These factors are a matter of discussion among the Member States which expressed their preferences. Table 7.2 summarizes the scenario after the first proposal.

From the table the question emerges of how the Commission proposed a reform which mixes the options presented in the early stages of the procedure. The combined quota and price cuts are coupled with direct payments to beet growers and the establishment of a restructuring fund sustaining the transformers. Member States reacted to such a mixed solution differently. On one hand the more competitive Member States in sugar production supported the proposal, even if some of them raised concerns about possible negative effects and applicability. Mediterranean Member States are, on the contrary, quite united in opposing the intervention, which is judged as being too early and unbalanced in its effects, especially in the distribution of sugar production within the EU.

Table 7.2   Member States reactions to Commission Guidelines

Member States' Reaction	Member States
In line with the proposed reform concerned about timing. Possible negative effect should be carefully addressed	Germany, France, UK
Agree in broad principles but expressed concerns over the feasibility of some of the proposed mechanisms	Austria, Romania, Belgium
In favour of a more radical reform, towards a fully liberalize sugar market	Denmark
Raised serious reservation. The proposed reform is too far reaching and unbalanced among Member States. Concentration of the production could result from the reform in contrast with the Lisbon strategy	Italy, Spain, Finland, Greece

The Member States supporting the guidelines of the reform are also in favour of the relative instruments, with some exceptions. Is this the case of the Netherlands, which supports the reform guidelines also asks, at the same time, for adjustments to the instruments proposed. More specifically it asks for price cut reductions and an increase in income percentage covered by the direct payments. The separation between northern and southern countries also applies to the instruments. In such a scenario the Commission presents it official proposal.

Member Stated positions towards specific instruments are summarized in Table 7.3. Due to the complexity of the positions expressed, we isolated the reaction of the Member States to the direct payment for beet growers. Member States supporting the guidelines of the reform are also in favour of the relative instruments, with some exceptions.

Table 7.3   Member States' positions between the Commission proposals – Specific Issues

Issue	Commission Proposal	Member States' Position	Member States
Prices	Price Reduction Abolition of Intervention Reference Price (400/450 EUR/t) Minimum Price for Beet growers	Support	Germany, Denmark, France
		Oppose	Italy, Netherlands, Greece, Spain, UK, Ireland
Quotas	Quota reduction A and B quotas merging Increase of iso-glucose quota	Support	Denmark, Germany
		Oppose	Poland, Portugal, Greece, Italy, Finland

Issue	Commission Proposal	Member States' Position	Member States
Market Balance Tool	Declassification abolition Private storage mechanism Quota carry-over	Support	Germany, France, UK
		Oppose	Italy, Greece, Spain
	Quota Transferability Conversion Scheme	Support	France, Netherland, Germany, UK
		Oppose	Italy, Greece, Spain
Income Support for Beet Growers	Income Compensation 60%	Support	Denmark, Germany, Belgium
		Oppose	Finland, France, Italy, Greece, Spain, Netherlands
	Full Decoupling	Support	Denmark, Germany, Belgium
		Oppose	Poland, Austria, Finland, Spain, Italy
	Regionalization of the aid	Support	Finland, Italy, Hungary
		Oppose	None Explicitly

The modifications that the Commission introduced to the first proposal are important if compared to the positions expressed by Member States during the discussion. In fact, the work done in this phase can be related to the necessity of gathering together the highest number of Member States in supporting its view over the sugar reform. More specifically, the Commission abandoned one of the most opposed instruments, namely the compulsory quota cut, simultaneously has better specified the restructuring fund in terms of aid and funding and has welcomed the call for a softer price cut coming from a number of Member States.

Commission behaviour can be explained through a strategic view. The Commission wants to prevent opposing Member States forming a blocking minority inside the Council, which will jeopardize the core elements of the reform together with its timing. Thus the Commission acted in the phase of the procedures it controls, the elaboration of the proposal, in order to understand Member States' positions and find the most-widely possible accepted compromise which encompasses Commission concerns and preferences. In other words the Commission acted in order to gather together Groups 2 and 3 of Member States which were not against the principles of the reform, reshaping the proposed instruments in order to build an 'acceptable' base of negotiation given the weight of Member States' position. In this way the Commission has not exerted its active role on the extreme Member States' positions, but instead on the Member States to closest to its point of view, building a solid base on which the Council can form an agreement which is not too far from the proposal.

In the next section we will briefly review the changes introduced to the proposal by the Presidency Compromise of November 2005 and show how it responds essentially to the specific position of the 'more status quo oriented' countries, not affecting the nature of the Commission proposal.

Presidency Compromise

Three measures have been added to the Commission's proposal in the direction of a more conservative approach, but all of them are temporary and do not affect the structure of the sugar market in the future. Firstly, the payment for beet growers has been added with a coupled component which covers 30 per cent of the income loss in addition to the 60 per cent covered by the decoupled aid. This is a temporary measure lasting until 2014.

Secondly, for a transitory period of eight years and for the Member States who suffer a severe quota reduction, the possibility of this sector receiving national coupled aid has been conceded. The national measures have to be authorized by the European Union.

Thirdly, there are some measures in favour of the stabilization of the EU sugar sector after the reform: those in favour of the machinery suppliers, the full time refineries and the increase in the supplementary quota for some Member States.

The point we want to highlight is that the reform passed practically unchanged by the Council, except for some transitional adjustments outlined in the previous paragraph. This is related to the intense discussion conducted in the earlier phases which facilitated mutual understanding between the Commission and those countries more in favour of a reform. At the same time, the Commission seemed reluctant to propose, in the intermediate stage of the procedure, measures which posed strong financial pressure on the EU budget, leaving, in some ways, this decision to the intergovernmental bargaining in the Council. The attention on such an aspect could have further strengthened the Commission's proposal and improved the chances of its being accepted.

Conclusive remarks

According to the questions tackled in the chapter, we can derive some interesting insights from this analysis.

First of all, the theoretical question we raised appears very useful in clarifying the bargaining situation in which we developed our analysis, not only for general and political issues but also in the evaluation of the modification of technical instruments. From this point of view Putnam's approach can serve as an important tool in the analytical and quantitative analysis.

Secondly, the choice of sugar as a case study to test our hypothesis about Commission's role in the EU decision-making process proved to be really effective for the variety of issues involved, the complexity of certain aspects, the relevance of the theme for EU institution and the relevance of the stakeholders in the sector.

As regards the relation between the nature of the issues in discussion (technical or political) and the phase of the procedure in which they are tackled, we can say that all of the issues are covered in each phase of the procedure. The Commission's proposal is considered as a 'unique piece of legislation' which contains many instruments, these instruments and/or their magnitude form the base for the intermediates and final compromises which characterize the development of the EU decision-making process.

The results of the analysis, even if preliminary and incomplete, showed that in the case of sugar, the main activity of the Commission has been developed in the definition of the proposal. We can thus identify this step of the procedure as the main institutional channel inside the consultation procedure in which the Commission can exert its active role in the political process which runs in parallel with the procedural and technical process.

Going into more depth in the sugar case study, Member States can be divided into three main groups. Two of them, namely group 2 and 3 are in favour of an intervention, whilst group 1 raised serious reservations about the proposed reform. The first two groups' positions are close to the Commission's proposal and also those who obtain a reform which encompasses their preferences. Group 1 has to accept the reform and try to find some sort of compensation inside the Council. The Commission played a relevant role of coordination in the specific case, accurately shaping the legislative proposal in order to gather the highest number of Member States. With this operation the Commission, in the case of sugar, succeeded in safeguarding the concerns and views expressed in the early stage of the procedure.

References

Putnam, R. 1988. 'Diplomacy and Domestics Politics: The Logic of Two Level Games'. *International Organization* 42.

Putnam, R., Evans P. and Jacobsen, H. 1993. *Double Edged Diplomacy*. University of California Press, Berkley.

Cavallo, A. Correani, L., Iagatti, M. and Sorrentino, A. 2007. E*U Agricultural Policy Bargaining and Domestic politics: An evolutionary Game Model: Global Conference 'Agricultural Policy Changes: Canada, EU and the World Trade Organization'* University of Victoria, Victoria B.C., 13-15 September 2007.

PART 3
The CAP Reform and Its Impact on EU Agriculture

Chapter 8

Implementation of Single Area Payment Scheme in the EU New Member States[1]

Sophia Davidova

Introduction

The potential impact of the 2003 CAP reform on the EU New Member States (NMS) depends on a range of factors, including the modalities of its implementation and the structure of the farming sectors in different countries. The EU acceding countries were given the choice of implementing the standard EU single payment scheme (SPS) or the simplified single area payment scheme (SAPS). Initially SAPS was supposed to be in force until 2008, but the CAP Health Check extended it until the end of 2013, acknowledging its efficiency (Council Regulation (EC) 73/2009). Those NMS which opted for the standard single farm payment (SFP)[2] had to implement a flat rate regional model. At the same time, NMS had to discontinue most of their pre-accession support measures though some of these were allowed to continue for a fixed transitional period (World Bank 2007).

Although the choice of ten out of twelve NMS was similar, namely to implement SAPS, the pre-accession policies, the structure of agricultural output, the distribution of farm sizes and the minimum threshold for eligibility for direct payments have created substantial differences in the distribution of support and the choice between de-coupled and coupled complementary national direct payments (CNDP), widely known as top-ups.

This chapter has two objectives: first, to present a comparative overview of the level and distribution of direct payments across seven NMS covering different sub-regions, i.e. the Baltics, Central and Eastern Europe, and the two most recent Member States from the Balkans – Bulgaria and Romania, and to discuss some reasons for the observed differences; and second, to provide some insights into the potential alterations of farmers' strategic plans as a result of the policy switch from the national pre-accession policies to CAP in order to enter into the debate

---

1   The author is grateful to N. Ivanova (UNWE), T. Medonos (VUZE), N. Levente and J. Popp (AKI), A. Miglavs (LSIAE), E. Stonkute and R. Zemeckis (LAEI), D. Milczarek and T. Wolek (UW), M. Luca (MAFRD), L. Luca (IAER), G. Blaas (VUEPP), L. Juvancic (ULJUB) who provided information and explanations about the national modalities of SAPS implementation.

2   Slovenia and Malta.

about the potential implications of CAP implementation for the structural change in NMS agriculture.

For the sake of simplification and the provision of a comparative cross-country overview, all direct payments coming from EAGGF Guarantee and CNDP, irrespective of whether the latter have been allocated in a decoupled way per hectare or were coupled to certain products, were summed up and expressed per hectare based on the area on which they have actually been allocated, per beneficiary based on the number of beneficiaries who have actually received the payments and per Annual Work Unit (AWU) taking into account the total number of AWU – family and non-family, regular and casual. Data were compiled from several EU publications, the OECD, as well as from national sources. The Central European Bank annual reference exchange rates were used whenever necessary to convert the national top-ups into euros.

Some differences with the amounts of single area payment (SAP) published elsewhere may have occurred due to: i) some preliminary data and forecasts from national sources concerning CNDP; ii) the difference in the accountancy procedure on a cash flow or an accrual basis – often CNDP are accounted for on a cash flow basis, while SAP coming from EAGGF Guarantee is accounted for on an accrual basis; iii) the attempt to use the area on which SAP has actually been allocated instead of the eligible area.

The chapter is structured as follows. The next section presents a comparative picture of the level and distribution of SAP. This is followed by a discussion based on results from survey work undertaken within the EU FP6 project 'Impact of Decoupling and Modulation on the Enlarged Union' (IDEMA) investigating how farmers plan to respond to SAPS implementation. In the last section some conclusions are drawn.

SAP levels and distribution: a comparative picture

The foreseen accession led to contradictory predictions concerning its impact on the structural change in agriculture of the acceding countries. Apart from budgetary considerations, what was a central issue in the pre-accession discussions was the extent to which the application of 'full' CAP support would be counter-productive and might impede the progress of the unfinished reform process in the acceding countries from Central and Eastern Europe (Ciaian and Swinnen 2006). This has been used as one of the arguments for the gradual phasing-in of direct payments. As a 'sweetener', the top-ups were introduced for a limited period while the payments funded by the EU budget would build up to the EU level. However, it was explicitly agreed that the total payments from EAGGF Guarantee and CNDP should not exceed the level in the EU-15.

The provision for CNDP introduced differences in the implementation of SAPS across NMS, but at the same time acknowledged the need for flexibility in order to address the diversity of the agricultural sectors in the NMS. NMS were allowed to top-up by 30 per cent the SAP funded by the EU or to use the pre-accession production coupled support and to increase its level by 10 percentage points on a product by product basis (for Slovenia, as a result of the negotiated incremental increases, these percentage points were allowed to reach 25 in 2007). Thus, the modalities concerning CNDP meant that countries with more distorted pre-accession support were given the opportunity to continue inducing intra-sectoral distortions.

Differences in SAP level per hectare

Figure 8.1 presents the level of SAP per hectare. As expected, due to the phasing-in process, there has been a trend towards an increase of the payment over time. However, this increase has not been uniformly manifested. For example, whilst during 2005-2007 payments per hectare in the Czech Republic increased by EUR44 and in Poland by EUR33, in Hungary this increase was only EUR8. In Hungary, CNDP remained below the ceiling due to budgetary constraints (OECD 2007).

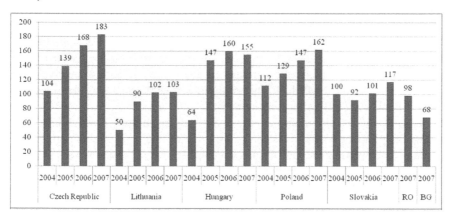

Figure 8.1   SAP per hectare 2004-2007 (EUR/ha)

Source: Own calculations based on data from national sources and Agriculture in the European Union – Statistical and economic information, 2009.

The first important source of differences in the level of SAP have stemmed from the different single payment ceilings and eligible land. Countries with higher production of CAP supported products were allocated relatively larger financial envelopes (World Bank 2007). For most of the analysed countries there is a neat correspondence between the pre-accession support and the single payment ceiling per hectare. The two exceptions are Romania, where in relation to the pre-accession support (the Romanian Government increased the level of domestic support in expectation of accession) the ceiling is relatively low, and Slovakia where the relationship is in the opposite direction.

In addition, some of the NMS have negotiated transitional arrangements which allowed them to give coupled payments to their farmers over and above CNDP. For example, during the period 2004-2008 Latvia was allowed to provide degressive state aid for fibre flax, milk and dairy, pigs, sheep and goats, and seeds (Official Journal 23/09/ 2003, Annex II). Thus, despite the 'common' rules, the pre-accession support and the negotiated transitional support arrangements have introduced substantial diversity in the level of direct payments that could be granted to farmers in the different NMS.

The extent to which individual countries have been able to grant CNDP has depended on the national budget constraints and on the decision to use (or otherwise) rural development funds to top-up the SAP. According to the Treaty of Accession, for a three year period rural development allocations could be used for top-ups. However, the amount from the EU rural development funding was capped at 20 per cent of the CNDP. For the three years following their date of accession all countries, with the exception of Hungary and the Czech Republic, used some rural development funds to increase Pillar I support (from 9 per cent in Latvia to the maximum allowed 20 per cent in Poland, Slovakia and Lithuania).

The option to allocate some of the CNDP in a coupled way has been another source of complexity and diversity. It has raised a debate about whether CNDP were supposed to maintain the incomes of some farmers at the pre-accession level or whether the NMS had more ambitious objectives about farm incomes (World Bank 2007). By allocating some of CNDP in a coupled way, NMS tried to minimise the redistribution of support from their sub-sectors which had been more highly supported pre-accession to those which had been less supported. The decision on how to allocate CNDP has varied widely across the NMS. For example, Bulgaria initially decided to allocate all CNDP on a per hectare basis and only to implement transitional soft fruit payments. During the first year of accession, Romania limited the coupled payments to fibre flax and hemp, hops, tobacco and a separate sugar payment (Cionga and Luca 2008). Slovakia has also employed a relatively simple scheme for CNDP allocation, whilst Hungary adopted one of the most complex systems (World Bank 2007). However, since the time of accession the policy has changed and in 2009, for example, Bulgaria introduced a coupled payment for sheep and goats, and in 2010 introduced two more coupled CNDP for suckler cows and a slaughter premium. In 2010 Romania decoupled all CNDP in the crop sector and left coupling only for bovine animals, and sheep and goats. Therefore post-accession, depending on their market situation and the efficiency of the sectoral farmers' lobbies, NMS have altered the ratio between coupled and decoupled top-ups and have changed the list of products benefiting from coupled payments.

In summary, the strategic national decisions on whether or not to alter the income distribution in comparison to the pre-accession period and whether or not to put more emphasis on Pillar I in comparison to Pillar II have acted as important sources of differences in the support levels and support distribution amongst NMS.

Differences in SAP level per Annual Work Unit and beneficiary

The state of technology and relative prices, together with the structure of agricultural output affect the labour intensity of production. The latter brings about additional differences amongst the NMS, particularly in the level of SAP per AWU (Table 8.1). In 2007, three of the seven studied NMS had labour intensity, measured in AWU to land ratio, of more than 0.1/ha – Bulgaria, Poland and Romania. It is not surprising that these countries have much lower amount of SAP per AWU. For example, in 2007 the payment in Poland was EUR1,005 whilst the level in the Czech Republic was EUR4,668. A more fundamental and important difference between the two countries is that whilst the labour intensity in the Czech Republic has gradually decreased post-accession, in Poland it has been stagnating at around 0.14 AWU per ha.

Table 8.1   SAP per AWU, beneficiary and SAP distribution, 2007

	Bulgaria	Czech Republic	Hungary	Lithuania	Poland	Romania	Slovakia
SAP per AWU (EUR)	354	4.468	1,925	1,478	1.005	414	2.365
SAP per beneficiary (EUR)	2796	30.349	3,921	1,254	1.565	747	14.101
Average size of beneficiary (ha)	41.2	166	25	12	10	8	121
SAPS beneficiaries as a share of total number of holdings (%)	16	54	32	92	61	31	22
Share of SAP smaller than EUR1,250 in the total amount of SAP (%)	na	1.6	13.2	43.2	51.4	na	2.5
Share of SAP larger than EUR100,000 in the total amount of SAP (%)	na	53.5	25.2	4.7	3.6	na	42.5
Share of beneficiaries receiving SAP smaller than EUR1250 (%)	na	40.3	76.7	93.0	90.4	na	74.2
Share of beneficiaries receiving SAP larger than EUR100,000	na	4.3	0.3	0.0	0.0	na	2.2

Source: Own calculations based on data from national sources; Agriculture in the European Union – Statistical and economic information, 2009.

The large variations in the farm structure and farm size distribution across the NMS have substantial implications for several aspects of SAPS implementation and its beneficiaries, including the amount of payment per beneficiary, the average size of a beneficiary, the share of holdings which have been included in the scheme, the segment of non-beneficiaries and the administrative costs brought about by SAPS implementation. Data on the average size of holdings and the share of large holdings above 50 hectares (ha) per country are presented in Table 8.2.

Table 8.2   Farm structure in New Member States, 2007

	Bulgaria	Czech Republic	Hungary	Lithuania	Poland	Romania	Slovakia
UAA/holding (ha)	6.2	89.3	6.8	11.5	6.5	3.5	28.1
Agricultural land per holding larger than 1 ESU (ha)	21.1	134.6	26	25	12.3	8.4	143
UAA in holdings larger than 50 ha %	81.9	92.7	74.7	46	24.3	40	92.9
Holdings larger than 50 ha in the total number of holdings (%)	1.3	16.7	1.9	3	1	0.4	4.2

Source: Agriculture in the European Union – Statistical and economic information, 2006; Eurostat, Farm structure survey 2007, Statistics in Focus for the respective countries.

Those NMS with more land use concentrated in large farms (most of them corporate farms, i.e. farm companies or producer co-operatives) are the Czech Republic and Slovakia. Hungary and Bulgaria occupy the middle ground, whilst Lithuania, Poland and Romania are characterised by small size farms almost across the board, although in the last two countries there are some very large farms as well (e.g. in Romania there are 38 holdings operating more than 5,000 ha each, Cionga and Luca (2008)). These structural differences affect the amount of SAP per beneficiary.

In the Czech Republic and Slovakia the amount per beneficiary is substantial, thus, it may generate a more important impact on farm incomes in comparison to the other NMS (Table 8.1). Moreover, the average amount disbursed per beneficiary in the Czech Republic has increased substantially, from EUR25.8 thousand in 2005 to EUR30.3 thousand in 2007, whilst during the same period the level in Slovakia has been almost flat. The differences with the other NMS are not only due to the fact that the Czech Republic and Slovakia have granted higher support per hectare, but mainly to the much larger average size of the beneficiaries, 166 ha in the Czech Republic and 121 ha in Slovakia, which is in stark contrast with the average size of a beneficiary in Romania – 8 ha.

The existing structural differences resulted in a widely different concentration of SAP beneficiaries. The SAP beneficiaries represent a relatively small segment of the existing farm holdings in Bulgaria and Slovakia, taking into consideration the number of holdings as reported by the Farm Structure Surveys including holdings smaller that 1 European Size Unit (ESU), whilst in Lithuania they cover 92 per cent of the holdings (Table 8.1). However, the concentration of SAP has been gradually decreasing together with the increasing farmers and administration learning curve for applications for payment, and processing and monitoring of the claims. For example, between 2005 and 2007 the share of beneficiaries in the Czech Republic increased from 44 per cent to 54 per cent, in Hungary from 28 per cent to 32 per cent and in Poland from 56 per cent to 61 per cent. Of course, there are many more very small farms that are not covered in the farm structure surveys and are not eligible for SAP as they are below the set thresholds. It could be argued that this particular segment exemplifies most explicitly the specificity of agriculture and rural areas in the NMS in comparison to the EU-15. For example, in Romania around 3 million household farms are not eligible for SAP as they do not fulfil the eligibility criteria – one hectare of land with min 0.3 ha of parcels (Giurca 2008). From this point of view, SAP in the NMS has a more skewed distribution than what appears as a result of analyses based on data from farm structure surveys.

An interesting comparison for illustrating the effect of concentrated or fragmented farm structure is to analyse the share of small payments of up to EUR1,250 per beneficiary and large payments of EUR100,000 and above per country. Whilst in 2007 small payments accounted for 43.2 per cent and 51.4 per cent of the total SAP in Lithuania and Poland respectively, their share was only 1.6 per cent and 2.5 per cent of the amounts in the Czech Republic and Slovakia (Table 8.1). Concerning large payments, they absorbed 53.5 per cent and 42.5 per cent of SAP amounts in the Czech Republic and Slovakia, but only 4.7 per cent and 3.6 per cent of those in Lithuania and Poland respectively. All of the above suggests that policy proposals to set payment limits, or to include eventually the NMS in a progressive modulation may have very different effects on different NMS and could generate substantially different political positions. The divergent interests of national farming communities and policy-makers that stem from the large farm structural variations among NMS may change, but only in the long term in parallel with the process of structural change.

Potential impact of SAPS implementation on farmers decisions[3]

Two of the above analysed NMS were chosen for further analysis owing to their largely different farm structure and SAP distribution; Slovakia and Lithuania. A sub-sample of holdings included in Farm Accountancy Data Network (FADN) was surveyed to understand whether farmers would change their strategic decisions as a result of SAP implementation. As the main interest of the research was to study the potential impact of SAPS, the respondents were advised that price changes should not be taken into consideration. The study focused on the strategic decisions of farmers, namely whether to exit or to remain in farming in a five-year and ten-year time horizon and, for those staying, on their plans about whether or not to increase their farm area within the next five years, to change the output mix or to diversify outside agriculture. Farmers were asked to state their intentions under different policy scenarios corresponding to a counterfactual continuation of the pre-accession policy, the actual SAPS implementation with the modalities of each country CNDP, and a hypothetical scenario of full-decoupling based on area payments and the elimination of CNDP. In Lithuania, 227 individual farms were surveyed and in Slovakia 154 farms. In addition, in Slovakia 152 corporate farms (101 producer cooperatives and 51 farming companies) were covered, since due to their size they are major SAP beneficiaries.

Farmers' plans are influenced by the changes in the level of support and the redistribution of support as a result of SAPS. During the pre-accession period, both countries implemented CAP-like coupled direct payments. Although the introduction of SAPS has increased the level of support, there were clear winners and losers from the policy change. In Slovakia, the main beneficiaries were crop farmers. The support per hectare of arable land has tripled as a result of accession, but livestock farmers have lost from the redistribution. Post-accession, only the suckler cow premium, and the sheep and goat premium were maintained (Blaas and Bozik 2007).[4]

In Lithuania since the first year of accession, SAPS has brought about an increase in payments for cereals and protein crops, as well as an introduction of new payments, e.g. for starch potatoes and grassland. However, there were two obvious losers from the policy change – the producers of flax for fibre and linseed. Payments for most livestock producers have also increased but there were variations

---

3   This part summarises results of research presented in Douarin et al (2007) IDEMA Deliverable 14.

4   Since 2007 in order to revitalise the livestock sector cattle, milk, sheep and goat were integrated in Large Cattle Units (LCU) and the support was distributed as payments for LCU (Blaas and Duricova 2008).

depending on the level of direct payments pre-accession and the stocking density. Overall, farmers who have benefited most from the policy change were the arable crop producers and the producers of previously unsupported crops.

As mentioned, the most important debate around the implementation of the CAP in NMS has been in relation to its effects on structural change. Although decoupled, per hectare payments might impede structural change as, first, they provide on average higher support and thus incentives to stay in farming, and second, they tend to increase land values (World Bank 2007). On the other hand, SAPS provides flexibility for farmers to leave their land out of production and keep it in good agricultural and environmental conditions (GAEC), and invest in non-agricultural activities that bring higher returns, thus developing alternative income sources and contributing to rural diversification. The results presented below can feed the debate about structural change and farm diversification under SAPS – at least in mid-term.

Farmers' intentions to respond to SAPS introduction

Individual farmers
In Lithuania, there is no evidence that farmers intend to change their exit decisions in response to the policy change (Table 8.3). However, they plan to grow much more under the CAP payments, particularly under the actually applied SAPS modalities with coupled top-ups, than under the pre-accession policies.

Table 8.3   Farmers intentions to stay in farming and grow in Lithuania and Slovakia (individual farms only)

Policy scenarios	Lithuania	Slovakia
Exit from agriculture within 5 years (%)		
Pre-accession policy	18.0	38.0
SAPS plus CNDP	17.0	18.0
Fully decoupled payments	16.0	24.0
Share of farmers intending to increase the cultivated area (% of those who will stay in agriculture within 5 years)		
Pre-accession policy	21.0	34.0
SAPS plus CNDP	47.0	51.0
Fully decoupled payments	31.0	26.0

Source: Douarin et al. (2007).

In Slovakia, the survey evidence supports studies arguing that the CAP introduction might impede structural change (Ciaian and Swinnen 2006). Under the CAP payments, either with or without top-ups, more farmers would like to stay in agriculture within the next five years and are willing to increase the size of their farms than under the pre-accession policy.

In Lithuania, under the continuation of the pre-accession policies, the main constraint to growth appeared to be farm leverage.[5] In general Lithuanian farms included in FADN have the lowest net worth amongst EU-25 (EC 2008). In our survey sample several farms had negative leverage. This constraint was no longer significant under SAPS. This is an important insight about the potential coupled effect of SAPS. Although SAP is decoupled (*ex ante*) it may still have (*ex post*) effect on farm investment and thus farm expansion (Latruffe et al. 2010). In the case of perfect credit markets, transfers through decoupled payments should not affect farm investment and production by changing liquidity constraints and farm creditworthiness. However, this is not the case under imperfect capital markets when farms are credit constrained. In NMS where imperfect credit markets constrained farmers' decisions prior to accession (see e.g. Latruffe 2005, Petrick 2004, Davis et al. 2003), SAPS is expected to improve the access to credit and stimulate farm investment.

In Lithuania, farm size has not indicated any statistically significant relationship either with exit or growth under any of the policy scenarios. The different SAP distribution in Slovakia and the existence of much larger holdings resulted in different considerations behind farmers' plans about how to respond to the policy change. In Slovakia, concerning the intentions to exit under SAPS (with or without top-ups) the only factor that had a significant influence was the farm size (measured in farm area) before the policy change. While under SAPS the larger farms indicated a lower probability to exit within five years, the farm size did not appear significant for exit decisions under the continuation of the pre-accession policy. This suggests that the policy change has provided incentives for large farmers to stay in agriculture.

Concerning the change in land use as a response to the new policy, there was no evidence that either Lithuanian or Slovakian farmers intended to keep land out of production in GAEC. Just the opposite, respondents indicated that they wanted to reduce the land left out of production. These intentions indicate that the increase in the level of payments in NMS has acted as an incentive towards more production irrespective of whether the payments were allocated in a fully or partially decoupled way. However, whilst in Lithuania SAPS without requirements for production has not provided incentives to invest outside agriculture, Slovakian farmers indicated their intentions to invest in diversification activities, especially under the hypothetical removal of the coupled element in CNDP.

---

5   Probit was employed to investigate the determinants of exit and growth under the different policy scenarios, including a set of characteristics affecting the utility derived by farmers from the different policy alternatives and a set of farmers' characteristics.

Corporate farms

As mentioned previously, owing to their importance in the farm structure in Slovakia, 152 corporate farms were surveyed. The respondents for the corporate farms were their managers or in some cases directors or accountants. Since the farms were large (on average 1,866 ha UAA) the picture concerning the size changes was more complex. The most intensive adjustments to farm size appeared to be stimulated by the SAPS actually implemented plus top-ups. This tendency is similar to the one observed for individual farms. Of the respondents, 34 per cent intended to increase their land area under SAP plus top-ups in comparison to only 9 per cent under the pre-accession policy. One important difference between the corporate and individual farms was the response to the opportunities for non-agricultural investment stemming from payments with no obligation to produce. If 10 per cent of corporate farms planned to diversify outside agriculture under the pre-accession policy, this percentage increased to 19 under SAP plus coupled top-ups and reached 21 under the fully decoupled scenario. This suggests that the corporate farms that are the core of competitive agriculture in some NMS might also turn into drivers of rural diversification.

To summarise, two main conclusions could be drawn as a result of the survey data analysis. First, as most farmers would like to increase their farm area there will be increasing competition for land and therefore an increase in land values. Second, since there are no clear signals that NMS farmers treat SAP as decoupled, there might be an increase in farm investment.

It is interesting to see whether such developments have been observed in practice in the first post-accession years. It is not possible to disentangle the effects of SAPS without an adequate modelling tool, thus the observed short-run changes reflect the simultaneous influence of a number of factors, including weather, farm prices and input costs. Most of all, it is difficult to disentangle the effects of SAPS from the general effect of accession. However, some anecdotal evidence supports the expectations formulated on the basis of the survey. There are reports that the availability of credit to farmers and farm investment have increased substantially (Blaas and Bozik 2007). For example, between 2003 and 2005 corporate farms in Slovakia doubled the investment per hectare, mainly investing in machinery and equipment.

Land values have increased almost everywhere. Land price statistics are still unsatisfactory with important gaps, but despite the differences in the reported levels Swinnen and Vranken (2008) confirm the general tendency towards increases in land sale and rental prices in NMS. Some commentators attribute the increase in land values more to the healthy general economic climate during the first post-accession years than to the effect of the SAPS.

Conclusions

This chapter has provided an overview of the level and distributional aspects of SAP across seven NMS and presented analytical results about the change (or lack of change) in farmers' strategic plans as a result of the implementation of SAPS in two NMS with contrasting farm structures and payment distribution – Lithuania and Slovakia. Several conclusions can be drawn from the analysis.

Despite the common rules, the level of support granted to farmers in NMS varies substantially. NMS with more distorted pre-accession support were given the opportunity to continue inducing intra-sectoral distortions. The structural differences and the strategic national decisions on whether or not to alter income distribution in comparison to the pre-accession period and whether or not to put more emphasis on Pillar I in comparison to Pillar II were one of the important sources of differences in the support levels and support distribution. In some countries, e.g. the Czech Republic and Slovakia, payments per beneficiary are substantial and thus have an important impact on farm incomes.

The distribution of support varies enormously from highly concentrated on large beneficiaries to much more clustered towards the small ones (e.g. Slovakia and the Czech Republic versus Lithuania). This suggests that policy proposals to set higher eligibility thresholds in order to increase administrative efficiency, to set payment limitations, or to include the NMS in a progressive modulation might have very different effects on farming in different NMS and could generate substantially different political positions.

The implementation of the SAP in the NMS means higher and more predictable payments. Therefore, it is not surprising that the results from the survey of farmers' plans indicate that SAPS increases the willingness to stay longer in agriculture and to operate larger farms. These farmers' intentions suggest that the implementation of SAPS might impede structural change. Moreover, due to the increased returns to agricultural activities under SAPS, farmers have not indicated any interest in leaving land out of production and keeping it under GAEC. As farmers want to grow, the implementation of the SAP might lead to greater agricultural land utilisation and an increase in the demand for land.

References

Agriculture in the European Union – statistical and economic information, 2009 available at: http://ec.europa.eu/agriculture/agrista/2009/table_en/index.htm [accessed 26 May 2010]

Blaas, G. and Duricova, I. 2008. *Report of the Slovak Republic on monitoring and evaluation of agricultural policies of the OECD Member States, 2007.* Bratislava: Research Institute of Agricultural and Food Economics.

Blaas, G. and Bozik, M. 2007. The state of agriculture in Slovakia – three years after EU accession, in *Changes in the Food Sector after the Enlargement of the*

EU, Warsaw: Institute of Agricultural and Food Economics-National Research Institute, 122-145.

Ciaian, P. and Swinnen, J. 2006. Land market imperfections and agricultural policy impacts in the new EU Member States: a partial equilibrium analysis. *American Journal of Agricultural Economics* 88(4), 799-815.

Cionga, C. and Luca, L. 2008. *Farm income support and investment incentives – underlying factors in agricultural sector's economic convergence.* Paper to the *Proceedings of the Romanian Conference on Growth, Competitiveness and Real Income Convergence edited by V. Lazea.* April 21st-22nd 2008, Bucharest.

Council Regulation (EC) No 73/2009. *Official Journal of the European Union*, L30/16 31 January 2009.

Davis, J., Buchenrieder. G. and Thomson, K. 2003. Rural credit for individual farmers: a survey analysis of rural financial markets, in *Romanian Agriculture and Transition toward the EU,* edited by S. Davidova and K. Thomson: Lexington Books, Lanham/Boulder/New York/Oxford, 45-56.

Douarin, E., Bailey, A., Davidova, S., Gorton, M. and Latruffe, L. 2007. *Structural, locational and human capital determinants of farmers' response to decoupled payments,* EU FP6 IDEMA, Deliverable 14 available at http://www.sli.lu.se/IDEMA/publications.asp [accessed 26 May 2010]

EC DG for Agriculture and Rural Development 2008. *EU farm economics overview report 2005*: Brussels.

Eurostat *Farm Structure Surveys 2005 and 2007, Statistics in Focus,* available at: http://epp.eurostat.ec.europa.eu/portal/page/portal/eurostat/home/ [accessed 27 May 2010]

Giurca, D. 2008. Semi-subsistence farming in Romania: prospects for catching-up in *Economic Growth and Convergence, Proceedings of the Romanian Conference on Growth, Competitiveness and Real Income Convergence*, edited by V. Lazea, Bucharest, April 21st-22nd 2008, .

Latruffe, L. 2005. The impact of credit market imperfections on farm investment in Poland. *Post-Communist Economies*, 17(3), 349-362.

Latruffe, L., Davidova, S., Douarin, E. and Gorton, M. 2010. Farm expansion in Lithuania after accession to the EU: the role of CAP payments in alleviating potential credit constraints. *Europe-Asia Studies*, 62(2), 351-365.

OECD 2007. *Agricultural policies in OECD countries: monitoring and evaluation*: Paris.

Official Journal of the European Union L236, 46. 23 September 2003 Act of accession 2003. Annex II agriculture.

Petrick, M. 2004. A microeconometric analysis of credit rationing in the Polish farm sector. *European Review of Agricultural Economics*, 31(1),77-101.

Swinnen, J. and Vranken, L. 2008. Review of transitional restrictions on acquisition of agricultural land in NMS. CEPS, available at: http://ec.europa.eu/internal_market/capital/docs/study_en.pdf [accessed 26 May 2010]

World Bank 2007. *EU8+2* Regular Economic Report, Part II: Special Topic, May.

Chapter 9

Assessing European Farmers' Intentions in the Light of the 2003 CAP Reform

Margarita Genius and Vangelis Tzouvelekas

Introduction[1]

The agricultural sector in the European Union is undergoing a big change at present. In effect, while the sector has traditionally been the greatest beneficiary of domestic support, it is trying to adapt to a new situation where farm payments are decoupled from farm production. The rationale behind the 2003 CAP reform was to put an end to the overproduction of agricultural products that was the result of the substantial support payments European farmers were receiving and to enhance the competitiveness of farmers. Under the lemma "farmers should produce what the markets demand" the reform established a single farm payment scheme whereas payments are not linked to production anymore (therefore eventually reducing the incentives to overproduce) but they are linked to area and historical payments. In fact, farmers could decide not to produce anything at all and still receive the single farm payment. In addition, the reform introduced the concept of cross-compliance making the receipt of farm payments contingent on "good practice" with respect to environmental, food safety and animal standards. The new reform shifted the weight from direct aids (Pillar I) towards rural development measures (Pillar II). Rural development measures evolve around three main axes targeting mainly the competitiveness of the agricultural sector, the environment and the quality of life in rural areas. It is expected that the new reform will bring about changes in the employment levels of rural areas. On the one hand, the adoption of decoupled payments –by reducing the incentives to overproduce- could lead to some farmers abandoning farming activity and therefore a reduction of hired labour in some farms. On the other hand, some of the rural development measures could act as an incentive to young farmers to get involved in agricultural production and other economic activities. It was not clear at the outset what the overall effect of CAP reform on employment levels and on farmers production decisions will be (see

1   The present paper is based on the study undertaken under the EU Carera program (contract no. SSPE-CT-2005-022653). We are indebted to our partners in Wageningen University, Corvinus University of Budapest and Aristotle University of Thessaloniki for administering the surveys and providing us with the statistical data on the three Regions.

Hennessy and Thorne 2005, Breen et al. 2005, Anton and Sckokai 2006, Gohin 2006).

The present study tries to shed some light on the above issue by trying to elicit the future plans of farmers in the EU from the information given by the farmers themselves. Three sets of surveys were conducted in the summer of 2007 in three different regions of the EU, namely the regions of Anatoliki Makedonia-Thraki in Greece, Flevoland in Holland and the Southern Great Plain in Hungary. The surveys focussed on farmers' future intentions with respect to their farming activity and targeted crops/livestocks were specific to each region. An econometric model that takes into account the sequential nature of the decisions to abandon, to change production volume and/or crop mix is estimated with the data.

The questionnaire also contemplates three different price scenarios for the price of agricultural products to take into account possible future price fluctuations.

The survey results show that when farmers are not given information about future prices, then a substantial proportion (38.1 per cent) will opt for abandoning farming activity in Greece, the corresponding figures being much lower for Holland (12.9 per cent) and for Hungary (9.2 per cent). Of those farmers who report they will not abandon, the majority will keep the same level of production and the same crop mix in the three countries. Farmers in Greece and Hungary are much more responsive to changes in prices than Dutch farmers. Small farms are more likely to abandon in the case of Greece and Hungary while the contrary holds for Holland. More specialized farms are more likely to abandon production in Greece and Hungary while in Holland they are less likely to do so. Moreover, once a farm decides to continue with farming, the more specialized it is, the less likely it is to increase production in the case of Greece and Hungary but the more likely it is to increase production in the case of Holland. One striking difference among the three countries is the level of information farmers claim to have about CAP reform, with Hungarian farmers being much less informed than the other two groups, and the Dutch farmers being the better informed ones.

The organization of the chapter is as follows: the next section briefly describes the three regions that were chosen for the study, section three offers a description of the survey and data, the econometric model applied to the data is developed in section four, the estimation results are presented in section five while section six concludes the chapter.

The Regions

The Region of Anatoliki Makedonia and Thraki

The region of Anatoliki Makedonia and Thraki comprises the eastern part of Greek Macedonia along with Greek Thrace. It has a land area of 1,403,400 ha (11 per cent of the total land area of Greece) and borders in the west with the region of Kentriki Makedonia, in the north with Bulgaria and in the east with Turkey.

According to the data published by the National Statistical Service of Greece (NSSG), the share of agriculture in the gross added value is calculated to be 20 per cent of the total regional gross added value, against 8.8 per cent nationally. The region produces 10 per cent of the output of the national rural sector, 4.4 per cent of the total national transportation and 3.5 per cent of the total national services sector (NSSG 2001). It also occupies the penultimate place in the classification of regions based on per capita production, with EUR10,200 in 2002, representing 79 per cent of the national average per capita production and 58.6 per cent of the EU-25 average. The primary sector occupies 96 per cent of the total regional land area. Forests and wooded areas cover 53 per cent of the region's total area, whilst the national average is 49.7 per cent. The main cultivated products are cereals (mostly wheat), tobacco, cotton, tomatoes, potatoes, olive oil and apples. Milk production is also important due to the regional development of livestock farming.

The agricultural sector is the leader in terms of employment although its share in total employment has been declining, going from 34.3 per cent in 2000 to 23.7 per cent in 2004 (NSSG 2005). Of some importance is the manufacturing sector whose share has gone from 11.9 per cent to 13.6 per cent in the same time period and mainly comprises food processing and tobacco industries.

The unemployment rate is higher than the national average as it was around 11 per cent in 2006 (NSSG 2006) and with a much higher incidence of female unemployment.

The Region of Flevoland

Flevoland is the youngest province of The Netherlands and is an area that has been reclaimed from the sea.

The most important sector that provides jobs is the business activities sector (21.1 per cent) followed by the trade sector (19 per cent), the health and welfare sector (11.9 per cent) and the manufacturing sector (10.6 per cent), while the agricultural sector provides 4.7 per cent of the jobs (Province Flevoland 2005).

According to the Central Bureau of Statistics data (CBS 2005), the arable acreage is mainly used to grow potatoes (26 per cent), arable vegetables, including onions (21.4 per cent) and cereals (19.2 per cent). The sugar beet contributes for 5 per cent to the total agricultural acreage and about 2 per cent of the agricultural production. Wheat crops cover 1,520,246 hectares of the agricultural area in Flevoland which is 6.9 per cent of the total acreage used for wheat in The Netherlands. It is worth noting that in Flevoland 7.8 per cent of the agricultural area is organic, while the national percentage is only 2.1 per cent.

The second largest type of agricultural holdings in Flevoland are grazing livestock farms (14.2 per cent of all holdings), including horses, cattle, sheep and goats. In Flevoland, dairy farming is the main livestock activity with a total of 46,759 dairy cows in 306 dairy farms representing 50.1 per cent of the total number of livestock farms.

In the 90's, the number of dairy cows in Flevoland decreased. However, the decline was lower than the national decline in dairy cows and therefore the share in the Dutch dairy stock increased from 1.5 per cent in 1990 to 1.7 per cent in 2000.

The participation of women in the total labor force is less than the participation of men in Flevoland and women account for a higher percentage in unemployment (almost 1 to 2).

The Region of the Southern Great Plain

The Southern Great Plain, the largest region in Hungary (19.9 per cent of Hungary's territory), is located in the south and south-east of Hungary. Most of the region's land (85 per cent) is suitable for agriculture and it is predominantly flat.

According to data provided by the Hungarian Central Statistical Office (HCSO 2005) the employment rate for people aged 15-64 in Hungary (56.9 per cent in 2005) corresponds to the average employment rate in the ten new member states. However, it is significantly lower than the rate of 64 per cent of the EU-15.

Although, the Southern Great Plain accounts for only 9 per cent of the total Gross Domestic Product (GDP) of Hungary, it accounts for 25 per cent of the agricultural GDP. Agriculture's share of the regional GDP was 9 per cent in 2002. The per capita GDP of the region is below the national average and up to 2003, the region had the lowest rate of growth in Hungary.

The Questionnaire, Data and Preliminary Results

The questionnaire was designed to elicit farmers' future intentions in the light of the recent CAP reform and the associated rural development measures. Emphasis was put on inferring the farmers' confidence level in their business and the possibility of future changes in their production activity in the light of the CAP reform. Future intentions about input use, labour use, size of business, investment levels and output diversification were addressed taking into account the respondent's socio-economic and demographic characteristics as well as the possibility of their succession in the business.

The questionnaire was divided in five parts. In the first part, information concerning the farming current activity level were collected including information on the location of the farm, the crop and livestock production, the ownership status, the use of labour and farmers' experience with CAP implementation in the past. An important part of this section refers to farmers responses to previous changes in agricultural policy in order to compare those with their intentions concerning the reform towards decoupling farm incomes. This part was excluded for Hungarian farmers as they did not have any previous experience with the CAP. The next section of the questionnaire dealt with farmers' level of information about the 2003 CAP reform as well as their sources of information. In part three the questionnaire focused on their perceptions about the 2003 reform. Specifically, farmers were asked to evaluate the anticipated changes in regional agricultural production, employment, off-farm occupation, women and young people involvement in farming business etc. Part four of the questionnaire was concerned with future intentions over a five-year time span. The focus of the exercise was to learn what the farmers' intentions were with respect to their continuing in the farming business and their future production plans both in terms of volume and crop mix. Each respondent was presented with a brief summary of the new SFP (Single Farm Payment) regime applying to his/her specific crops and country and then was asked to state his/her future intentions in the light of the new regime and under three different scenarios corresponding to no price information, a 10 per cent price decrease and a 10 per cent increase. For Greece the chosen crops were cotton and tobacco together with sheep breeding, sectors in which the region of Anatoliki Makedonia and Thraki is specialized. For the Netherlands, sugar beet, wheat and dairy farming were chosen as the most indicative activities in Flevoland. Finally, in the Southern Great Plain the sectors investigated were corn, fresh vegetables and pig production.

In addition, farmers were asked about the likelihood of introducing new crops, increasing their off-farm activities, increasing their capital investment, changing the amount of labour used in the farm as well as the possible uses of the SFP payments between investment and leisure. Finally, in the fifth section some important socio-economic characteristics were collected like the age and education of the head of the household, the number of family members, their experience in farming and off-farm income sources.

The initial sample of farmers was randomly selected using the information provided by the local agricultural directorates and it was 176 for Anatoliki Makedonia and Thraki, 191 for Flevoland and 225 for Southern Central Plains. Some of the collected questionnaires were incomplete as farmers were not able or willing to answer all the included questions. Therefore the final sample sizes used in the present study were 160 for Greece, 85 for Holland and 153 for Hungary. The big difference in the response rates among the three regions is due to the fact that in the case of Greece and Hungary data was collected through face-to-face interviews while in the case of Holland, telephone interviews was the chosen method. The average age of the respondents was 48 for Greece, 50 for Holland and 51 for Hungary while the average level of specialization differed across countries with the Herfindahl index being 0.76 for Greece, 0.4 for Holland and 0.46 for Hungary.

The analysis of farmers' responses by country reveals some interesting patterns. In the case of Greece, although almost one third of the farmers declare having participated in some structural program in the past, less than one third considers they have at least a fair level of information about the 2003 CAP reform, only 34 per cent have a fair knowledge of the requirements for direct farm supports while more than 30 per cent declare they are not familiar with the terms "single farm payment" and "cross-compliance".

This lack of information can be explained from the sources of information that farmers declared. The main source consisted of private agricultural extension agents that regularly visit their farms (78.4 per cent), other farmers (77.5 per cent) and from various media like TV or newspapers (73.3 per cent) and only 1.7 per cent mentioned the local authorities. The above response pattern reveals the lack or weakness of an organized information campaign from both local and central governmental authorities as well as from farm cooperatives. With regard to their perceptions about the future changes in their region initiated by the recent CAP reform, 89.4 per cent of the questioned farmers agree that the agricultural production in their region will decrease after the implementation of the new regime. With regard to the crop they cultivate the share believing production will decrease is also high (86.2 per cent). Only 29.3 per cent of the respondents believe that the new policy will increase job opportunities outside farming, whereas 39.4 per cent strongly disagree this will be so. Most respondents strongly disagree (83.1 per cent) that employment levels will increase in their specific farming activity in the next five years. Finally, only 2.5 per cent think that farm income arising from the specific crops that they cultivate will increase as a result of the decoupled farm payments. Almost 70 per cent of the farmers responded they will use the single farm payment for investment, around 20 per cent still don't know how they will use the SFP and the rest will use it for leisure. Of those farmers planning to invest the single farm payment, the majority intends to invest it in the farm.

On the other hand, for Holland only 1.2 per cent of the farmers have participated in a previous EU structural program, while all of them declare to be familiar with the term "single farm payment" and 93 per cent are familiar with the term "cross-compliance". Although only 28 per cent have a fair amount of information about the CAP reform, as many as 63 per cent know the requirements for direct farm supports. The opinion of the farmers about the decrease of agricultural production, the creation of new jobs outside the agricultural sector and the decrease of agricultural production for the specific farm activity, caused by the CAP reform, is not unambiguous. Most farmers disagree with the statements that the CAP reform will create new jobs in their specific farming activity (71.8 per cent) and that the income of farmers in their specific farming activity will increase (55.3 per cent). More than half of the farmers will use their single farm payment for investments (54.1 per cent), mainly inside the farms (97.8 per cent of those investing the SFP). Only 3.5 per cent of the farmers will use single farm payment for leisure. The remaining farmers do not know yet for what they will use the single farm payment.

In the case of Hungary, only 12 per cent of respondents declare to have at least a fair amount of information about the CAP reform and the corresponding figure for knowing the requirements for direct farm supports is 28 per cent. In addition 82 per cent of the farmers are not familiar with the term "cross-compliance" and 32 per cent with the term "single farm payment". Concerning their perceptions about the effects of the CAP reform, these are not very positive since 39.9 per cent believe agricultural production will decrease, while only 15.7 per cent think there will be new jobs created outside of agriculture. Almost half of the farmers will use their single farm payment for investment (47.1 per cent) while the rest still don't know, and all of the former are planning to invest inside the farms.

Table 9.1 below, shows the percentage of farmers, for the three regions, who declared that sometime in the next five years they will either abandon farming activity, or change their level or production or change their crop mix, under the three alternative price scenarios. As it appears on the table, Greek and Hungarian farmers appear to be very reactive to different price scenarios, with 62.5 per cent and 28.1 per cent, respectively, declaring their intention to abandon if prices were to decrease by 10 per cent, while the corresponding percentages under the scenario of a price increase are only 8.1 per cent and 5.9 per cent. If we consider the age distribution of respondents choosing the abandon option it turns out that for Holland, older farmers are a majority across scenarios while in the case of Greece and Hungary the percentage of young farmers greatly varies across scenarios. As far as changing their crop mix, the majority of farmers would keep the same mix, except for Greece under the scenario with increasing prices.

Table 9.1   **Future Production and Crop Mix Plans Under Three Scenarios for Future Prices.[a]**

	Abandon	Decrease Prod.	Same Prod.	Increase Prod.	Change Mix	Keep Mix
GREECE						
No info. Fut. prices	38.1	3.8	36.3	21.9	20.6	41.3
Decr. 10%	62.5	6.3	16.3	15.0	10.6	26.9
Incr. 10%	8.1	0.0	29.4	62.5	51.3	40.6
HOLLAND						
No info. Fut. prices	12.9	0.0	51.8	35.3	32.9	54.1
Decr. 10%	18.8	1.2	49.4	30.6	29.4	51.8
Incr. 10%	4.1	0.0	50.6	35.3	28.2	57.6
HUNGARY						
No info. Fut. prices	9.2	3.9	54.2	32.7	29.4	61.4
Decr. 10%	28.1	6.5	43.8	21.6	28.1	43.8
Incr. 10%	5.9	2.6	55.6	35.9	26.1	68.0

[a] The three scenarios are 1. No info: no information is given about future prices, 2. Decr. 10%: a 10% decrease in prices, 3. Incr. 10%: a 10% increase in prices

In order to further analyse the factors that affect the different choices of the farmers: abandon, change production level, change crop/livestock mix, we develop an econometric model, where the farmers face a sequential choice under the first scenario only. In the first step they choose whether to abandon or not and then those who choose to stay in business, are faced with the simultaneous choice of the production level and crop mix.

The Econometric Model

Each choice described above can be represented by an equation linking the "propensity towards a choice" or latent variable Y* to a set of characteristics of the farmer denoted by X.

First Equation: to abandon or not

$$Y_{1i}^* = X_{1i}\beta_1 + \varepsilon_1 \qquad (1)$$

Second Equation: acreage (livestock size) decision

$$Y_{2i}^* = X_{2i}\beta_2 + \varepsilon_2 \qquad (2)$$

Third Equation: crop (livestock) mix decision

$$Y_{3i}^* = X_{3i}\beta_3 + \varepsilon_3 \qquad (3)$$

In equations (1)-(3), Y_j^* (j=1,2,3) are the usual latent variables governing each decision and ε_j are stochastic terms representing possible factors that affect the farmers' decision but are not observed by the researcher. Since the latent variables are not observed we define the following three observable dichotomous variables:

$$Y_{1i} \begin{cases} 0, & Y_{1i}^* < 0 \\ 1, & Y_{1i}^* \geq 0 \end{cases} \qquad (4)$$

to represent whether a farmer plans to continue ($Y_{1i} = 0$) or abandon ($Y_{1i} = 1$),

$$Y_{2i} \begin{cases} 0, & Y_{2i}^* < 0 \\ 1, & Y_{2i}^* \geq 0 \end{cases} \qquad (5)$$

to represent whether a farmer plans to increase acreage/size ($Y_{2i} = 1$) or not ($Y_{2i} = 0$) and Y_{2i} is observed only when $Y_{1i} = 0$. The last observed variable gives us information about whether a farmer is planning to change the crop/livestock mix ($Y_{3i} = 0$) or not ($Y_{3i} = 1$) and is defined analogously as,

$$Y_{3i} \begin{cases} 0, & Y_{3i}^* < 0 \\ 1, & Y_{3i}^* \geq 0 \end{cases} \tag{6}$$

where once again the latter is observed only for respondents who answered "not abandon".

In order to allow for correlations among the three decisions, the three errors terms $\varepsilon_1, \varepsilon_2, \varepsilon_3$ are assumed to follow a trivariate normal distribution with zero means, unit variances and correlations $\rho_{12}, \rho_{13}, \rho_{23}$. The log-likelihood corresponding to equations (1) to (6) is given by:

$$
\begin{aligned}
logL_n = & \sum_n y_{1i} \Phi(x_{1i}\beta_1) \\
& + \sum_{n_1} (1 - y_{1i}) \{ \, y_{2i} y_{3i} P(\varepsilon_1 < -x_{1i}\beta_1, \varepsilon_2 \geq -x_{2i}\beta_2, \varepsilon_3 \geq -x_{3i}\beta_3) \\
& + y_{2i}(1 - y_{3i}) P(\varepsilon_1 < -x_{1i}\beta_1, \varepsilon_2 \geq -x_{2i}\beta_2, \varepsilon_3 < -x_{3i}\beta_3) \\
& + (1 - y_{2i}) y_{3i} P(\varepsilon_1 < -x_{1i}\beta_1, \varepsilon_2 < -x_{2i}\beta_2, \varepsilon_3 \geq -x_{3i}\beta_3) \\
& + (1 - y_{2i})(1 - y_{3i}) P(\varepsilon_1 < -x_{1i}\beta_1, \varepsilon_2 < -x_{2i}\beta2, \varepsilon_3 < -x_{3i}\beta_3) \, \}
\end{aligned}
\tag{7}
$$

Note that due to the sequential nature of the model, the first summation is taken over all respondents and the second over the respondents who do not abandon. Thus the first line of equation (7) describes the probability of abandoning, the second line the probability of not abandoning and increasing acreage/livestock size and keeping the same mix and so on. The computation of expression (7) involves the evaluation of trivariate integrals and therefore the GHK algorithm (Hajivassiliou et al. 1996) will be used to simulate the log-likelihood with 100 replications.

Estimation results

The econometric model presented in the previous section was estimated for the three regions separately, the likelihood ratio test was used to select the set of explanatory variables included in the estimated model and also to test correlations among the three equations. After several attempts and using different variables as explanatory ones in the trivariate probit model we ended up with the specifications presented in Table 9.2a, 9.2b and 9.2c while Table 9.3 gives a description of the variables used in the estimation. The signs of the coefficient estimates give us information about the direction –but not about the magnitude- of the effects of explanatory variables on the three different probabilities: to abandon, to increase production and to keep the same mix. With respect to the decision of abandoning the estimation results show that as expected, the closer a farmer is to retiring the more likely he/she is to abandon. The level of satisfaction with the current situation of farming business affects negatively the probability to abandon for Greece and Hungary. The higher the specialization of the farm the higher the probability that farmer will exit farming in the case of Greece and Hungary indicating the significant risks that farmers perceive about the future course of the sector in the light of CAP changes. It is the foremost important factor influencing the probability to abandon in Greece. For Holland the opposite holds with respect to specialization. In the case of Greece the level of information about the CAP reform and the previous experience with CAP structural programs lessens the adverse perceptions as it reduces the probability to exit the sector, while the latter factor contributes positively as well in the decision to increase production. However, small farms seems to be more vulnerable to changes as they have less opportunities to survive exhibiting a higher probability to abandon farming both in Greece and Hungary.

This is also supported for Greece by the parameter estimate of FFARMINC which is negative and statistically significant indicating that farms with high profitability (mainly of large size) are having a lower probability to abandon. Once again, we get an opposite effect in the case of Holland in the case of small farms. Finally, for Greece the age of the head of the household increases the probability of abandoning but the experience of the farmer (as measured by tenure) does not. The more experienced the farmer is, the higher is the possibility to adjust himself into the new environment and thus the less the probability to exit the business.

Table 9.2a   Estimation results: GREECE[a]

Parameter	Abandon		Production		Mix	
	Estimate	StdErr	Estimate	StdErr	Estimate	StdErr
CONSTANT	-2.2850	0.8656	2.6266	1.0173	0.1788	0.7774
PROBRET	0.0126	0.0043	-	-	-	-
SATISF	-0.7439	0.3145	-	-	-	-
PARSTRUC	-0.4429	0.3159	0.6971	0.3266	-	-
INFCAP	-0.6858	0.3606	-	-	-0.8499	0.3445
HDAGE	0.0274	0.0184	-0.0615	0.0190	0.0471	0.0213
FFARMINC	-0.0335	0.0164	-	-	0.0054	0.0054
SPEC	2.2843	0.6434	-1.4841	0.6322	-	-
TENURE	-0.0300	0.0191	-	-	-0.0253	0.0228
SIZLO	0.5820	0.4122	1.0850	0.8353	-	-
PINVT	-	-	0.0150	0.0046	-0.0069	0.0044
DEDU1	-	-	-0.5916	0.4244	-	-
DEDU2	-	-	-0.3506	0.4499	-	-

Table 9.2a   continued

Parameter	Abandon		Production		Mix	
	Estimate	StdErr	Estimate	StdErr	Estimate	StdErr
DCOTTON	-	-			-0.8443	0.3457
ρ_{12}	-0.5065	0.4347				
ρ_{13}	0.8108	0.2751				
$Ln(\theta)$	-156.31					

[a]   *Res*ults from trivariate probit model.

Table 9.2b   Estimation results: HOLLAND[a]

Parameter	Abandon		Production		Mix	
	Estimate	StdErr	Estimate	StdErr	Estimate	StdErr
CONSTANT	2.6375	3.1744	-1.1515	0.6177	1.3400	0.6519
PROBRET	0.0807	0.0424	-	-	-	-
FAMSIZ	-3.2670	1.6918	-	-	-	-
PCTRENTL	-3.8939	2.1398	-	-	-	-
TENURE	0.0714	0.0806	-0.0376	0.0211	-0.0191	0.0173
SPEC	-3.5698	2.5280	2.0523	0.7538	-	-
SIZLO	-1.7106	1.2766	-1.1398	0.6114	-	-
INFCAP	-	-	-	-	-0.2723	0.3615
PINVT	-	-	0.0213	0.0064	-0.0125	0.0054
DEDU2			-0.1402	0.1496	0.4811	0.3883
DWHEAT	-	-			1.1241	0.4471
DSUGAR	-	-	-	-	-1.3833	0.4859
$Ln(\theta)$	--79.645	-	-	-	-	

[a]   *The* likelihood ratio statistic for the null hypothesis $\rho_{12} = \rho_{13} = \rho_{23} = 0$ is 0.76 and therefore we can not reject the null hypothesis. The results above are therefore from three univariate probits.

Table 9.2c   Estimation results: HUNGARY[a]

Parameter	Abandon		Production		Mix	
	Estimate	StdErr	Estimate	StdErr	Estimate	StdErr
CONSTANT	-3.5764	1.8246	-0.3616	0.3370	0.6255	0.7377
PROBRET	0.0161	0.0053	-	-	-0.0088	0.0037
HDAGE	0.0542	0.0244	-	-	-	-
PCTRENTL	0.6428	0.4722	-	-	-	-
SATISF	-0.8448	0.4324	-	-	-	-
IMPAG	-1.0238	0.4330	-	-		
FAMSIZ	-0.4315	0.2073	-	-	-0.1845	0.1021
SPEC	1.5851	0.9222	-1.1347	0.6934	1.6809	0.7412
SIZLO	0.8420	0.4956	-0.5885	0.4464	-	-
OPCAP	-	-	-0.8405	0.2763	-	-
POFF	-	-	0.0096	0.0063	-	-
PINVT	-	-	0.0177	0.0039	0.0002	0.0003
TENURE	-	-			0.0362	0.0224
DCORN	-	-			-0.5721	0.4181
ρ_{23}	-0.3705	0.1578				
$Ln(\theta)$	-168.76					

a Results from trivariate probit model.

Although small farms are more likely to exit, we have in the case of Greece that if they stay in business they are also more likely to increase production, while the opposite holds for Holland and Hungary. Also, conditional on staying, more specialized firms are more likely to increase production in Holland, while the opposite holds in Greece and Hungary. More educated farmers are more likely to increase production in both Greece and Holland than less educated ones.

As far as the decision to change the crop mix is concerned, the results show that more informed farmers are more likely to change their crop mix in both Greece and Holland. On the other hand cotton growers in Greece, sugar beet growers in Holland and corn growers in Hungary are more likely to change their crop mix than farmers who do not grow those crops in their respective samples.

Table 9.3 **Description of variables used in the estimation of the model: scenario no information on prices**

Variable	Description	Sample means		
Explanatory		GR	HO	HU
PROBRET	Probability the farmer will retire	24.750	26.694	20.686
SATISF	Dummy for satisfaction current business	0.55		0.751
PARSTRUC	Dummy for part in any prev. EU str. program	0.33		
INFCAP	Dummy for level of knowledge about CAP reform	0.29	0.29	
HDAGE	Age of household head in years	47.956		51.294
FFARMINC	Family farm income in this EUR[a]	15.407		
SPEC	Specialization index[b]	0.780	0.412	0.467
TENURE	Years in farming[c]	19.288	21.059	12.726
SIZLO	Dummy indicating small farms[d]	0.175	0.424	0.170
PINVT	Probability of increasing investment	24.300		
DEDU1	Dummy for up to primary school	0.606		
DEDU2	Dummy for up to secondary school	0.263	0.765	
DCOTTON	Dummy for growing cotton	0.475		
FAMSIZ	Family size		2.635	3.582
PCTRENTL	Percent of land that is rented		0.509	0.435
DWHEAT	Dummy for growing wheat		0.494	
DSUGAR	Dummy for growing sugar beet		0.635	
IMPAG	Dummy for import. of activity in region econ.			0.667
OPCAP	Dummy for belief agr. prod. of their act. will fall			0.353
POFF	Probability of increasing off-farm labour			11.013
DCORN	Dummy for growing corn			0.811

[a] The original variable in the questionnaire was measured in euros; [b] For crops the acreage of each crop was used for the index, for livestock the number of heads was used, while for farms involved in both activities we opted for using the minimum of the two indices; [c] The original variable in the questionnaire was the actual year the respondent became the main decision maker; [d] Small farms are those whose size is below the 20 per cent quartile, where size is computed in terms of total acreage for crops and number of heads for livestock.

Conclusions

The 2003 CAP reform represents a substantial change with respect to the way the EU faces the agricultural sector. By decoupling farm payments and shifting agricultural policy towards rural development measures it is expected that the agricultural sector will undergo a structural reorganization whereas farmers whose existence depended in the past on direct supports and not on market conditions, will adapt to the new situation and become more market oriented. Therefore the new regime could in principle encourage some farmers to either abandon farming activity in the immediate years following the application of the reform, or decrease their levels of production, or switch to other crops. On the other hand, the rural development measures by targeting the development of rural areas as the main objective could lead to an increase of the employment opportunities in rural areas. The final effect that these two forces can have on the employment levels in rural areas is not clear at the outset. In order to assess what farmers intend to do with their farming activity we have conducted surveys for Greece, Holland and Hungary. The main results worth highlighting from our analysis of the collected data are summarized in what follows.

Those farmers who intend to abandon farming in the next five years, are mostly older farmers in the case of Holland and only in the event of crop prices decreasing do some young farmers decide to leave farming. In the case of Greek farmers though, those who intend to abandon farming are young irrespective of the future price scenario presented to them, while the percentage of young farmers intending to abandon farming greatly changes across future prices scenarios. For Hungarian farmers we find that if future crop prices increase then young farmers are not very likely to abandon farming. So overall young farmers are very susceptible to market price changes in Hungary and Greece and their future actions depend greatly on what will happen to world prices. Although the biggest percentage of farmers declaring they will abandon farming occurs in the case of Greece, it is also the case that the biggest percentage of farmers who already know how they will use the SFP occurs in the sample for that country as well. Indeed, more than half of the farmers in Hungary do not know yet how they will use the SFP, while the equivalent percentages are a little bit over 40 for Holland and around 20 for Greece. However it is the case that most (in Holland) and all (in Hungary) farmers who will invest the SFP will invest it within the farm, while this is not the case for Greece. Therefore, it could be the case that farmers in Greece feel greater uncertainty about the future and try to diversify their investments.

If we examine the factors that affect the probabilities of abandoning farming, to increase acreage/livestock size and keeping the same mix for the three countries the following conclusions can be derived from the analysis. In the case of Hungary and Greece it is small farms that are more likely to abandon while in the case of Holland the opposite occurs. However for those farmers who intend to stay in business, the smaller ones are more likely to increase production than the bigger ones in Greece while the opposite holds for Hungary. When it comes to the effect of specialization then again we have different effects for Holland and the other two countries. Indeed, more specialized farms are more likely to abandon production in Greece and Hungary while in Holland they are less likely to abandon production. Moreover, once a farm decides to continue with farming, the more specialized it is, the less likely it is to increase production in the case of Greece and Hungary but the more likely it is to increase production in the case of Holland. Therefore policy implications differ for the two groups of countries. If the aim is to prevent farmers from abandoning farming then structural programmes should be devised that promote alternative cultivation and decrease the risk of monoculture in Greece and Hungary but the contrary holds for Holland.

Our results also indicate that the level of information received is very important to reduce farmers' uncertainty about the future and that the more informed a farmer is, the more willing he/she will be to change his/her crop mix in the case of Holland and Greece. Therefore, policies that increase farmers' level of information could prove useful if farmers are to switch crops. On an ending note we should emphasize that the evolution of crop prices in world markets could be after all the most important factor dictating farmers intentions in the light of the new policy regime.

References

Anton, J. and Sckokai, P. 2006. The Challenge of Decoupling Agricultural Support. *Eurochoices*, 5(3), 13-18.

Breen, J.P., Hennessy, T.C and Thorne, F.S. 2005. The effect of decoupling on the decision to produce: An Irish case study. *Food Policy*, 30(2), 129-144.

CBS, Statistics Netherlands, 2005. The Hague. Available at http://www.cbs.nl.

Gohin, A. 2006. Assessing CAP Reform: Sensitivity of Modelling Decoupled Policies. *Journal of Agricultural Economics*, 57(3), 415-440.

Hajivassiliou, V. D., McFadden D., and Ruud P. 1996. Simulation of Multivariate Normal Rectangle Probabilities and their Derivatives Theoretical and Computational Results. *Journal of Econometrics*, 72(1-2) , 85-134.

Hennessy, T.C. and Thorne F. S. 2005. How decoupled are decoupled payments? The Evidence from Ireland. *Eurochoices*, 4(3), 30-34.

HCSO, Hungarian Central Statistical Office, Statistical yearbook of Hungary, 2003, 2004, 2005.

NSSG, National Statistical Service of Greece, 2001, 2005. National Accounts, Athens.

Province Flevoland, 2005. Available at http://provincie.flevoland.nl/flevoland-in-beeld-en-cij/Feiten-en-cijfers/.

Chapter 10

Assessing the Impact of the 'Health Check' in an Italian Region: An Application of the RegMAS Model

Antonello Lobianco and Roberto Esposti[1]

Introduction

With the 2003 Reform, the Common Agricultural Policy (CAP) underwent a major regime change, with a substantial migration from coupled payments and market intervention (and distorting) measures to farm-specific decoupled support based, at least in Italy, on historical payments. During 2008, further modifications of that Reform, the so-called Health Check (HC), were proposed by the European Commission (EC) for implementation in 2009.

Farm-based modelling approaches allowing for a direct representation of such changes in the CAP regime, therefore seem better suited than partial or general equilibrium models (like ESIM, FAPRI/AGMEMOD or GTAP) to analyse their impacts (Heckelei and Britz 2005). In particular, mathematical programming, and more specifically, Positive Mathematical Programming (PMP) models, are widely used within the scope of agricultural political analysis (Paris 1991, Arfini 2000). However, modelling representative farmers, they miss the interaction between them that is instead considered in Agent-based models (AMB).

RegMAS (Regional Multi Agent Simulator) is an open-source spatially explicit multi-agent model framework, developed in C++ language specifically designed for long-term simulations of effects of agricultural policies on farm structures, incomes, land use, etc.

More specifically, RegMAS conceives rural social systems (and in particular agricultural ones) as complex evolving systems, made of a heterogeneous set of `agents" (that is, farmers) whose behaviour is generated by a profit-maximisation, Mixed-Integer linear Programming (MIP) problem; they compete in the land-market and use purchased resources to increase their competitiveness (mainly through scale effects).

1   Authorship may be attributed as follows: sections 2, 3 and 5 to Lobianco, sections 1, 4 and 6 to Esposti. The corresponding author wishes to thank the IAMO team for their support and training on agent-based modelling.

Differently from similar models, the spatial dimension is initialised from real land-use data, using satellite information, and plots are explicitly modelled in the agents' problem as individual resources. As common in GIS, spatial information is organised in layers to facilitate its usage within the model. This approach allows very detailed analysis along the spatial dimension, as farmers' decisions can be based on individual plot properties and result of farmers' activity can be directly observed and, for example, evaluated on an environmental point of view.

While Lobianco and Esposti (2010) detail the RegMAS framework internal algorithms, in this chapter we apply it by evaluating the effect of the CAP reform known as the Health Check.

In particular, we are interested to observe how measures specifically designed to maintain a neutral aggregated offset, such as the regionalisation (which was admitted in the 2003 Reform but then adopted by very few countries), may shift public support across different types of farmers and areas, eventually generating aggregate modifications on the whole area. The focus here is on the effect of new 'parameters' applied to the political instruments introduced in the 2003 Fischler reform, such as the Single Farm Payment (SFP) passing from the historical based to an area-based flat payment and stronger modulation.

The chapter is structured as follows. Section 2 introduces the methodological approach underlying RegMAS. The case-study region is then presented in Section 3, together with the steps required to derive a ``virtual" region on which the simulations are eventually ran. Section 4 illustrates the hypothetical policy scenarios under which results are generated.

From these simulations we obtain a large amount of information, including the status of individual farms, environmental effects (soil use, land abandonment, agents and objects location), as well as aggregate results. Nonetheless, to better emphasise the possible impact of HC on the case-study area, we prefer to report and discuss some selected, mostly aggregated, evidence (5). Section 6 concludes.

Methodological approach

The use of spatial explicit Agent-Based Models (ABM) within the specific agricultural domain was pioneered by Balmann (1997) with the Agricultural Policy Simulator (AgriPoliS) model.

ABM allows the representation of social systems as the result of individually-acting agents. When they are applied to agriculture, they can simulate, at the micro-level, the fundamental behaviours of individual farmers, without the need to aggregate them in 'representative' agents. Furthermore ABM can catch the iterations of the heterogeneous farms when competing over common finite resources, e.g. land.

Boero (2006) and Parker (2003) have reviewed several ABM involving land use changes in various scientific areas, including agricultural economics, natural resource management, and urban simulation, but this section will briefly describes

AgriPoliS as RegMAS borrows many concepts from it, *in primis* the utilisation of a profit-maximisation algorithm to derive farmers behaviours.

In AgriPoliS agents are mainly farmers.[2] They have their own goals; in AgriPoliS, the farmer's objective is the maximisation of household income. To achieve this objective, farmers solve a MIP problem that, in some aspects, is specific to each farmer. Outside the linear programming problem, they can also decide to rent other agricultural plots or to release rented land.

Using a mixed integer linear programming approach to simulate each agent behaviour on one hand is very flexible, as it can cover the whole range of farm activities, from growing specific crops to investing in new machinery or hiring new labour units. Furthermore, it is easy to add new regional-specific activities.

On the other hand, however, linear programming techniques require a long calibration phase to assure a balanced choice of farm activities, avoiding unrealistic outcomes.

Each farmer in the model is a real farmer whose data are taken from farm-level datasets (in Europe, FADN) and explicitly associated to a spatial location. Due to privacy-protection regulations, however, researchers do not normally have access to the real farm localisation. Therefore, farms have to be randomly distributed along the virtual region. Space (i.e. location) is important in the model because it influences transport costs and indirectly makes the farmers interact each other, e.g. by competing for the same land plots.

AgriPolis, as it takes into account many aspects of a real farm, is a very complex model, with lot of code dedicated to cover specific aspects (e.g. quota markets, generational changes, multi-years investments). A detailed description of AgriPolis can be found in Happe et al. (2004) or in Kellermann et al. (2007). While Happe et al. (2004) focuses on the methodological advantage of using ABM in agriculture as compared with other instruments as partial and general equilibrium models on one side and individual farm-level models on the other, Kellermann et al. (2007) details the latest implementation of AgriPoliS (2.0). In addition to these two papers, Sahrbacher et al. (2005) describes AgriPoliS implementation over several case-study regions and Lobianco (2007) presents an adaptation of AgriPoliS for the Mediterranean regions, further adding some general background on agent-based modelling and to its motivations.

As AgriPoliS, RegMAS is spatially explicit, a characteristic that cannot be neglected when modelling the agricultural sector. For example the spatial heterogeneity allows the model to associate on each plot a different rental price and investigate possible land abandonment phenomena even when the land is, on average, profitable.

Differently from AgriPoliS, the spatial dimension is initialised from real land-use data, using satellite information, and plots are explicitly modelled in the decision matrix as individual resources.

---

2   Other agents in the model perform some specific tasks, e.g. managing land or coordinating product markets.

As a further distinction, RegMAS has been designed from the ground-up to explicitly consider farmers as one of several possible types of agent. In RegMAS farmers have sensitivity to the overall environment, including extra-agricultural variables. On a technical point, 'farmer' agents derive from a more general type of 'spatial' agents that in turn derive from a 'base' type. Each agent type has its own 'manager' agent that dialogue with a 'Super Agent Manager'. The former are a sort of interface 'agent side' while the latter implements the same interface on the program core side. In this way the model core does not need to know anything about agents' internal logic. While this approach allows for rapid development of different agent types (only specific characteristics need to be modelled) at the current RegMAS development stage only farmer agents are fully implemented.

They autonomously make their decisions solving a profit-maximisation mixed integer linear problem, where activities, coefficients, gross margins and capacities are initially exogenous.

The model is able to adjust year upon year the capacities of the individual farmers and modify the coefficients and gross margins according to several algorithms (Lobianco and Esposti 2010).

A first algorithm consists in changing the farmer capacities according to the investments acquired in previous years and the rented (or released) land. This will allow farmer-agents to evolve during simulation as their production decisions will depend on the current capacities level while scale-effects are expected to emerge in the simulation because of the integer programming.

Land is allocated to the farmers according to a bid system (Kellermann et al. 2008).

A further algorithm (and the novelty of this approach) is appointed to dynamically build the problem matrix according to the individual plots owned or rented by the farmer, adding for each plot all the compatible spatial explicit activities, changing at the same time their coefficients and gross margins according to the plots' unique characteristics. For example, gross margins are adjusted according to the plot altitude. This GIS-alike functionality allows a full linkage between the social and the geophysical parts of the model.

The case-study region

Our simulations are carried out on a hilly region of central Italy, Colli Esini (Marche region), including 24 LAU2 municipalities and approximately 50,000 UAA, hosting in 2001 around 6,000 farms. Their main characteristic is to have a well-established homogeneous agricultural area on the east and a more heterogeneous, mixed agro-forestry area on the south-west.

Actually, the computer simulation is run on a 'virtual' region based on this region and more specifically built upon the following datasets:

Quantitative regional data

Aggregated data of the region, normally available from the Census.

Individual farmer detailed data

Individual farmers are used in the model as 'bricks' to build a simulation region and the crucial information here becomes the individual farms production factors. In order to obtain satisfactory congruence between the real and the simulated region, a basket in the magnitude of tens of farmers' data is often necessary.[3]

Technical coefficients and prices

Technical coefficients, production prices and factor prices are needed to link the activities pool with the resource pool and to establish the objective function.

Land use map

As RegMAS is fully spatial explicit, it requires a detailed map of land uses (in Europe this is available from the Corine Land Cover project).[4]

The specificness of this virtual region (and its differences with a real one) is the fact of being composed uniquely of ``typical" farms, while still having its aggregated values as close as possible to the region under study.

Typical farms are a subset of all the farms in the region for which detailed data is available (e.g. because member of the FADN network). These are weighed with a scaling coefficient that minimise the difference between the simulation region and the real one (Eq. 1).

$$\min \sum_{k=1}^{K} \left(\frac{\sum_{n=1}^{N} \left(FADN_{n,k} * UC_n \right)}{REGIO_k} - 1 \right)^2 sub\, UC_n \geq 0 \forall n \qquad (1)$$

Where:

Indices	Variables:
$n = \{1,..,N\}$ Individual farms	$FADN_{n,k}$ FADN data
$K = \{1,..,K\}$ Characteristics	$REGIO_k$ Regional aggregated data
	UC_n ``upscaling" coefficient

---

3   The exact number depends on three parameters: (1) the number of elements that should be compared between the real and the simulated regions, (2) how good the typical farms reflect the total of the farms in that region and (3) the statistical discrepancy that the user is willing to accept.

4   http://dataservice.eea.europa.eu/dataservice/metadetails.asp?id=950

This procedure is called 'upscaling' and it is well documented in Kellermann et al. (2007), while a practical implementation is discussed in Sahrbacher et al. (2005).[5]

The upscaling can be conveniently obtained using the quadratic solver embedded in Excel.

Scenarios

Simulations discussed in this chapter start from 2001 in order to include the reference period. The period covered is 2001-2003 for most activities,[6] over those years, the model 'collects' the subsides received by each farm, then automatically calculates the single-farm payment (SFP) due to any individual farmer and finally assigns the SFP to farmers.

In more detail, the model keeps track, for each farmer, of three vectors: the **dRights, dYears** and **dHa**.

The **dRights** are the average entitlements that a farmer 'owns' for the decoupled payment, differentiated by each specific production activity. It is already averaged by the number of years of the reference period. In a similar way the **dHa** are the average hectares that have generated the entitlements for the specific activity. Finally **dYears** are the years for which these averages have been calculated.

Using an activity-specific flag to indicate the reference period, every year the model updates the entitlements for each agent and each activity:

$$dRights_t = (dRight_{t-1}*dYears_{t-1}+newRight_t)/(dYears_{t-1}+1)$$

$$dHa_t = (dHa_{t-1}*dYears_{t-1}+newHa_t)/(dYears_{t-1}+1) \qquad (2)$$

$$dYears_t = dYears_{t-1}+1$$

where *newRight* is the coupled premium obtained by the farmer on the specific activity for that year (only if the activity flag is in 'registration' mode for that year). In this way different products may have different reference periods, even if not continuous.

When due the model assigns back the entitlements to each farmer in terms of SFP. Starting from version 1.1, RegMAS could distinguish between history-based SFP (Eq. 3) and area-based SFP (Eq. 4):

$$dpaymentI = \sum_{i=1}^{N} dRights_i * dRateCoef_i \qquad (3)$$

---

5   Both papers refer to the preparation of a simulation region for AgriPoliS, but the methodology can be equally applied to RegMAS.

6   The exception is the olive oil sector, where, due to its higher yield fluctuation, the reference period is extended to 2004.

$$dPayment = \left(\frac{\sum_{i=1}^{N} \sum_{y=1}^{A} dRights_{i,y} * dRateCoef_i}{\sum_{i=1}^{N} \sum_{y=1}^{A} dHa_{i,y}} \right) * \sum_{i=1}^{N} dHa_i \qquad (4)$$

where N are the activities; $dRateCoef_i$ counts for eventual partial decoupling and A are the number of agents in the model. Please note that the farmer can still benefit for a given year/activity of a mixed of coupled and decoupled premium.

This farm-based modelling approach allows for a very detailed implementation of the various policy instruments that can barely be achieved with conventional equilibrium models. Beside macro-economic and general, policy-specific parameters (e.g., modulation), RegMAS allows for dynamically setting activities' gross margin, matrix coefficient or decoupling entitlements along the temporal dimension.

We used such flexibility to build the two following scenarios:

Decoupling scenario (dec)

In this scenario, the introduction of historically-based SFP starts in 2005, the modulation on payments over EUR5,000 rises from 3 per cent in 2005 to 5 per cent in 2007. All major payments are decoupled but quality premiums remain (for durum wheat and ex Art. 69) and these are treated in the model as coupled subsides.

This scenario approximately matches the actual implementation of the 2003 Fischler CAP reform, including the Italian national decisions in terms of decoupling options and Art. 69.

Health Check scenario (hc)

The hc scenario is equal to the dec scenario till 2008, but from 2009 onward it assumes the following changes:

Modulation

Starting from 2009 it becomes much stronger, arriving in 2012 to a maximum of 22 per cent for payments over EUR300,000. Furthermore, payments below EUR250are totally dropped;

Set aside

Mandatory set-aside minimum share (10 per cent) is abolished from 2009;

Regionalisation

From 2010 the SFP calculation changes following area-based implementation (also known as 'regionalisation') where the unit-value of the subsides are averaged. Our implementation of the regionalisation does not allow the redistribution of the subsides to farmers without eligible land;

Full decoupling

Since Italy opted for full decoupling in 2003, the only novelty is the decoupling of the specific durum wheat payment (EUR40) starting from 2010, on the base of the 2005-2008 reference period.

While the durum wheat payment has been decoupled, the other quality payments, ex Art. 68, have been maintained. This scenario is aimed at implementing the Health Check reform, as know by the preparatory legislation acts by the EU Commission.[7]

Selected results

Tables 10.1 and 10.2 present the outcomes on the simulated region under the dec and hc scenarios, when we run the model till 2015, showing first the overall results and then results subdivided by farm size classes.[8]

In particular the number of farms seems only marginally influenced by the contingent policy option. During historical periods (1990-2003) the yearly abandonment rate in Italy has been 2.32 per cent (Eurostat), while we report slightly higher rates in our region for the period 2008-2015 in the two dec and hc scenarios (respectively 3.30 per cent and 3.35 per cent). While the differences between the two scenarios seems small, it increases when we look down by farm size. The smallest group of farm seems much more influenced by the hc scenario (Figure 10.1). This is likely an outcome of the suppression of smaller payments (EUR250). In fact, while these small payments represent only 0.68 per cent of the total support (referring to 2008) if we consider only

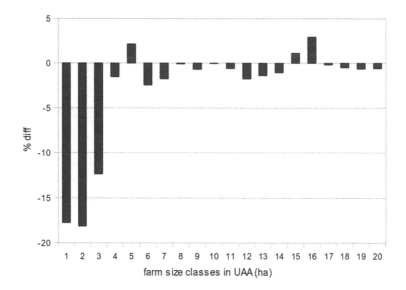

Figure 10.1   Number of active farms by size class, % differences between hc vs dec (2015)

7   COM (2008) 306, Proposal for a Council Regulation, 20 May 2008
8   These simulations have been conducted with Version 1.3 of RegMAS software. Readers can replicate them downloading RegMAS at http://www.regmas.org.

farms up to three hectares they represent 22.36 per cent of the support involving 68.15 per cent of those farms.

The effect of modulation on farmers becomes evident when we look at the farm profits with CAP payments. While the net farm profit, without considering the CAP support, increases along all classes, probably due to a higher production freedom following the drop of mandatory set-aside and further decoupling of durum wheat, the farm profit when the CAP support is included, strongly reduce the gains.

Adapting production to more intensive crops (e.g. substituting set aside areas) requires also more work, subtracting it from off-farm activities that in this region are particularly important. Consequently total incomes, composed of farm profits plus off-farm activities, seems at the end to remain steady between the two scenarios.

Table 10.1   Main results

	dec_2015	hc_2015	% var dec 2015-2008	% var hc 2015-2008	% var 2015 2015-2008
All Farms					
number of farms (n)	4,304	4,288	-20.9	-21.2	-0.4
average size (UAA ha/farm)	10.97	11.01	23.5	24.0	0.3
quitted farms (n)	1420	1436			
abandoned land (%)	3.25	3.27			0.5
farm profits (EUR/ farm)	10,981	11386	12.1	16.2	3.7
farm profits w/CAP (EUR/farm)	16,068	16,202	15.7	16.6	0.8
incomes (EUR/farm)	20,942	20,982	7.6	7.8	0.2
off-farm labour (h/farm)	975	956	-12.6	-14.3	-1.9
total agr labour (AWU)	2,884	2,928	-12.5	-11.2	36,647

Table 10.2   Main results by farm size class

	dec_ 2015	hc_2015	% var dec 2015-2008	% var hc 2015-2008	% var 2015 2015-2008
Small farms - [0-3] ha					
number of farms (n)	405	355	-73.9	-77.2	-12.35
average size (UAA ha/farm)	2.2	2.3	16.4	19.0	2.22
farm profits (EUR/ farm)	4,069	4,196	-9.2	-6.3	3.14

	dec_ 2015	hc_2015	% var dec 2015-2008	% var hc 2015-2008	% var 2015 2015-2008
farm profits w/CAP (EUR/farm)	4,726	4,582	-3.2	-6.1	-3.04
incomes (EUR/farm)	9,679	9,484	-17.4	-19.1	-2.01
off-farm labour (h/farm)	991	980	-27.6	-28.4	-1.03
Middle farms - [4-15] ha					
number of farms (n)	3,004	3,092	-2.4	0.4	2.93
average size (UAA ha/farm)	7.0	7.1	0.7	2.6	1.89
farm profits (EUR/farm)	8,963	9,119	-5.5	-3.9	1.74
farm profits w/CAP (EUR/farm)	11,960	11,936	-4.5	-4.7	-0.20
incomes (EUR/farm)	16,608	16,566	-4.7	-4.9	-0.25
off-farm labour (h/farm)	929	926	-5.1	-5.5	-0.38
Large farms - [>16] ha					
number of farms (n)	895	824	10.4	1.6	-7.93
average size (UAA ha/farm)	28.4	29.0	-4.4	-2.3	2.24
farm profits (EUR/farm)	20,884	22,781	-1.3	7.7	9.09
farm profits w/CAP (EUR/farm)	34,987	36,789	-3.8	1.2	5.15
incomes (EUR/farm)	40,586	42,170	-3.6	0.2	3.90
off-farm labour (h/farm)	1,120	1,076	-2.4	-6.2	-3.91

Regionalisation redistributive effects

The regionalisation of the SFP is expected to introduce significant redistributional effects between farmers. However these effects are interconnected with the other policy changes that the Health Check introduced.

Comparing the two scenarios on 2015[9] and considering all the subsides (still-coupled payments+SFP), the number of farms that 'lose money' compared to those that 'win money' is slightly smaller (46.43 per cent against 51.31 per cent). However the average gain (EUR647) is much higher than the average loss (EUR1,146). While there are exceptional cases losing over EUR10,000 or gaining

___

9   This results take into consideration only those farms that are still in the model on both scenarios. There is however a limited number of farms that reach 2015 only in one scenario.

over EUR5,000, the 92.4 per cent of farms is within the ± 2,000 range and 47.24 per cent of them are within the EUR500 range. Figure 10.2 shows quite clearly that the distribution is asymmetric, especially at its tails, where the left tail is much stronger than the right one, due to the modulation.

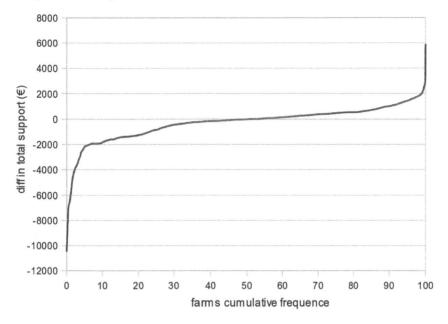

Figure 10.2   Support level along farms, differences between hc vs dec (2015)

Territorial consequences

While we used a very conservative coefficient to establish altitude influence over the gross margin (2 per cent every 100 meters), we can nevertheless report that the majority of abandoned plots (that is, those plots that are either unrented or unused by the tenants) tend to be in the hilliest part of the region (Table 10.3). An important role seems to be played by the fragmentation that this area has with non-agricultural areas, increasing the average distance and so the transport costs compared with homogeneous agricultural areas in the east part of the region. On these areas the land freed by small farms that, especially in the hc scenario, quit agricultural production, may be too far away to be used by remaining farms, leading to land abandonment.

As our simulations do not take into account the increase in producer prices that happened over the last few years, the fact if this increase could slow down the farm quitting phenomena and so the resulting localised land abandonment is still an open question.

Table 10.3   Abandoned plots (2015)

	dec		hc		hc-dec	CV
	ab. plots	%ab. rate	ab. plots	%ab. rate	% diff	
0-200m	769	2.312	774	2.327	0.65	0.028
200-400m	679	4.995	678	4.987	-0.15	0.010
400-900m	135	7.508	139	7.731	2.96	28

Conclusions

We used RegMAS, an open-source, spatially explicit agent based modelling framework, to asses the possible impacts of the Health Check (regionalisation and further modulation, in particular) on the heterogeneous structures, farmer incomes and land use of a central Italian region (Marche). RegMAS allows economical agents (that is, farmers) to contemplate spatial-explicit information within the formulation of their behaviours (in our case, income maximisation) and to asses the economic, social as well as environmental outcomes of these behaviours on whole area. Our results seem to indicate that the Health Check, while increasing the farm profit net of CAP support, may slightly reduce the overall farmers incomes, also through a reduction of the off-farm labour, and that these effects may be greater on small and large farms compared with middle-size ones. Allocation of land freed by quitting farms depends on distance from neighbouring farmers, and in some internal areas this land may eventually be abandoned.

References

Arfini, F. 2000. I modelli di programmazione matematica per l'analisi della politica agricola comune. INEA seminar Valutare gli effetti della Politica Agricola Comune, Rome, 24 October 2000. Available at: http://web.archive. org/web/20041024164241/http://www.inea.it/opaue/pac/arfini.PDF.

Balmann, A. 1997. Farm-based modelling of regional structural change: A cellular automata approach. *European Review of Agricultural Economics* 24(1-2), 85-108. http://dx.doi.org/10.1093/erae/24.1-2.85 doi:10.1093/erae/24.1-2.85.

Boero, R. 2006. The spatial dimension and social simulations: A review of three books. *JASSS, Journal of Artificial Societies and Social Simulation*. Available at: http://jasss.soc.surrey.ac.uk/9/4/reviews/boero.html.

Bousquet, F., Bakam, I., Proton, H. and Page, C.L. 1998. Cormas: Common-pool resources and multi-agent systems, in T*asks and Methods in Applied Artificial Intelligence* edited by A.P. del Pobil and J.M. Ali', Vol. 1416 of *Lecture Notes in Computer Science*, Springer, pp.826-837.

Brady, M. and Kellermann, K. 2005. Methodology for assessing the regional environmental impacts of decoupling: A focus on landscape values. IDEMA

Assessing the Impact of the 'Health Check' in an Italian Region 149

working paper 11. Available at: http://www.sli.lu.se/IDEMA/WPs/IDEMA_ deliverable_11.pdf.

Castella, J., Boissau, S., Trung, T. and Quang, D. 2005. Agrarian transition and lowland-upland interactions in mountain areas in northern vietnam: application of multi-agent simulation model', *Agricultural systems*. Available at: http:// www.sciencedirect.com/science/article/B6T3W-4F29HSM-1/2/5d018e2f9b9 40709e3857c26d8d1f86d.

Happe, K., Balmann, A. and Kellermann, K. 2004. The agricultural policy simulator (agripolis) - an agent-based model to study structural change in agriculture', IAMO discussion paper 71. Available at http://www.iamo.de/dok/dp71.pdf.

Heckelei, T. and Britz, W. 2005. Models based on positive mathematical programming: state of the art and further extensions, in *Modelling agricultural policies: state of the art and new challenges* edited by F. Arfini. Proceddings of the 89th European seminar of the European Association of Agricultural Economics', Monte Università Parma, pp.48-73.

Kellermann, K., Happe, K., Sahrbacher, C. and Brady, M. 2007. Agripolis 2.0 - documentation of the extended model', IDEMA working paper 20. Available at: http://www.sli.lu.se/IDEMA/WPs/IDEMA_deliverable_20.pdf.

Kellermann, K., Sahrbacher, C. and Balmann, A. 2008. Land markets in agent based models of structural change, in *Modelling Agricultural and Rural Development Policies*. 107th EAAE Seminar, Sevilla'. Available at: http:// agecon.lib.umn.edu/cgi-bin/pdf_view.pl?paperid=29447&ftype=.pdf.

Lobianco, A. 2007. The effects of decoupling on two Italian regions. An agent-based model., PhD thesis, Università Della Tuscia. Available at: http:// associazionebartola.univpm.it/pubblicazioni/phdstudies/phdstudies2.pdf.

Lobianco, A. and Esposti, R. 2010. The Regional Multi-Agent Simulator (RegMAS): An open-source spatially explicit model to assess the impact of agricultural policies. *Computers and Electronics in Agriculture* 72(1). Available at: http://dx.doi.org/10.1016/j.compag.2010.02.006

Makhorin, A. 2007. Gnu linear programming kit. reference manual'. Available at: http://www.gnu.org/software/glpk/.

Paris, Q. 1991. *An economic interpretation of Linear Programming*, Iowa State University Press. Also available in Italian with title Programmazione lineare. Un'interpretazione economica.

Parker, D.C. 2003. Multi-agent systems for the simulation of land-use and land-cover change: A review', *Annals of the Association of American Geographers* 93(2), 314-337. Available at: http://www.blackwell-synergy. com/doi/abs/10.1111/1467-8306.9302004, http://dx.doi.org/10.1111/1467-8306.9302004 doi:10.1111/1467-8306.9302004.

Sahrbacher, C., Schnicke, H., Happe, K. and Graubner, M. 2005. Adaptation of agent-based model agripolis to 11 study regions in the enlarged european union', IDEMA working paper 10. Available at: http://www.sli.lu.se/IDEMA/ WPs/IDEMA_deliverable_10.pdf.

Chapter 11

Impact of the CAP Reform on the Spanish Agricultural Sector

Lucinio Júdez, Rosario de Andrés, Miguel Ibáñez, José María de Miguel, José Luis Miguel, and Elvira Urzainqui

Introduction

The purpose of the CAP health check conducted by the EU Commission, among others, is to assess the 2003 CAP reform and propose modifications to enhance CAP effectiveness, (see Commission Staff Working Document SEC (2008) 1885). These tasks have been undertaken in a scenario of substantial rises in food prices due to the expansion of agro-energy crops and the increase in the world-wide demand for cereals.

In that context, this chapter aims to evaluate the impact of the decoupling measures adopted by Spain in 2006 taking into account recent measures proposed or studied for possible future proposals. More specifically, the modifications studied are: abolition of the 10 per cent set-aside requisite to qualify for compensatory payments for CAP crops, the increase in the milk quota and recent provisions for the cotton and sugar beet sub-sectors.

The farm types defined in the Spanish Farm Accountancy Data Network (FADN) are used to perform a static comparative analysis between the results of the positive mathematical programming (PMP) model PROMAPA (PROgramación Matemática para el Análisis de Políticas Agrarias) for the base year 2002 and the findings for a simulated year in which a new price scenario is established and the decoupling scheme measures are assumed to be in effect.

The chapter is structured as follows: section 2 discusses the interest aroused by and recent developments in positive mathematical programming for the analysis of agricultural policy, section 3 contains a brief description of PROMAPA and sections 4, 5 and 6 respectively describe the farm types, prices and agricultural policy scenarios considered. Finally, section 7 analyses the results obtained.

Positive Mathematical Programming and Agricultural Policy

Mathematical programming and in particular linear programming has been and continues to be a widely used technique in the context of agricultural economics.

Despite this extensive use, however, considerable criticism has been levelled against linear programming. Specifically, to obtain solutions that accurately reflect reality, it is felt that certain – usually arbitrary – constraints must be included.

One way of avoiding this problem is to use PMP, as devised by Howit (1995). Briefly, by estimating the coefficients of the target function for a non-linear programming model, this technique can calibrate the model so that it reproduces the situation existing in a base year for the unit modelled (farm or region). The calibration method proposed by Howit was subsequently enhanced by including entropy maximization in the procedure (Paris and Howit 1998).

The suitability of PMP for formulating and evaluating the Common Agricultural Policy (CAP) has driven further development of this technique, as can be seen in the recent revisions by Heckelei and Britz (2005) and Henry de Frahan et al. (2007).

Variations designed to correct some of the shortcomings of the Howit and Paris calibration procedure have also been proposed. Such variations have been published, among others, by Júdez et al. (1998, 2001), who propose to perform calibration without running the first stage of PMP, Gohin and Chantreuil (1999) introduce a procedure for processing marginal activities, Helming et al. (2001) include supply elasticities obtained exogenously to calibrate the target function coefficients, Röhm and Dabber (2003) propose a method for linking different variants of the same crop, and more recently Severini and Cortignani (2008) and Júdez et al. (2008) suggest procedures for including activities in PMP that are not present in the base year. In addition to these proposals in which calibration is achieved for each unit modelled with the data for a single year, in Heckelei and Wolf (2003) and in Buysse et al. (2007) calibration is replaced with econometric estimation procedures using datasets.

In parallel with the theoretical developments around calibration, PMP has been applied in a fair number of cases lately to analyse the effects of agricultural policy (essentially Common Agricultural Policy measures) on the agricultural sector. In addition to the above papers, in which the authors illustrate their calibration proposals with applications, others have been published by Arfini and Paris (1995), Heckelei and Britz (1999), Barkaoui and Butault (1999, 2003), Röm and Dabbert (1999), CAPRI (2000), Paris et al. (2000), Osterburg et al. (2001), Arriaza and Gómez-Limón (2003), Júdez et al. (2003), Buysse et al. (2004), Buysse and Van Huylenbroeck (2005), Offermann et al. (2005), Blanco and Iglesias (2005), Adenauer et al. (2006) and Kuepker and Klainhauss (2006).

Finally, for several years now a number of European teams have been developing PMP models at the farm level, using national and European FADN. Some of these models, which are often used by national and/or Community officials as a tool for analysing the impact of agricultural policy are: FARMIS (FAL- Germany),

SEPALE (Ghent University, CAE Brussels, Catholic University of Louvain – Belgium) and CAPRI (Bonn University -Germany).[16] The PROMAPA model pursues the same line of research as the foregoing studies.

Brief Description of the PROMAPA Model

PROMAPA is a PMP representative farm model, designed to study the impact of change in agricultural policies on the Spanish agricultural sector.[2] Model calibration can be performed with several procedures. Exogenous supply elasticities are used in this study.

The activities covered by the model included some 15 non-irrigated and around 25 irrigated crops, as well as dairy cattle, rearing cattle, and dairy and non-dairy sheep. Livestock feed is endogenous, whether produced on the farm or purchased to meet the energy and protein needs of different livestock categories; intake capacity was taken into account as well. The activities associated with the agricultural policy tools implemented in the model included the mandatory set-aside in irrigated and non-irrigated land requisite to receiving direct payments for COP crops, several premiums for livestock (dairy and rearing cattle and sheep), several types of (coupled and decoupled) direct payments for crops in the context of the Single Payment System, modulation and crop and livestock quotas.

The primary source of the data to feed the model was the Spanish FADN, although information provided by experts was likewise used, especially to determine unit costs for crops and different categories of livestock and to establish livestock feeding needs.

Farm Types

The farm types considered were the mean types listed in the Spanish FADN in 2002 by autonomous community for each of the farm sizes in the TFs most affected by the CAP reform. A total of 140 farm types, covering 188,310 farms nationwide, were included.

---

1   The AROPAJ model developed in France applying linear programming, which also uses information from the farm accountancy data networks, is used to analyse agricultural and agro-environmental policies (see Jayet et al. 2000).

2   The model is being developed by the Departamento de Estadística y Métodos de Gestión en Agricultura, ETSIA (UPM) and the Instituto de Economía, Geografía y Demografía, CCHS (CSIC).

Price Scenario

The prices of 2007 were adopted to reflect the price increase with respect to the base year.

Table 11.1   Price variations 2002-2007

Rice	-1.34
Sugar beet	-47.87
Cotton	44.69
Potato.	49.88
Chick pea	-9.63
Vetch	31.34
Common wheat	56.82
Durum wheat	150.85
Barley	55.33
Rye	45.58
Oats	25.36
Maize	49.27
Sunflower	50.84
Veal (7-18 days)	-17.00
Veal (6 months)	-11.85
Cow's milk	23.42
Sheep's milk	0.21
Lamb	15.58
Dehydrated alfalfa	10.40
Concentrated feed, dairy cow	10.46
Concentrated feed, rearing cattle	7.43
Concentrated feed, sheep	17.29

The variation in prices from 2002 to 2007 according to data published by the Spanish Ministry of the Natural, Rural and Maritime Environments are shown in Table 11.1. In sugar beet the price variation is between the minimum base year price and the minimum price in place under sub-sector reform.

Agricultural Policy Scenarios

The base year measures considered were the Agenda 2000 arrangements in effect in 2002, while in the main scenario simulated, partial decoupling measures adopted in Spain in 2006 were included, with the following modifications:

- For sugar beet, according to the new proposal for the sub-sector, a coupled payment of EUR8.78/t and a decoupled payment of EUR12.83/t were assumed. The sugar quota was reduced by 50 per cent.
- For cotton, the new measures entailed a coupled payment of EUR1,551/ha, while the decoupled payment was the same as in the base year EUR1,351/ ha. The maximum farming area eligible for guaranteed coupled payments

was lowered from 70,000 to 48,000 ha.
- The compulsory 10 per cent land set-aside was eliminated.
- The dairy quota was increased by 2 per cent.
- Furthermore, for the full decoupling scheme simulation, the decoupled measures were defined to be the sum of the coupled and decoupled measures in the main partial decoupling scenario.

Results

The following assumptions were made to obtain the results: i) the reduction,due to modulation, for direct payments totalling over EUR5,000 was set at 5 per cent; ii) the decoupled aid for each farm type was established on the basis of farming area and livestock numbers in the base year. That is, the base year replaced the reference period (mean for calendar years 2000, 2001 and 2002); and iii) as a result, the land set aside in 2006 was the same as in the base year.

The results for the 140 farm types were obtained with GAMS. The analyses in the following sections concerns the weighted sum of the results for each farm type. The weighting coefficient was the number of farms represented by each type nation-wide.

Impact of Price Variations and of the New Agricultural Policy Measures

The effect of prices and the new agricultural policy measures on crop distribution and on gross margin and payments are given in Table 11.2 as variations with respect to the base year.

In scenario 1, the agricultural policy measures were the ones in effect in 2006. The variations in the results for this scenario with respect to those of the base year 2002 scenario (Agenda 2000) were due to the agricultural policy adopted in 2006 and the increase in prices between 2002 and 2007.

Scenario 2 differed from scenario 1 only in the elimination of the compulsory 10 per cent set-aside. That is, the land set aside under both the 2002 and the 2006 measures was available for farming in scenario 2.

In the main scenario, the primary object of the present analysis, the new measures referred to above for cotton, beet and the dairy sector were included, and the mandatory set-aide was eliminated.

Table 11.2   Variation (in %) in the results for simulated scenarios compared to the base year

	SCENARIO 1 2006 measures Compulsory set-aside: 10%	SCENARIO 2 2006 measures Compulsory set-aside: 0%	MAIN SCENARIO 2006 + new measures Compulsory set-aside: 0%
Cereals (ha) (except rice)	2.31	12.22	12.71
Rice (ha)	-0.35	-0.34	-0.34
Oilseed (ha)	-19.12	-2.12	-1.62
Grain legumes (ha)	-40.02	-26.64	-26.64
Sugar beet (ha)	-3.66	-1.55	-50.00
Cotton (ha)	-19.00	-19.00	-34.28
Gross margin (EUR, real terms)	2.22	5.39	4.41
Payments (EUR, nom. terms)	9.99	10.08	15.83

The following may be deduced from the analysis of the variations shown in the table.

Cereals. Despite the partial decoupling of the compensatory payments for cereals, the steep rise in prices led to a 2.5 per cent increase in the farming area, even with the mandatory set-aside in effect (scenario 1). When it was not (scenario 2), the area increased by approximately 12 per cent. The rest of the new measures studied had no significant effect on this rise.

Rice. The slight price decline between 2002 and 2007 barely impacted the farming area for this crop, despite the competition from other crops with steep price rises. This was due to the substantial rise in payments for rice between the base year and 2006.

Oilseed. The partial decoupling of oilseed payments and their price made them much less profitable than the cereals that competed with them for farming area. Nonetheless, the amount of area yielded to the latter was much smaller when the set-aside, still mandatory in 2006, was recovered for farming. Moreover, with the introduction of the new cotton and beet policies (main scenario), oilseed occupied part of the farming area formerly devoted to those crops.

Grain legumes. The Spanish decision to fully decouple payments for grain legumes, in conjunction with their price, which was lower in 2007 than the price paid for the cereals with which they compete, led to a substantial reduction in the farming area used to grow these crops. This decline was smaller, although nonetheless significant, when the 10 per cent set-aside was released for farming. The new cotton and beet policies had no impact on the grain legume farming area, because in the farm types studied, grain legumes are non-irrigated, whereas cotton and beet are irrigated crops.

Sugar Beet. Despite partial decoupling, the substantial price rise in the cereals studied made them more profitable under the 2006 measures than sugar beet on the farms where the two types of crops competed. While this led to a decline in sugar beet farming area, approximately half of the loss was recovered when the 10 per cent set-aside was released for farming. The 50 per cent reduction in sugar beet farming area in the main scenario was due to the lower sugar quotas established in the new proposal for the sub-sector. Be it said in this regard that without this constraint the simulations showed that the sugar beet area would be about 30 per cent of its area in the base year.

Cotton. Under the 2006 measures, the farming area for this crop came to approximately the total eligible (70,000 ha) for coupled aid, regardless of whether a 10 per cent or a 0 per cent mandatory set-aside was used. When the recently proposed measures were assumed to be in effect, the area devoted to cotton declined to 56,758 ha (65.72 per cent of the base year figure), which is more than would be eligible for coupled payments (48,000 ha), despite the penalization per hectare applied for exceeding that ceiling.

Gross margin. In scenario 1 the gross margin rose by approximately 2 per cent compared to base year 2002. This increase was essentially due to the steep rise in prices (a simulation with the 2006 partial decoupling measures and base year prices showed a 2.8 per cent decline in gross margin).

The recovery of set-aside land for farming (scenario 2) led to a three percentage point increase in gross margin. When all the new measures considered in this study, i.e., a 2 per cent increase in the dairy quota and new cotton and sugar beet policies, in conjunction with the cultivation of mandatory set-aside land, were implemented, the gross margin was just slightly over two points higher than in scenario 1. The reason for this dip compared to scenario 2 is that the new measures for cotton and beet partially offset the increase in gross margin induced by the growth in farming area and the dairy quota.

Payments. The 2006 measures led to higher payments than in the base year due to the increase in certain types of aid (for rice for instance) and the institution of new measures (such as for cotton and dairy products). The enlargement of the farming area with the elimination of the mandatory set-aside (scenario 2) did not, logically, lead to higher payments: on the one hand, the penalizations per ha applied to coupled payments for exceeding the eligible farming area kept the total sum unchanged despite increases in the amount of farming area that would initially qualify for payments. On the other hand, the decoupled payments could not grow either, for they were limited to the amounts payable for the area eligible for such aid in the base year.

Payments were higher under the new measures as a result of the new provisions for beet and the higher dairy quota.

Livestock. The variations with respect to the base year were similar in the three scenarios. Table 11.3 gives the results for the main scenario.

Table 11.3   Variations in livestock numbers with respect to the base year, in per cent

	All farm holdings	Farm holdings in northern Spain	Proportion of total farm holdings in northern Spain
Suckler cows	-7.53	-0.66	40.08
Dairy cows	0.61	0.45	79.28
Dairy sheep	-16.13	-1.90	6.25
Non-dairy sheep	-23.82	-14.66	4.19

Suckler cows. The decline in the selling price of livestock and the increase in purchased feed prices were the chief reasons for the 7.5 per cent decline in the number of suckler cows. This type of cattle was also adversely affected, albeit to a lesser degree, by the 7 per cent decrease in payments in Spain, further to Article 69.

Dairy cows. Despite the decline in the selling price of weaned animals and the rise in the price of purchased feed, the upward trend in milk prices, the coupled payments for dairy farmers and the possibility of increasing the milk quota led to growth in dairy livestock numbers, although the increase was smaller than allowed under the 2 per cent rise in the quota.

Sheep. A sizeable proportion of the sheep-raising farm types considered in this chapter are heavily dependent on purchased feed. The rising price of such feed and the high payment decoupling rate for this type of livestock (nearly 50 per cent of the total) led to a considerable decline in the herd size, which was less steep in the case of dairy sheep.

Regional variations. Substantial regional variations were observed for suckler cows and sheep in the 140 farm types studied. These differences are illustrated in Table 11.3 which shows that the number of suckler cows varied very little in northern Spain, which accounted for approximately 40 per cent of the total number of cows in all the farm types studied. Similarly, the variation observed for dairy sheep in northern Spain was less than 2 per cent, while the figure for non-dairy sheep was nearly 50 per cent lower than for the farms considered as a whole. This smaller decline in livestock numbers in what is known as humid Spain was due to the fact that the abundant pasture land in that region makes the activity less dependent on purchased feed.

Effects of a Possible Adoption of the Full Decoupling Scheme

Table 11.4 shows how the change from the present partial to possible full decoupling would affect the main crop groups and certain economic indicators.

As the table shows, full decoupling only affected rice, sugar beet and cotton. In all three cases, this was due to the fact that the prices considered did not make these crops more profitable than others with which they compete when crop-coupled payments were decoupled. In the case of sugar beet, the farming area dipped to below the allowed maximum 50 per cent of the base year area. The cotton farming area came to around 55,000 ha, higher than the 48,000 ha for which coupled payments are guaranteed under the partial decoupling scheme.

Table 11.4   **Variations (in %) in farming area and economic indicators stemming from the change from partial to full decoupling**

Cereals (except rice) (ha)	**0.04**
Rice (ha)	-4.46
Oilseed (ha)	0.14
Grain legumes (ha)	0.00
Sugar beet (1)	-10.17
Cotton (ha)	-2.72
Potato (ha)	-0.50
Gross margin (EUR)	0.13
Payments (EUR)	2.62

The change from partial to full decoupling went hand-in-hand with a decline in farming activity, translating into a larger number of non-farmed hectares and a substantial downturn in the numbers of cattle and sheep. This decline did not affect all the regions of Spain to the same degree, however, for as Table 11.5 shows, hypothetical full decoupling had little impact on livestock in northern Spain.

Table 11.5   Variations (in %) in livestock numbers stemming from the change from partial to full decoupling

	All farms	Farms in northern Spain
Suckler cows	-19.65	-0.75
Dairy cows	-0.21	-0.23
Dairy sheep	-8.16	0.36
Non-dairy sheep	-18.34	-1.26
Total L.U.	-9.44	-0.35

Note, finally, that despite full decoupling, payments would be higher. This is because activity was lower in certain sub-sectors (suckler cows, sheep and cotton) under partial decoupling than in the base year. As a result, when payments were wholly decoupled, they were associated with a higher level of activity than when only partially decoupled. Thanks in part to this increase in direct payments, the total gross margin for all the farms as a whole was similar under the two decoupling schemes.

Conclusions

The following conclusions can be drawn from the results obtained for the scenario in which new measures for cotton, sugar beet and dairy products were incorporated and the mandatory set-aside was eliminated:

- Even with a guaranteed minimum price, the optimum sugar beet farming area would be less than allowed under the present quota, although higher than 50 per cent of that quota, as provided in the new reform for this crop.
- Despite the penalization applied to coupled payments for exceeding the 48,000-ha ceiling, the high price for cotton assumed in the simulated scenario would lead to a farming area for this crop 15 per cent above that limit.
- Under the price conditions simulated, dairy farms would not exhaust the 2 per cent rise in the quota, for the assumed increase in milk prices over the base year would be partially offset by the decline of nearly 20 per cent

envisaged in the selling price of weaned animals.

- The recovery of the 10 per cent mandatory set-aside for farming and the substantial rise in cereal prices would raise the amount of farming area devoted to these crops, which would occupy a sizeable portion of the recovered area. The magnitude of the rise shown in the model may possibly be greater than the increase that would be obtained if farms not represented in the Spanish FADN were included.
- The change from the present partial to full decoupling would not prompt any substantial variations in farming area for the chief crops or in the results for farms taken as a whole. Sheep and rearing cattle would be affected, however, with substantial declines (in addition to the downturn recorded under the partial decoupling scheme). Nonetheless, not all regions would be affected to the same extent. Before any possible full decoupling scheme is adopted, a detailed study should be conducted of its effects on sheep and rearing cattle sub-sectors in the various autonomous communities.

References

Adenáuer, M., Britz, W., Gocht, A., Gömann, H., Cristoiu, A. and Ratinger, T. 2006. Modelling Impacts of Decoupled Premiums: Building-Up a Farm Type Layer within the EU-Wide Regionalised CAPRI Model: 93rd EAAE Seminar. Prague, 22-23 September 2006.

Arfini, F. and Paris Q. 1995. A Positive Mathematical Programming Model for Regional Analysis of Agricultural Policies. Proceedings of the 40th Seminar on The Regional Dimension In Agricultural Economics And Policies. 26-28 June 1995.

Arriaza, M. and Gómez-Limón, J.A. 2003. Comparative performance of selected mathematical programming models. Agricultural Systems, 77, 155-171.

Barkaoui, A. and Butault, J.P. 1999. Positive Mathematical Programming and Cereals And Oilseeds Supply within EU Under Agenda 2000: 9th European Congress of Agricultural Economics, Warsaw, August 1999.

Barkaoui, A. and Butault, J.P. 2003. Révision a Mi-Parcours De La PAC: Simulations de L'Effect du Découplage sur L'Offre Dans Les Régions Françaises. Document de Travail. INRA-ESR Nancy.

Blanco, A. and Iglesias, E. 2005. Modelling New EU Agricultural Policies: Global Guidelines, Local Strategies: 89th EAAE Seminar .3-5 February. Parma.

Buysse, J., Fernagut, B., Harmignie, O., Henry De Frahan, B., Lauwers, L., Polomé, P., Van Huylenbroeck, G. and Van Meensel, J. 2004. Modelling the impact of sugar reform on Belgian Agriculture: International Conference on Policy Modelling. Paris, 30 June- 2 July, 2004.

Buysse, J. and Van Huylenbroeck, G. 2005. Impact of alternative implementations of the Agenda 2000 Mid Term Review: 11th Congress of the EAAE. Copenhagen, Denmark, August 24-27, 2005.

Buysse, J., Van Huylenbroeck, G. and Lauwers, L. 2007. Normative, Positive and Econometric Mathematical Programming as Tools for Incorporation of Multifunctionality in Agricultural Policy Modelling. Agriculture, Ecosystems & Environment, 120, 70-8.

Commission of the EU. 2008. Commission Staff Working Document SEC 1885.

CAPRI 2000. Common Agricultural Policy Regional Impact (Fair3-CT96-1849). Final Consolidated Report. Institute for Agricultural Policy, University of Bonn, Department of Economics, University College Galway, Institut Agronomique Meditérranéen de Montpellier, Departamento de Economia, Sociologia y Politica Agraria, Universidad Politécnica de Valencia, Università degli Studi di Bologna, Dipartamento di Protezione e Valorizzazione Agro-Alimentare (DIPROVAL), Sezione Economía.

Gohin, A. and Chantreuil, F. 1999. La Programmation Mathématique Positive dans les Modèles D´Exploitation Agricole: Principes et Importance du Calibrage. Cahiers dÉconomie et Sociologie Rurales, 52, 59-78.

Heckelei, T. and Britz, W. 1999. Maximum entropy specification of PMP. CAPRI Working Paper 08199.

Heckelei, T. and Wolff, H. 2003. Estimation of Constrained Optimisation Models for Agricultural Supply Analysis Based on Generalised Maximum Entropy in European. Review of Agricultural Economics, 30(1), 27-50.

Heckelei, T. and Britz, W. 2005. Models Based on Positive Mathematical Programming: State of the Art and Futher Extensions. Plenary paper to the 89th EAAE Seminar, 3-5 February. Parma, 48-73.

Helming, J.F.M., Peeters, L. and Veendendaal, P.J.J. 2001. Assessing the Consequences of Environmental Policy Scenarios in Flemish Agriculture in Agricultural Sector Modelling and Policy Information Systems edited by T. Heckelei, H.P. Witzke, and W. Henrichsmeyer. Vank Verlag, Kiel.

Henry de Frahan, B., Buysse, J., Polomé, P., Fernagut, B., Harmignie, O., Lauwers, L., Van Huylenbroeck, G. and Van Meensel, J. 2007. Positive mathematical programming for agricultural and environmental policy analysis: review and practice, in Management of Natural Resources: A Handbook of Operations Research Models, Algorithms and Implementations edited by A. Weintraub, T. Bjorndal, R. Epstein, and C. Romero. Springer, International Series in Operations Research and Management Science (Series Editor: F.S Hillier. Stanford University).

Jayet, P.A., Marzochi, E., Hofstetter, A. and Donati, M. 2000. Modular Approach of Agricultural Supply Modelling Through Linear Programming. EUROTOOLS, Working Paper, Series n.26.

Júdez, L., Martínez, S. and Fuentes-Pila, J. 1998. Positive Mathematical Programming Revisited. Trabajo presentado al meeting sobre EUROTOOLS (Proyecto FAIR 5-CT97-3403).Madrid. 16-17 Octubre 1998.

Júdez, L., Chaya, C., Martínez, S. and González, A.A. 2001. Effects of the Measures Envisaged in 'Agenda 2000' on Arable Crop Producers and Beef and Veal Producers: An Application of Positive Mathematical Programming to Representative Farms of a Spanish Region. Agricultural Systems. 67, 121-138.

Júdez, L., de Miguel, J.M., Piniés, M., Legorburu, I.G. and Miguel, J.L. 2003. Effects of Mid-Term Review Measures on some Representative Spanish Farm: An Application of the PROMAPA Model:80th EAAE Seminar. Ghent, 24-26 September 2004.

Júdez, L., de Andrés, R., Ibáñez, M. and Urzainqui, E. 2008. A Method for Including in PMP Models Activities Non –Existent in the Baseline Situation: XIIth Congress EAAE. Ghent, Belgium, 26-29 August .

Kuepker, B. and Kleinhanss, W. 2006. Comparative Assessment of National and Alternative Decoupling Schemes in EU Member States: 93rd EAAE Seminar. Prague 22-23 September 2006.

Offermann, F., Kleinhanss, W., Huettel, S. and Kuepker, B. 2005. Assessing The 2003 CAP Reform Impacts on German Agriculture Using the Farm Group Model FARMIS, in Modelling Agricultural Policies: State Of The Art And New Challenges edited by Arfini F. Proceedings of the 89th EAAE Seminar. Parma, Italy.

Osterburg, B., Offermann, F. and Kleinhanss, W. 2001. A sector consistent farm group model for German agriculture, in Agricultural Sector Modelling and Policy Information Systems edited by Heckelei, T., Witzke, H.P. and Henrichsmeyer, W. Kiel, Wissenschaftsverlag, Vauk 2001.

Paris, Q. and Howitt, R.E. 1998. An Analysis of Ill-posed Production Problems Using Maximum Entropy. American Journal of Agricultural Economics. 80(1), 124-138.

Paris, Q., Montresor, E., Arfini, F. and Mazzocchi, M. 2000. An Integrated Multi-Phase Model for Evaluating Agricultural Policies Through Positive Information: Proceedings of the 65th EAAE Seminar on Agricultural Sector Modelling and Policy Information Systems. Bonn University, Vauk Verlag Kiel. March 29-31, 2000.

Röhm, O. and Dabbert, S. 1999. Modelling regional production and Income Effects in Countryside Stewardship: Farmers, Policies and Markets edited by G. Huylenbroeck, M. Witby. Pergamon, Amsterdam.

Röhm, O. and Dabbert, S. 2003. Integrating Agri-Environmental Programs into Regional Production Models: An Extension of Positive Mathematical Programming. American Journal of Agricultural Economics, 85(1), 254-265.

Severini, S. and Cortignani, R. 2008. Introducing Deficit Irrigation Crop Techniques Derived by Crop Growth Models into a Positive Mathematical Programming Model: Paper at the XIIth Congress EAAE. Ghent, Belgium, 26-29 August 2008

PART 4
The Impact of CAP Reform on Specific Sectors

Chapter 12

The Abrogation of the Set-Aside and the Increase in Cereal Prices: Can They Revert the Decline of Cereal Production Generated by Decoupling?

Simone Severini and Stefano Valle

Introduction

The decoupling of direct payments, caused by the introduction of the Single Payment Scheme (SPS), has generated an incentive for farmers to decrease the production of cereals, oilseeds and protein crops (COP) and (because of the reform of sugar CMO and fruit and vegetable CMO), sugar beet and vegetables. In some cases, this has also provided a strong enough incentive for farmers to leave some of the available land idle.

However, in the period 2006-08, cereal prices increased sharply under the pressure of growing world demand and the EU Commission abrogated the set-aside requirement, allowing cultivation on idle land. In this way, the Commission intended to allow EU farmers to take advantage of the new market conditions and to stabilise the cereal market.

This chapter aims, firstly, at assessing how much the abrogation of the set-aside requirement can be effective in increasing cereal production. This is not a trivial question, given that the introduction of the SPS has also resulted in some land that was previously being cultivated (i.e. not set-aside) not being cultivated at all. Under this circumstance, the abrogation of set-aside may not result in an increase in production. The second aim of the chapter is to evaluate to what extent increases in cereal prices could foster cereal production and reduce the amount of uncultivated land. The third objective is to analyse how much the impact of the abrogation of set-aside requirement could differ under different cereal price levels.

The analysis has been carried out on a sample of FADN farms located in two regions of Italy (Emilia Romagna and Veneto) by means of the Positive Mathematical Programming (PMP) models. The sample has been stratified to take into account the different behaviour of farmers running small, medium and large farms. The performed simulations refer to the introduction of the SPS and reforms of other CMOs, the abrogation of the set-aside requirement and increases in cereal prices with and without the abrogation of the set-aside requirement.

The following paragraph reviews the evolution of cereal prices in the recent past; paragraph 3 provides information on the programming models, the selected farm samples, and the simulation scenarios. Paragraph 4 describes simulation results while the last paragraph draws some conclusions on these results.

Background

The regulation (EC) n. 73/2009, apart from completing the decoupling of direct payments and reinforcing modulation, abrogates the set-aside requirement.

The rationale for the set-aside abrogation derives from the increasing demand for cereals (for food, animal feeding and bio-fuels) that has occured over the last few years (Tangermann 2008).

The evolution of the world market for cereals has had an impact on cereal prices in Italy, too. Figure 12.1 depicts the evolution of price indexes for all crop products and for cereal in the years following the Fischler reform (ISTAT).

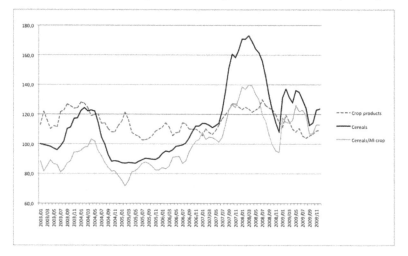

Figure 12.1   The evolution of price indexes for cereals and all crop products in Italy (2003-2009)

Source: elaboration on ISTAT data

Cereal prices, in comparison with the price index for all crop products, rose steadily from 2005 up to the beginning of 2008. The ratio among the price indexes for cereals and for all crop products between 2003 and 2007 (yearly averages) increased by 28.9 per cent. However, the data shows that the increase in cereal prices was even stronger in the period September 2007 – August 2008: in fact, from 2003 up to this period, the price index ratio increased by 49.1 per cent. Note that average cereal price level in 2009 was approximately at the 2007 average price.

There is no consensus on the level at which cereal prices will stay in the future. Therefore, the simulations performed by means of the farm PMP models consider the hypothesis of relative increases of cereal prices of between 30 and 50 per cent from baseline levels i.e. similar in magnitude to the ones observed in the recent past.

Material and methods

The analysis is conducted by means of the Positive Mathematical Programming (PMP) farm models. The PMP was developed in the 90's (Paris 1993, Howitt 1995, Arfini and Paris 1995, Paris and Howitt 1998) and it has been largely utilized for agricultural policy analysis in Europe (Heckelei et al. 1997, 2000; Paris et al. 2000, Arfini et al. 2003, Buysse et al. 2005a, 2005b).

Despite this large utilization of PMP models, it is important to highlight some limits of this methodology in order to be able to evaluate the obtained results correctly. These models refer to short-term conditions, thus they give quite conservative results especially in the case of big changes in the market or policy framework. Another limit is that these models represent only the evolution of the supply structure and, for this reason, it is necessary to assume exogenous price conditions. Furthermore, it is not possible to include activities that are not part of the baseline situation in the models, even if the Paris and Arfini approach permits the consideration of all activities that are found in a homogeneous group of farms. In the empirical analysis presented here, the introduction of the SPS represents a relevant change therefore it is likely that the utilised approach underestimates the size of the real adjustment shown by the farms in the post-reform periods.[1]

These models have been developed by using the approach proposed by Paris and Arfini (2000) and also used by Severini and Valle (2007). Therefore, the programming models used for the simulations include cost functions recovered by using the Maximum Entropy approach (Golan et al. 1996, Paris et al. 1998) in their objective functions. This methodology derives from the theories of mathematical programming duality and of production cost.

Differently from the Paris and Arfini approach, in this model the set-aside constraint is also considered in the first step of the PMP. The measures foreseen by the SPS and sugar and fruit and vegetable CMO reforms have been introduced in the models on the basis of pre-reform data (2003).The simulation scenarios considering the set-aside abrogation and the increase of cereal prices have been analysed using this last version (i.e. post-reform) of the model.

---

1   The analysis has been carried out when data referring to the years after the introduction of SPS was still not available.

For each representative macro-farm (sum of a group of homogenous farms), the model has the following quadratic objective function:[2]

$$(1) \qquad \max_{x,h,\mathit{hsp},\mathit{hgaec},\mathit{mod} \geq 0} \pi = \left(p'\underline{x} + s'\underline{h} + \text{uev}\,\underline{\mathit{hsp}} - \tfrac{1}{2}\underline{x}'\,Q\underline{x} - \text{cgaec}\,\underline{\mathit{hgaec}} - \underline{\mathit{mod}} \right)$$

where:

π	farm gross margin (EUR);
p	vector of average prices of the j activities (j x 1 dimension) (EUR/t);[3]
x	vector of the productions (j x 1) (t);
s	vector of direct payments coupled with land use (jx1) (EUR/ha);
h	vector of lands cultivated with different activities (j x 1) (ha);
uev	unitary entitlement value (EUR/ha);
hsp	land considered for the single payment (ha);
Q	matrix of the quadratic cost function coefficients (j x j) (EUR, EUR/t, EUR/t^2);
cgaec	unitary cost for good agricultural and environmental practices (EUR/ha);
hgaec	uncultivated land following good agricultural and environmental practices (ha);
mod	modulated aids (EUR).

The maximization of function (1) is subject to the following constraints:

(2) $A \cdot x + \underline{\mathit{hsa}} \leq LAND$ land availability

(3) $D \cdot x = \underline{h}$ balance between productions and cultivated areas

(4) $\underline{\mathit{hsa}} \geq \overline{\mathit{hsa}}$ set-aside

(5) $\underline{\mathit{hsp}} \leq \text{Heleg} \cdot \underline{h}$ availability of eligible land

(6) $\underline{\mathit{hsp}} \leq EntAv$ entitlement availability

(7) $\underline{\mathit{aid}} \leq \underline{\mathit{aid1}} + \underline{\mathit{aid2}}$ payments below and above modulation threshold

---

2   Symbols in **bold** refer to vectors or matrixes; other symbols refer to scalars; model variables are underlined.

3   The models consider 20 crops for Veneto Region (j = 1, ..., 20: wheat, durum wheat, maize, barley, millet, rapeseed, sunflower, soybean, pea, sugar beet, alfalfa, grassland, tomato, salad, garlic, watermelon, melon, shallot, asparagus, pumpkin, set-aside). For the Emilia Romagna Region the crops are 26 (j = 1, ..., 26: wheat, durum wheat, maize, barley, sorghum, rice, sunflower, soybean, sugar beet, alfalfa, grassland, grass meadow, potato, tomato, salad, garlic, watermelon, shallot, carrot, cabbage, onion, bean, fennel, bean, celery, courgette, set-aside).

(8) $\underline{aid1} \leq tresh$ modulation threshold

(9) $\underline{mod} \geq \underline{aid2} \cdot \mathrm{mod}\, r$ definition of modulated payments

(10) $\underline{x_{beet}} \cdot YSUG \leq Q_{SUG}$ sugar quota

where:

A	vector of crop yields inverse (j x 1) (ha/t);
hsa	set aside land (scalar) (ha);
LAND	availability of land for crops (scalar) (ha);
D	matrix where yields inverse are inserted in the diagonal (j x j) (ha/t);
hsa	set aside land according to the number of withdrawal entitlements (scalar) (ha);
Heleg	vector which identifies crop eligibility to be associate with entitlements (j x 1);
EntAv	available entitlements (scalar) (ha);
aid	pre-modulation direct payments (EUR);
aid1	payments below the threshold and not subject to modulation (EUR);
aid2	payments above the threshold and subject to modulation (EUR);
modr	modulation rate (0,05);
x_{beet}	amount of sugar beets produced by the macro-farm (t);
YSUG	output in sugar from a unit of sugar beet (t sugar/t sugar beet); [4]
Q_{SUG}	quota of sugar assigned to the macro-farm (t).

Models have been developed on a sample of 133 cereal and sugar beet producing FADN farms of three study areas located in Emilia Romagna (hilly and plain areas) and Veneto. The sample has been chosen considering only farms belonging to Type of Farming 13 (specialist cereals, oilseeds and protein crops) and 14 (general field cropping). The sample has been stratified to take into account of likely different behaviour of farmers running small, medium and large farms. Therefore, a total of nine farm models have been developed and used for simulations. Only land cultivated with field crops has been considered. Permanent crops and livestock have been excluded, where present, since they are not the object of short term planning.

The sample farms are strongly specialised in COP crops and sugar beet and they show a low level of crop pattern differentiation. This is especially true in the case of the Veneto model where four crops (wheat, maize, soybean and sugar beet) use about 90 per cent of the UAA.[5] In these farms about 70 per cent of the land is used for COP crops, while 20 per cent is used for sugar beet. Forage and vegetable

---

4   The results refer to the sum of all models of each region.
5   The Utilized Agricultural Area figures refer only the land devoted to non-permanent crops.

crops represent only 0.7 per cent of the cultivated area. The main cereals are maize (27 per cent) and wheat (16 per cent). The land cultivated with soybean (27 per cent) is also quite relevant.

The Emilia Romagna model farms (hilly and plain areas) are less specialised because more relevance is given to vegetable and forage crops. Sugar beet uses 14 per cent of the cultivated land (less than in the Veneto model). Indeed, on these farms rice, forage and vegetable crops cover a larger area and the main cereals are maize and wheat.

Economic data also show this high specialisation of the farms: in Veneto about 61 per cent of revenues is generated by COP crops, while 38 per cent of it is generated by sugar beet. In Emilia Romagna, on the other hand, due to rice cultivation, these values are respectively 65 per cent and 23 per cent.

In the Veneto sample, direct payments represent 26 per cent of the total income, while variable costs represent 35 per cent. In Emilia Romagna payments are less influent (22 per cent); indeed, on these farms there are higher unitary revenues as well as higher unitary variable costs. For this reason the Veneto farms have a higher land unitary gross margin, whereas the farms in Emilia Romagna have a higher unitary production value.

The models have been calibrated using 2003 data (*Baseline*). The first simulation has simulated the impact of the introduction of the Single Payment Scheme considering the decoupling of direct payments and other measures such as modulation and cross-compliance on uncultivated land (*MTR* scenario). Product and factor prices have been kept at the original 2003 level to account only for the change of the system of payment. Furthermore, to account for the reform of Common Market Organizations for sugar and fruit and vegetables, an additional scenario is considered (*Post-Reform* scenario). This scenario considers a decrease of sugar beet prices and the introduction of compensatory direct payments as expected after the transitional period of the reform. To account for the reform of CMO fruit and vegetable this scenario considers a decrease of the price of tomato for transformation, the introduction of the compensatory direct payment for this crop and the abrogation of Art. 51 constraint as expected at the end of the transition period (*Post Reform* scenario).

The post reform results (*MTR* and *Post Reform* scenarios) have been compared to the 2003 condition to underline the effect of the considered reforms. Particular emphasis is given to changes in crop patterns to show their effect in terms of decreasing cereal and COP production and of leaving some land uncultivated. The *Post Reform* scenario results have provided a new baseline to show the impact of the two factors considered in this chapter: the abrogation of set-aside requirement (Without *Set-aside* scenario) and the increase of cereal prices. Regarding cereal prices, two relative price increases have been considered: 30 per cent and 50 per cent (*Cereal Price Increases: 30 per cent and 50 per cent* scenarios). The impact of these increases has been considered with and without the set-aside requirement. In this way it is possible to show the impact of the abrogation of set-aside under different cereal price conditions.

Simulation results[6]

Evolution from Baseline to post-reform conditions

In the Emilia Romagna models, the MTR reform determines a strong decrease of COP crops, especially cereals, oilseeds and protein crops (Table 12.1). Sugar beet (within the available quota) and forage crops slightly increases but around 10 per cent of the UAA remains uncultivated.

Also, in the Veneto models the MTR reform determines a strong decrease of COP crops. The area not cultivated with COP is substituted by sugar beet, forage and vegetable crops. In this case, no land is left uncultivated.

In both farm groups, the CMOs sugar and fruit & vegetable reforms lead to a cut of the area cultivated with sugar beet (-22.5 per cent in Emilia Romagna and -15.7 per cent in Veneto). This decrease allows a little recovery of the COP: in comparison with the baseline situation in the Emilia Romagna models, the decline is reduced to less than 5 per cent (in particular rice and soybean), while in the Veneto models there is an increase of the COP crops (+2.3 per cent). In the Emilia Romagna models it is possible to note a reduction of the vegetable crops, in particular tomato for processing.

In Emilia Romagna, the reduction of the area cultivated with sugar beet and tomato determines a decrease of the uncultivated lands from 10 to 5 per cent of the UAA. On the contrary, in Veneto the reduction of sugar beet leads to the introduction of a small amount of uncultivated land. In general terms, it can be stated that the increased availability of titles (about +21 per cent in Emilia Romagna and +26 per cent in Veneto) and their greater value, orient the choices of farmers towards cereals, oilseeds and protein crops.

Table 12.1   Crop patterns under post-reform scenarios

	Emilia Romangna			Veneto		
	(ha)	Var. % on Baseline		(ha)	Var. % on Baseline	
	Baseline	MTR^	Post Reform^	Baseline	MTR^	Post Reform^
COP	11,359	-15.03	-4.4	5,591	-7.8	2.3
- cereals	*10,003*	-14.9	-5.1	3,411	*-8*	*3*
- oilseeds & prot. crops	*1,357*	-18.3	0.7	2,180	*-10.7*	*1.1*
Forage crops	897	7.4	22.1	47,000	155.4	160.1
Sugar beet	2,170	7.5	-22.5	1,590	22.3	-15.7
Vegetables	394	1.2	-15	8,000	139	49.4
Field cropping land	14,820	-10.2	-5.7	7,940	-	-0.6

6   The results refer to the sum of all models of each region.

Set-aside	1,040	0	0	604,000	-	-
Uncultivated land *(% on UAA)*	0	9.5	5.4	0	0	0.6
- uncult. with GAEC (ha)^^	-	0	24	-	-	-

^ MTR scenario includes SPS measures and the CMO rice reform. Post-Reform scenario includes also the reform of sugar and fruit and vegetables CMOs.

^^ GAEC = Good Agricultural and Environmental Conditions.

Source: elaboration on FADN data (2003). Estimates carried out by PMP models.

In the Emilia Romagna models, the considered reforms determine a clear contraction of both production value and production costs (Table 12.2). This is due to the reduction of the areas cultivated with cereals, sugar beet and vegetable crops, to the decrease in the sugar beet price due to the CMO sugar reform and, above all, to the decision to leave some land uncultivated. On the other hand, the value of direct payments increases. Therefore, the impact on the farm gross margins is quite small. However, it is important to highlight the relevant increase of the gross margin net of both direct payments and sustained price of sugar beet (GM net DA&P),[7] which testifies a market re-orientation of the considered farms.

Also in the Veneto models both production value and costs decrease, but less than in Emilia Romagna since there are no uncultivated areas. The value of direct payments increases but, differently than in Emilia Romagna, the impact on farm gross margins is severely negative because of the high production costs. Finally, there is a lower market re-orientation in comparison with that observed in the Emilian models.

In conclusion, the analysis suggests that, in the considered relatively homogeneous productive systems, farm behaviour consists in minimizing the production costs and in taking complete advantage of the available entitlements. This also results in a process of extensification proven by the decreasing value of the unitary total production value observed in the post reform scenarios.

---

7   This gross margin does not account for the direct payments and for that portion of sugar beet revenues (and also rice in the case of Emilia Romagna) derived from price policies. This can be considered a rough estimation of the social profitability of the farming activities as indicated by the Policy Analysis Matrix (Monke and Pearson 1989).

Table 12.2   Farm economic results under post-reform scenarios

	Emilia Romagna			Veneto		
	(.000 EUR)	Var. % on Baseline		(.000 EUR)	Var. % on Baseline	
	Baseline	MTR^	Post Reform^	Baseline	MTR^	Post Reform^
Total production value	19,605	-10.3	-18	9,024	5.7	-16.4
- cereals (%)	60.6	54.6	66.6	41.1	36.8	50.9
- sugar beet (%)	23.0	27.5	13.0	38.3	44.2	23.2
Total direct payments	5,688	19.9	43.1	3,131	-2.8	27.8
– coupled pay.s (%)	100	23.7	18.3	100	8.3	10.4
Modulation (% on tot. pay.s)	0	4.73	4.75	0	4.63	4.69
Net total direst payments	5,688	14.3	36.3	3,131	-7,3	21.9
Total costs	11,353	-15.7	-13.8	4,282	9.3	-4.2
Gross Margin (GM)	13,939	4.2	0.8	7,872	-1.4	-7.8
GM net of direct payments	8,252	-2.8	-23.7	4,742	2.5	-27.5
GM net of DA&P^^	5,869	3.7	7.2	3,361	-5.6	2.4
Unit. Prod. Value (EUR/ha UAA)	1,236	1,109	1,014	1,151	1,217	962
Tot. pay.s/Prod. Value & Tot. dir. pay.s (%)	22.5	27.9	33.6	25.8	24.2	34.7
Total pay.s/GM (%)	40.8	47	57.9	39.8	39.2	55.2

^ MTR scenario includes SPS measures and the CMO rice reform. Post-Reform scenario includes also the reform of sugar and fruit and vegetables CMOs.

^^ GM net of DP&P: gross margin net of direct payments and price support for rice and sugar beet.

Source: elaboration on FADN data (2003). Estimates carried out by PMP models.

Abrogation of set-aside under baseline price conditions

The exclusion of the set-aside constraint (*Without SA* scenario) causes different effects in the models of the two studied regions.

In the Emilia Romagna models, this makes about 1,000 hectares of land available. Under the price conditions of the pre-reform situation, only a small amount of this area is actually cultivated. Indeed, the total area cultivated increases only a little more than one per cent (Table 12.3). This increase regards exclusively cereals, including rice, oilseed crops (soybean) and sugar beet. The other land released from the set-aside remains uncultivated, even if respecting the good agricultural and environmental conditions (GAEC) required by conditionality. The economic impact is positive, but extremely limited. Indeed, gross production and coupled direct payments slightly increase but, at the same time, production costs increase (Table 12.4).

On the contrary, in the Veneto models, all land released by set-aside is reutilized so that the cultivated area increases by about 8 per cent (Table 12.3). This determines a consistent raise of both production value and production costs, which leads to a slight increase of gross margins (Table 12.4).

Table 12.3   Set-aside and cereal price simulation results. Crop patterns

	Emilia Romagna						Veneto					
	(ha)	Var. % on post-reform scenario					(ha)	Var. % on post-reform scenario				
			Cereal price increases						Cereal price increases			
			With SA		Without SA				With SA		Without SA	
	Post Reform^	W/o SA	30%	50%	30%	50%	Post Reform^^	W/o SA	30%	50%	30%	50%
COP	10,857	1.2	9.1	12.8	12.0	20.4	5,718	8.8	3.2	4.9	11.7	13.3
- cereals	9,491	1.5	13.1	19.7	16.1	26.9	3,514	6.7	14.7	23.9	21.7	31.1
- oil. & proteic	1,366	-0.4	-18.5	-35.2	-16.7	-24.8	2,204	12.2	-15.3	-25.4	-4.2	-15.1
Forage crops	1,095	-0.2	-12.1	-21.7	-10.4	-16.4	123	31.1	-9.6	-17.6	24.6	18.8
Sugar beet	1,682	1.1	-9.7	-17.6	-6.9	-11.9	1,343	4.4	-9.6	-16.4	-4.6	-11.1
Vegetables	335	-0.1	-3.4	-5.8	-2.5	-4.0	12	13.9	33.9	37.4	93	127
Field cropping land	13,968	1.1	4.9	6.0	7.6	13.0	7,194	8.4	0.6	0.6	9	9
Set aside	1,040	-100	-	-	-100	-100	604	-100	-	-	-100	-100
Uncultivated land (% on UAA)	5.4	11.0	1.1	0.1	5.2	0.4	0.6	0.6	-	-	-	-
- uncult. with GAEC (ha)^^	24	867	-	-	-	-	-	-	-	-	-	-

^ Post-reform scenario includes SPS measures, the CMO rice reform and the reform of sugar and fruit and vegetables CMOs.

^ ^ GAEC = Good Agricultural and Ecological Conditions.

Source: elaboration on FADN data (2003). Estimates carried out by PMP models.

On the whole, removing the set-aside constraint causes, in both regions, extremely limited changes in economic results. Emilia Romagna's models seem to have more difficulties in adjusting than the Veneto models.

Cereal price increases

In the Emilia Romagna models, a 30 per cent increase in the cereal prices determines the immediate cultivation of a larger share of the area than in the post reform conditions was left uncultivated (Table 12.4). This area is utilized to expand cereals cultivation. Part of the land previously cultivated with forage, oil seed, vegetable crops and sugar beet is also assigned to this destination. A 50 per cent increase in cereal prices determins almost the total disappearance of uncultivated areas (with the exception of compulsory set-aside) and the area cultivated with cereals significantly overcomes the pre-reform levels.

In the Veneto models, the increase in cereal prices leads to an expansion of these crops, which decreases the area of land devoted to oil seed, forage crops and sugar beet (Table 12.3).

Table 12.4 Set-aside and cereal price simulation results. Economic results

	Emilia Romagna						Veneto					
	(.000 EUR)	Var. % on post-reform scenario					(.000 EUR)	Var. % on post-reform scenario				
			Cereal price increases						Cereal price increases			
			With SA		Without SA				With SA		Without SA	
	Post Reform	W/o SA	30%	50%	30%	50%	Post Reform	W/o SA	30%	50%	30%	50%
Total prod. value	16,075	0.8	28.2	47.3	31.1	55.9	7,541	7.3	19.9	34.9	29.0	45.2
– cereals (%)	66.6	66.9	76.5	81.3	76.6	81.3	50.9	50.1	63.6	70.7	62.4	69.2
– sugar beet (%)	13.0	13.0	9.2	7.3	9.2	7.3	23.2	22.5	17.5	14.4	17.1	14.2
Total direct payments	8,138	0.3	1.6	2.2	2.1	3.5	4,002	0.8	-0.1	-0.2	0.6	0.5
– coupled pay.s (%)	18.3	18.5	19.5	20.0	20.0	21.0	10.4	11.1	10.3	10.2	11.0	10.9
Modulation (% on tot. pay.s)	4.75	4.75	4.75	4.75	4.75	4.76	4.69	4.69	4.69	4.69	4.69	4.69
Net total direct payments	7,752	0.3	1.5	2.2	2.1	3.5	3,815	0.8	-0.1	-0.2	0.6	0.5
Total costs	9,783	1.2	13.1	19.3	16.1	31.3	4,101	12.6	6.3	11.2	20.4	26.4
Gross Margin (GM)	14,044	0.3	24.0	41.8	25.5	44.1	7,255	0.9	17.1	29.8	19.0	32.4
GM net of direct pay.s	6,293	0.2	51.7	90.7	54.4	94.2	3,440	1.1	36.2	63.2	39.3	67.7
Unitary Production value (EUR/ha UAA)	1,014	1,022	1,299	1,493	1,329	1,580	962	1,032	1,153	1,297	1,240	1,397
Tot. pay.s/Prod. value & total dir. pay.s (%)	33.6	33.5	28.6	26.0	28.3	25.2	34.7	33.3	30.7	28.2	29.3	26.9
Total pay.s/GM (%)	57.9	58.0	47.4	41.8	47.1	41.6	55.2	55.1	47.0	42.4	46.7	41.9

^ Post-Reform scenario includes SPS measure, the CMO rice reform and the reform of sugar and fruit and vegetables CMOs.

Source: elaboration on FADN data (2003). Estimates carried out by PMP models.

The cereal prices raise has a strong positive impact on the farm economic results. In the Emilia Romagna farms, in comparison with the post-reform condition, the total production value increases by 28 and 47 per cent and the gross margins by 24 and 42 per cent. In the Veneto models these increases are smaller (20 and 35 per cent and 17 and 30 per cent), since in this region almost all land was already being cultivated (Table 12.4).

In both models the gross margin increases are smaller than those of the revenues because of the steady raise of production costs. The gross income net of direct payments remarkably increases since these latters account for a smaller share of the whole revenues and of the gross margins if compared with the post-reform condition.

Increases in cereal prices without the set-aside requirement

The effect of the abrogation of the set-aside constraint is more relevant if cereals prices increase too. In the Emilia Romagna models, the area assigned to cereals increases by about 16 and 27 per cent (while maintaining set- aside by 13 and 20 per cent) and the total cultivated land reaches 95 and 99.6 per cent of the UAA (Table 12.3). In the case of a 30 per cent cereal price increase, most of the set-aside area becomes cultivated, while with a 50 per cent increase all post-reform uncultivated land is utilized. In the Veneto models, the area assigned to cereals increases by about 22 and 31 per cent.

The abrogation of set-aside permits to take advantage of the positive price evolution. Indeed, the gross margins increase, due to the cereal price raise, is much more consistent if associated with the release of the set-aside constraint (Table 12.4). While the abrogation of set-aside maintaining the old cereal price condition generates a 0.3 per cent increase in Emilia Romagna and a 0.9 per cent increase in Veneto, in the case of higher cereal prices the increases are more consistent: about 1.7 and 2.5 per cent (Table 12.4).

Conclusions

The introduction of the SPS has generated an incentive to decrease the production of cereals, oilseeds and protein crops and, in some cases, to leave a share of the available land uncultivated. In the years following the reform, cereal prices increased sharply under perculiar world market conditions. Under these circumstances, the EU abrogated the set-aside requirement to allow the cultivation on idle land.

This analysis has shown that, in the farms that were taken into consideration, the combined effect of increasing cereal prices and set-aside abrogation generates a relevant increase of cereal production and farm economic performances. However, the role of these two factors is quite different. First of all, the abrogation of set-aside under the cereal price levels found in the pre-reform period is generating only a negligible impact in this respect. Secondly, the considered increases in cereal prices provide (even maintaining the set-aside requirement) a strong incentive for farmers to use all available land, to increase cereal production and to improve the economic results of their farms. Indeed, the considered price increases seem able to significantly modify the economic environment in which farmers operate. Thirdly, the set-aside abrogation *per se* is able to foster cereal production and to improve farm economic results only under the scenarios where high cereal prices are taken into consideration. Under these circumstances, the idle land has a productive use and the increase of cereal production generates large economic benefits.

To summarise, the analysis suggests that the abrogation of the set-aside constraint is very appropriate under high cereal price conditions. This allows cereal-producing farmers to take advantage of the new market conditions, to respond to the increasing demand for cereal in the international markets and to eliminate the economic distortions caused by the set-aside requirement. However, the analysis has shown that, at least in the considered farms, the abrogation of the set- aside requirement under the low cereal prices found in the period immediately following the introduction of the SPS and nowadays, could generate only a negligible impact on land use, cereal production and farm economic results.

The increases in cereal prices could be more effective than the abrogation of the set-aside requirement in increasing cereal production. The combination of both considered factors is expected to reverse the decline of cereal production experienced in the considered farms after the introduction of the SPS, even if the magnitude of this effect is strongly affected by the level of cereal prices. In particular, under the current situation of low cereal prices, it seems that the possibility to crop the land that has been set-aside could not have a relevant impact on farmers' choices and the economic performances of their farms.

References

Arfini, F. and Paris, Q. 1995. A positive mathematical programming model for regional analysis of agricultural policies, in *The Regional Dimension in Agricultural Economics and Policies* edited by F. Sotte. *Proceedings of the 40th EAAE Seminar*, June 26-28, Ancona, pp. 17-35.

Arfini, F., Donati, M. and Paris, Q. 2003. *A National PMP Model for Policy Evaluation in Agriculture Using Micro Data and Administrative Information.* Paper to the *International Conference on Agricultural policy reform and the WTO: where are we heading?* Capri (Italy), June 23-26, 2003.

Buysse, J., Fernagut, B., Harmigni, O., Henry de Frahan, B., Lauwers, L., Polomé, P., Van Huylenbroeck, G., and Van Meensel, J. 2005a. *Quota in Agricultural Positive Mathematical Programming Models,* in *Modelling Agricultural*

Policies: State of the Art and New Challenges, edited by F. Arfini. *Proceedings of the 89ᵗʰ EAAE Seminar,* February 3-5 2005, Parma, pp. 233-250.

Buysse, J., Fernagut, B., Harmigni, O., Henry de Frahan, B., Lauwers, L., Polomé, P., Van Huylenbroeck, G. and Van Meensel, J. 2005b. *Farm-based modelling of the EU sugar reform: impact on Belgian sugar beet suppliers,* European Review of Agricultural Economics, received November 2004, final version received December 2006 (in print).

Golan, A., Judge, G. and Miller, D. 1996. *Maximum Entropy Econometrics: Robust Estimation With Limited Data,* John Wiley, New York.

Heckelei, T., and Britz, W. 1997. *Pre-study for a medium simulation and forecast model of the agricultural sector for the EU,* CAPRI Working Papers, University of Bonn.

Heckelei, T., and Britz, W. 2000. Concept and explorative application of an EU-wide, regional Agricultural Sector Model (CAPRI-Project) in *Agricultural Sector Modelling and Policy Information Systems* edited by T. Heckelei, H.P. Witzke and W. Henrichsmeyer. *Proceedings of the 65ᵗʰ EAAE Seminar*, March 29-31, 2000, Kiel, Wissenschaftsverlag Vauk, pp. 281-290.

Howitt, R.E. 1995. Positive mathematical programming. *American Journal of Agricultural Economics*, 77, 329-342.

ISTAT – www.istat.it.

Monke, E. A. and Pearson, S. R. 1989. *The Policy Analysis Matrix for Agricultural Development.* Cornell University Press, Ithaca and London.

Paris, Q. 1993. *PQP, PMP, Parametric Programming, and Comparative Statics,* Department of Agricultural Economics, University of California, Davis, *Lecture Notes*.

Paris, Q. and Arfini, F. 2000. Funzioni di costo di frontiera, auto-selezione, rischio di prezzo, PMP e Agenda 2000, *Rivista di Economia Agraria*, 3(2), 211-242.

Paris, Q. and Howitt, R.E. 1998. An Analysis of Ill-posed production problems using maximum entropy. *American Journal of Agricultural Economics*, 80(1), 124-138.

Severini, S. and Valle, S. 2007. La riforma dell'OCM zucchero in Italia: una valutazione di impatto sui produttori bieticola di Veneto ed Emilia Romagna, *Rivista di Economia Agraria*, n.1. Available at: http://www.inea.it/public/it/pubblicazioni.php?action=8&id=298

Tangermann, S. 2008. *Policies, markets and food prices: who/what to blame? What's ahead:* Paper to the XIIᵗʰ of the EAAE Congress: People, Food and Environments: Global Trends and European Strategies. August 26-29, Ghent (Belgium).

Chapter 13

Policy Impact on Production Structure and Income Risk on Polish Dairy Farms

Adam Wąs and Agata Malak-Rawlikowska

Introduction

The adoption of European Common Market principles has been the main driving force behind dairy sector restructuring in Poland after EU accession, strongly affecting the income situation of farmers (Wilkin et al. 2007). Today further policy changes may be expected. Dependence on policy related transfers (market price support and direct income support) means that farm incomes are increasingly exposed to price and income risk related to Common Agricultural Policy (CAP) reforms. The most recent studies of the dairy sector in Poland focused predominantly on the issue of dairy farms' efficiency and the relationship between profitability and cow herd size (Parzonko 2006). Relatively little attention, however, has been devoted to investigations of dairy farms' income risk in changing policy conditions. Available studies on farmers' income and revenue risk, although providing valuable insights, do not analyse the problem in the Polish context (Moschini et al. 2001) nor the dairy sector in particular (Majewski et al. 2008). To fill this gap the present study examines the impact of changes of key policy factors on the income risk of Polish dairy farms in the context of the years 2013 and 2018.

Methods

In the long term, policy changes influence not only farm incomes but also production patterns. A deterministic linear programming farm model was used for estimation of changes in production patterns, while a stochastic simulation farm model was applied for examining income risk.

The research based on the Polish FADN and pre-FADN datasets was conducted for dairy farm type TF41, according to the FADN typology (FADN 2006a). There were two economic size clusters analysed: 8-16, 16-40, ESU. For a number of policy scenarios optimal production structure has been generated with the use of the LP farm model. The objective function in the LP model was maximization of Net Farm Income, calculated in line with the FADN income derivation scheme. In the optimization model apart from FADN data disaggregated parameters based on farm surveys, normative data and expert estimations were used.

Production patterns for policy scenarios obtained from optimal model solutions have been applied as one of the entry parameters in the Monte Carlo simulation model. Remaining parameters of the model, inter alia distributions of yields, prices and correlations were derived from data series from farm accountancy systems (a sample of 285 farms) for years 1998-2004 and general statistics.

For future policy scenarios appropriate adjustments of parameters based on historical data have been introduced. Future yields were extrapolated from long term trends while assumptions regarding input prices were based on specialists' judgment. Future prices of agricultural commodities were taken after Majewski et al. from FAPRI/OECD price predictions (Majewski et al. 2008).

Scenarios

The milk quota system and direct payments are presumed to be the most important policy factors determining incomes of dairy farms. Thus scenarios developed for the analyses assume mainly changes in those areas from 2004 (base year) situation up to the year 2018.

The following EU agricultural policy scenarios were considered:

- Base 2004 – historic reference scenario;
- CAP 2013 – reflection of continuation of all existing policies including implementation of the already agreed reforms (Luxembourg 2003) with minor changes assumed (10 per cent mandatory modulation of direct payments, gradual increase of the milk quota by 1 per cent annually since 2008 [EU Commission 2008];
- LIB 2013 – full liberalization of agricultural policy, withdrawal of all market and direct support and regulatory measures. EU farm prices equal world market prices for 2013;
- CAP 2018 – further decrease of market price support level, 20 per cent of mandatory modulation of direct payments, gradual increase of the milk quota by 1 per cent annually;
- LIB 2018 – withdrawal of all market and direct support and regulation measures. EU farm prices equal world market prices as predicted for 2018.

For each scenario two sets of results have been calculated. In the first set (FIX) cropping structure as observed in 2004 was fixed in all policy scenarios, while in the second (OPT) cropping structures were optimized in the LP model. Comparison of both variants gave an overview on the impact of adjustments in cropping structures on income level and its volatility. No investment activities which would lead to farm growth were considered in the model. Both crop and animal production were optimized for the base year farm resources. Such an assumption was made in order to ensure comparability of FADN size clusters within all scenarios.

Basic assumptions regarding model parameters are presented in Table 13.1.

Table 13.1   Indices of changes of the key model parameters for policy scenarios considered (BASE 2004 = 100)

Scenario	BASE 2004	CAP 2013	LIB 2013	CAP 2018	LIB 2018
Milk quota [2004 =100]	100	105	No quota	110	No quota
Milk quota lease	Not allowed	10	No quota	10	No quota
price [% of milk price]	100	100	0	100	0
Sugar quota	100	125	125	141	141
Maximum milk yield increase	100	114	114	123	123
Average yield increase in crop production					
Input prices*					
Fertilizers & pesticides	100	120	115	130	115
Seeds	100	125	120	140	125
Labour	100	150	150	180	180
Concentrates	100	120	115	130	125
Veterinary services	100	120	115	130	125
Fuel	100	120	120	130	130
Agricultural commodity prices**					
Wheat	100	99	93	99	93
Barley	100	102	97	101	95
Other cereals	100	96	90	94	89
Corn (grain) & Proteins	100	95	90	93	88
Oilseeds	100	99	94	100	95
Potatoes	100	97	97	103	103
Sugar beets	100	56	43	56	43
Milk	100	83	68	84	69
Beef	100	108	62	109	63
Pork	100	108	97	112	101

*own assumptions

**own assumptions based on E. Majewski et al 2008.

The analysis of historical data reveals low rates of yield improvements in Poland which can be attributed to a variety of unfavourable (both financial and structural) conditions related with the economic transformation. Relatively low current yields and general improvement in economic conditions due to EU accession suggest that growth rates above those calculated from historical trends should be applied in most cases due to a likely catching-up process. This explains relatively high coefficients of yield growth assumed for modelling.

Characteristics of the analysed farm types

There were two dairy farm types modelled. The type 1 represents 8-16 ESU cluster, with its 11.8 ESU and 22.1 ha of UAA, 16 dairy cows, 4.6 LU of other cattle, and 0.6 LU of pigs. The type 2 is a larger farm of 22.2 ESU, which represents the 16-40 ESU cluster and possesses 38.5 ha of UAA, 28.1 cows, 9.3 LU of other cattle and 0.5 pigs. Both farms have a similar cropping structure dominated by cereals (84 per cent in type 1 and 85 per cent in type 2). However, the smaller farm has a higher share of the fodder crops on arable land (7.4 per cent vs 5.3 per cent).

The estimation of standard deviation in the base period, which is a basic measure of variability of yields and prices in the simulation model, created some difficulties related mainly to available sources of data. Data from two different sources have been merged in order to achieve a minimum length of required time series for the estimation: FADN for the period 2002-2004 and the Farm Survey[1] for the years 1997-2001 after adjusting to the FADN standard. For a given farm type (activity, size) all observations have been pooled across years (1997-2004) and standard deviations were estimated for the whole set of variables. Both data bases were merged for our estimations in the following way:

- all farms from the Farm Survey which represent farm types selected for simulations;
- randomly drawn 10 per cent of the FADN farm population;
- As a result, the total number of farms in the "merged" data base varied in consecutive years. Simulation model parameters has been estimated based on 285 farm records, 171 for 8-16 ESU and 114 for 16-40 ESU farm size cluster.

Models

Optimization model

In order to simulate the effects of different policy scenarios, a linear programming farm model has been used to optimize the production structure of two FADN farm types. The model has been constructed in an Excel spreadsheet and solved with the Solver function. The farm model uses over 80 decision variables and up to 200 constraints. Net farm income was the objective function in the model.

---

1   Farm Survey conducted by the Institute of Agricultural and Food Economics in Warsaw. Polish FADN, which have been established very recently, provides data for the years 2002-2004 only, but for a large sample of farms 12000 in the year 2004. The Farm Survey, which is not fully compatible with FADN, provides historical data for a long period, but for a much smaller population of farms (about 1000 on average in the period considered).

A set of balances has been incorporated into the model to secure internal integrity of the results. The most important are the balances of stands for animals with farm buildings available and the balance of agricultural land in which full utilization of land is assumed with rotational ties for crops. An animal feed nutrient balance is obtained by optimization of the farm produced fodder area and the calculated necessary supply of purchased concentrates.

All parameters were introduced into the model in a disaggregated form including the farm enterprises with associated yields and input requirements, product prices, input costs, cost of land lease and production quotas, services, seasonal and permanent employment and other financial burdens of the farms.

Simulation model

The level and volatility of farm income was estimated with the use of the Monte Carlo simulation method in a farm model constructed for the @Risk package. The main parameters of the base model which were calculated from historical data are as follows (Majewski et al. 2007b):

- Means of structural variables to describe the farm types (e.g. size of activities, yield, prices, inputs or costs);
  - for the Base scenario, calculated from FADN data base for the years 2002-2004;
  - for future scenarios production structure obtained from the optimization model, while prices and yields assumed based on available forecasts.
- Standard Deviation for selected variables:
- Cross correlations:
  - farm related (input-output, input-input) from historical farm data;
  - market related (price-price, price-yield; yield-yield) from national statistics data.

Due to data limitations, input-output correlations for crop production were not included in the model.

Most of the farm activities in the model were described by the parameters of the distributions (standard deviation) of yields and prices. Similarly, the standard deviation was estimated for selected cost variables (energy, fertilizers, pesticides, seeds, purchased and farm produced feed for animals). Other variables of the model (e.g. fixed costs) were introduced as constant values specific for each farm type.

For simplification, a normal distribution for all variables was assumed. The distribution was truncated on the left side at 0 for yields and for prices at the values, optionally, of $x - 2\sigma$ or 0 or the intervention price, depending on which was the highest.

The simulation model was solved with 5,000 replications to ensure stability of results.

Calibration

To obtain consistency between both models a calibration procedure has been applied. In the initial run of the LP model the production structure was fixed at the 2004 level. A number of technological parameters (feeding balance parameters, input levels etc.) were adjusted adequately in order to generate results for all activities considered fully consistent with FADN averages.

After such calibration, the LP model has been used to produce results for both FIXED and OPT variants.

In the next step production structure obtained from LP model was applied in the simulation model. The initial run of the model was made to check whether the mean farm income simulated in the model approximates the level farm income from the optimization model. Due to difference in nature of deterministic and stochastic models in this comparison, all input volatility in the simulation model was set to 0. After confirming the consistency between both models, volatility parameters were applied according to scenario assumptions to get final results.

Results

Comparison of model results across scenarios cereals the direction of possible adjustments in the cropping structure and to point out the impacts of policy changes.

Production structure and net farm income

In both farms certain adjustments of the production pattern to the given scenario conditions has been observed. The difference is significant, especially in case of the historic and optimal production structure in 2004 (Figure 13.1). In the optimal solution for the base year, the model increased significantly the area of fodder crops compared to the initial, real cropping structure. This is because of substantial changes of economic conditions (mainly prices of agricultural commodities and eligibility of fodder crops for area payments) after Polish accession to the European Union in the year 2004. The historic cropping structure was decided by farmers before the accession, when price – cost relations were different.

There are more changes in the cropping structure observed in model solutions for the 2013 and 2018 scenarios which are influenced by varying prices and costs, but also yield increases are assumed. The main difference in comparison to the Base scenario is the removal of fodder crops grown on arable land after an optimal diet for cattle is composed of fodder from permanent grasslands and concentrates. The share of wheat and barley, the most profitable of all cereals, is increased to the maximum level allowed by constraints. In the CAP scenarios the model takes more of other cereals than in the Liberal scenario, which favours rape oil seed rape.

On the larger farm 16-40 ESU (Figure 13.1) the pattern of changes in the share of fodder crops is similar to what was observed in the model solutions for the 8-16 ESU farm type. There are no visible differences in the crop production structure under CAP and LIB 2013 and 2018 scenarios.

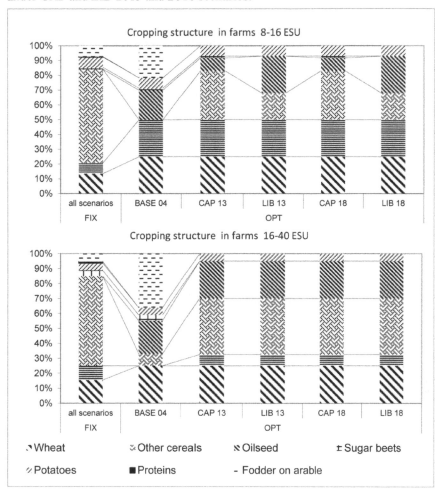

Figure 13.1 Cropping structure

Source: own calculations basing on the models

In both the 2013 CAP and LIB scenarios, the model increases milk production (Table 13.2) to the maximum possible resulting from the number of cows which can be kept on existing stands and the maximum yield of milk.

In the 2018 optimal solutions for the 16-40 ESU farm, the model does not utilize full milk production potential, setting the milk yield 15 per cent below the increased maximum. In the CAP 18 model solution the number of cows is slightly reduced.

Changes in the cattle herd structure in both farm types and removal of pigs from the larger farm indicate that the model tends to increase the specialization level in milk production.

Table 13.2   Animal production and economic performance in considered scenarios

	Farm size [ESU]	BASE 2004	CAP 2013	LIB 2013	CAP 2018	LIB 2018
Animal production						
Cows [heads]	8-16	15.9	17.1	17.1	17.1	17.1
	16-40	28.1	31.2	31.2	29.9	31.2
Milk production [th. litres]	8-16	68.56	92.2	92.2	104.3	104.3
	16-40	136.2	180.5	180.8	172.2	180.8
Milk quota lease [th. litres]	8-16	-	20.2	-	28.9	-
	16-40	-	37.7	-	22.4	-
Other cattle [LU]	8-16	4.6	3.4	3.4	3.4	3.4
	16-40	9.3	6.2	6.2	7.5	6.2
Pigs [LU]	8-16	1.3	1.5	1.5	1.5	1.5
	16-40	1	-	-	-	-
Average Net Farm Income [PLN]						
FIX* structure	8-16	33,452	24,295	-5,453	21,243	-2,752
	16-40	72,432	55,356	242	48,504	4,837
OPT** structure	8-16	38,299	29,881	-1,727	25,498	1,645
	16-40	87,740	64,631	9,186	56,318	14,271
Direct Payments						
Share of payments in Farm Income [%]	8-16	28.5%	55.8%	-	58.2%	-
	16-40	21.6%	45%	-	45.9%	-

*FIX – observed production structure

**OPT – optimal production structure form LP model

Source: own calculations basing on the models

Optimization model results prove that policy changes considered in the analysis deteriorate from base year 2004 farm incomes (Table 13.2). This applies to both farm types and all scenarios. Even an increase of payments in the year 2013 (Table 13.2), due to phasing-in[2] does not protect farm incomes from dropping.

---

2   Increase of direct payments from the level 25 per cent of the rates negotiated with the EU Commission (plus so called top-up – about 30 per cent paid from the national budget) to 100per cent in the year 2013.

A substantial difference in the level of farm incomes between the CAP and LIB scenarios is to a large extent because of direct payments which in both analysed years constitute about 50 per cent of income. Under the liberal scenario the complete withdrawal of payments and assumed decreases of prices of most of the commodities turns the incomes into negative values, except LIB 18 scenario and the 16-40 ESU farm. This is an indication that worsening of farming conditions for milk producers may inevitably lead to increases of scale of production, which was not considered in the analysis.

It is noteworthy that, albeit somewhat obvious, after optimization in the OPT scenarios the Net Farm Income is generally higher (by 14 – 22 per cent) than in solutions without optimization, with fixed production structure on the base year level (FIX). It shows that there are opportunities for farmers to improve their financial results by adjusting the production structure to actual policy situations.

Volatility of income

Volatility of farm income measured by standard deviation is similar in CAP and LIB scenarios. The optimization of the production structure does not change the value of the Standard Deviation significantly (Table 13.3). It means that the policy changes do not influence the range of income variability. However, they have a strong impact on the coefficient of volatility.

Table 13.3 Income risk and income volatility in considered scenarios

Farm structure	Farm size [ESU]	BASE 2004	CAP 2013	LIB 2013	CAP 2018	LIB 2018
Standard Deviation						
FIX*	8-16	16 552	20 976	19 842	24 595	23 729
	16-40	31 229	35 688	34 254	38 735	38 769
OPT**	8-16	16 648	21 880	20 839	26 084	25 437
	16-40	16 552	38 887	37 734	40 838	41 847
Coefficient of volatility						
FIX	8-16	49%	86%	-364%	116%	-862%
	16-40	43%	64%	14140%	80%	802%
OPT	8-16	43%	73%	-1207%	102%	1547%
	16-40	33%	60%	411%	73%	293%
Value at Risk 0						
FIX	8-16	1.88%	12.6%	61.2%	19.2%	54.5%
	16-40	0.89%	6.1%	49.9%	10.4%	44.9%
OPT	8-16	0.87%	8.8%	53.5%	16.6%	47.7%
	16-40	0.00%	4.9%	40.6%	8.4%	37.2%

*FIX – observed production structure

**OPT – optimal production structure form LP model

Source: own calculations basing on the model.

The risk of low incomes has been measured using a percentage of farms with the level of farm income below zero (Value at risk 0 – Table 13.3). The obvious result is, that in liberal scenarios as compared to the more protective CAP environment, farms are strongly exposed to risk due to incomes decreases. No market protection in the LIB 2013 and LIB 2018 scenarios is a serious threat to the farms' financial stability. Even 40-53 per cent of income observations fall into the below 0 category in the LIB 2013 scenario. The difference between CAP and LIB could be attributed to Common Agricultural Policy protection which stabilizes the market and lowers the income risk. Optimization of the production structure reduces risk. It means farmers in a more liberal policy environment and exposed to a greater risk should pay more attention to adjusting farm organization to market and economic conditions.

The range of farm income calculated as a difference between 5 per cent and 95 per cent percentiles was also considered. However CAP scenarios results showed lower risk of gaining lost, the range of income is higher when compared with appropriate liberal scenarios. It could be explained by a higher absolute income value and greater impact of yields volatility at the higher commodity prices.

Very likely liberalization of the agricultural policy will foster ongoing concentration processes in the milk production sector in Poland. As the modelling results show, incomes in smaller farms are noticeably lower, and the risk of low incomes is much higher.

Conclusions

Introduction of a less protective CAP or the complete liberalization of the agricultural policy would inevitably lead to decreasing farm incomes on dairy farms in Poland.

Abandonment of a milk quota and weakening of CAP direct support would also affect significantly increases of the risk of lower incomes. The dairy sector, which benefits from the milk quota regulation and CAP payments, loses much more in the liberal scenario. Full liberalization causes financial threats to all farmers. More radical policy changes, however, would dramatically worsen the financial situation of smaller scale milk producers, very likely driving a large number of dairy farmers out of business.

The most recent years in Poland are marked by the rapid concentration of production in a cluster of commercial farms that are growing in size. A strong liberalization of agricultural policies, together with a simultaneous reduction of production limits, would speed up significantly the process of structural changes in the Polish dairy sector.

Results of the optimization model also show that adjusting production structures to the more liberal policy environment will have a significant role as a tool for improving incomes and reducing income risk.

References

EU Commission. 2008. Health check and future perspectives of the CAP: Challenges for agriculture. Paper to the EAAE Congress, Brussels.

FADN Poland. 2006a. Standard results obtained by the farms participating in Polish FADN in 2005. [Online]. Available at: http://www.fadn.pl/mediacatalog/documents/wyniki_stand_ogolne_2005r.pdf

FADN Poland. 2006b. Standard results obtained by the individual farms having the accountancy system in 2005. [Online]. Available at: http://www.fadn.pl/mediacatalog/documents/wyniki_stand_indywidualne_2005r.pdf

FAPRI. 2007. Baseline 2007 Outlook for EU and Irish Agriculture. FAPRI.

Majewski, E., Dalton, G. and Wąs, A. 2007a. Anticipated impacts of decoupling on the pattern of production in Poland. Paper to the 93rd Seminar of the EAAE Impacts of decoupling and cross compliance on Agriculture in the EU. Czech Republic, 22-23 September 2006.

Majewski, E. and Wąs, A. 2007b. Farm income risk assessment for selected farm types in Poland - implications of future policy reforms: Paper to the IFMA Congress: A Vibrant Rural Economy – The Challenge for Balance, 2007.

Majewski, E., van Asseldonk, M., Meuwissen, M., Berg, E. and Huirne R. 2008. Economic impact of prospective risk management instruments under alternative policy scenarios. Paper to the 108 EAAE Seminar, Warsaw, 8-9 February 2008.

Moschini, G. and Hennessy, D.A. 2001. Uncertainty, risk aversion, and risk management for Agricultural producers, in Handbook of Agricultural Economic, vol. 1A Agricultural Production, edited by B.L. Gardner and G.C. Rausser. Elsevier.

OECD 2008. Working Party on Agricultural Policies and Markets, The OECD-FAO Agricultural Outlook 2008 – 2017, OECD 2008.

Parzonko, A. 2006. Efektywność ekonomiczna gospodarstw ukierunkowanych na produkcję mleka w Polsce i innych krajach europejskich. Paper to the Conference organised by Katedra Ekonomiki i Organizacji Gospodarstw Rolniczych, Wydział Ekonomiczno–Rolniczy, SGGW, 2 February 2006, Warsaw.

Wilkin, J., Milczarek, D., Malak-Rawlikowska, A. and Fałkowski, J. 2007. The Dairy Sector in Poland, Regoverning Markets Agrifood Sector Study, IIED, London.

Chapter 14

An Impact Assessment of the CAP Reform Health Check on the Italian Tomato Sector

Filippo Arfini, Michele Donati, Gaetana Petriccione and Roberto Solazzo

Introduction[1]

The fruit and vegetable CMO was broadly revised in 2007, when Regulation (EC) No. 1182/2007 entered into force. The new CAP perspectives of this sector in Italy are characterized by a progressive subsidy decoupling that will be completed in 2011 for the tomato sector and by 2012 for other processed fruit and vegetables. The Italian Ministry of Agriculture has adopted the option of the transitional period, preserving coupled payments, before applying the full decoupling process as stated by the same regulation. The reform has been particularly relevant for the processed tomato sector, where the subsidies made up about 50 per cent of the entire producers' revenue. Decoupling 50 per cent of the product payments in the first period and the 100 per cent at the end of the transitional payments means a large portion of the specific crop revenue moves from the product to the producer with the risk of reducing the marginal convenience of the crop. Such risk has become a serious concern for the tomato industry because the traditional raw material basin could reduce the supply to such a degree that the demand from the processing plants might not be completely satisfied. As it is known, these concerns significantly increased the prices paid by the industries to the producers, but it was said increase was insufficient to guarantee that all the traditional producers maintain their previous production level.

In 2008 the European Commission adopted a series of regulation proposals in order to prepare the actual CAP to the European agricultural support after 2013. The set of documents, called CAP's Health Check (HC), defines new options for Member States concerning the decoupling mechanism, reinforces the role of the modulation in transferring more financial resources from the first to the second pillar and defines a new frame for applying art. 69. In particular the HC reintroduces the possibility for the Member States that had adopted decoupling according historical criteria to implement a single farm payment based on the regionalisation principle or, in other words, on flat rate equivalents for all the farmers in a certain region.

---

1   This chapter is the result of the joint work of the authors. Nevertheless, paragraphs 1, 4 and 6 should be attributed to F. Arfini, paragraph 2 to R. Solazzo, paragraph 3 to G. Petriccione and paragraph 5 to M. Donati.

The expected change in the CAP mechanisms would affect all the agricultural sectors and every single farm payment. Regionalisation would bring less money to the sectors and farmers that currently benefit from large financial transfers (tomato sector, milk producers, etc.). At the same time much more money would go to the sectors that were not historically beneficiaries of CAP subsidies (i.e. fresh vegetables) and to farmers located in marginal areas (mountains). The question we pose here is whether or not regionalisation will produce changes in farmers' behaviours and the economic results of farms.

With particular reference to the fruit and vegetable sector, it is interesting to examine how the producers will be affected by the HC, both in terms of land allocation and economic results. The objective of this chapter is to evaluate the effects of the HC proposals on fruit and vegetable producers to identify their potential reactions to regionalisation. The assessment will be carried out on a sample of fruit and vegetable farms collected by the Italian FADN. The analysis focuses on the effects of HC on those producers, but the emphasis will remain on the tomato sector, the most important in term of CAP payments made as part of fruit and vegetable interventions. In order to evaluate its impact on farming behaviours, the analysis will be developed by implementing a simulation model based on the positive mathematical programming (PMP) methodology (Paris and Howitt 1998).

This work is articulated as follows: the first section gives an overview of the tomato sector and its trend under the CAP reforms; the second section presents the main contents of the HC proposal with particular attention to the changes affecting the fruit and vegetable sector; the third section describes the PMP model; the fourth section evaluates several pertinent policy scenarios with respect to a group of farms collected by the Italian FADN; the last section concludes with some remarks.

Approaching the Health Check

The production of the Italian fruit and vegetable sector amounted to EUR11,049 million in 2006, making up 25 per cent of total domestic agricultural production. Most of that production is concentrated in southern Italy which produces 56 per cent of the total sector value. In the last few years the fruit and vegetable sector has taken on an even more important role in Italian agriculture, showing a positive trend (Figure 14.1).

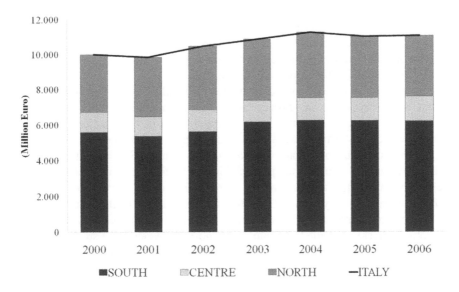

Figure 14.1 Value of fruit and vegetable production – Italy (2000-2006)

Source: our processing on ISTAT data.

Territorial analysis underlines the conspicuous heterogeneity of the Italian fruit and vegetable sector: in southern Italy this sector makes up almost 40 per cent of all agricultural production; in central and northern Italy this percentage is lower, at 21.4 per cent and at 16.2 per cent respectively. Over 50 per cent of total fruit and vegetable production by monetary value comes from four regions, three of which are situated in southern Italy (Sicilia, Campania and Puglia) and the other one (Emilia-Romagna) in northern Italy. On the whole, vegetables are more prevalent than fruit, exceeding 80 per cent in some regions, such as Puglia and Campania. These two regions comprise almost 30 per cent of all vegetable production.

About 35 per cent of total national production is made up of organized farming. As a result of the CMO reform in 1996, in Italy there has been an important organizational process involving fruit and vegetable production. Nevertheless it rose, especially after the passing of Regulation (EC) No 2699/2000,[2] but was not evenly spread out geographically. In 2006 more than 70 per cent of fruit and vegetable production in northern Italy was organized and the average value of the marketed production exceeded EUR40 million for each producer organization. In southern Italy organized production represented about 20 per cent of the total sector value of the area, while the average value of the marketed production was under EUR7 million per producer organization. An explanation of these

---

2   As is noted, this regulation capped the amount of Community financial assistance at 4.1 per cent of the value of the marketed production by each producer organization.

remarkable differences can be found in the deeply rooted co-operative culture that has influenced, especially in North-eastern Italy, the growth of a system of associations that favours a different economic aspect of these organizations and the development of suitable competitive strategies.

Major differences exist not only by territorial dimension but also by the production processes used. Processed tomato production is almost entirely organized and is one of the main agricultural sectors in Italy. It occupies over 90,000 hectares of surface area and its national production amounts to about 5 million tons, of which more than 50 per cent comes from the South. Eighty per cent of Italian processed tomatoes are produced in Emilia-Romagna and Lombardia in the North and Puglia (the Foggia area alone makes up 30 per cent of national production by monetary value), Campania and, to a lesser degree, Basilicata in the South. The North has larger farms and higher levels of mechanization than the national average, while the South is composed prevalently of smaller farms. Another important difference between the North and the South is the industrial process: production in the North is characterized by the prevalence of large co-operatives that process their own tomatoes whereas in the South there are many small private industries.

Campania (in particular the Salerno province) is Italy's main tomato industry hub. In fact 90 per cent of the tomatoes from Puglia stay in that region to be processed. The industrial facilities located there process nearly 50 per cent of the country's entire tomato production. Emilia-Romagna represents the other important tomato industry hub, processing nearly 30 per cent of the country's total tomato production, most of which are grown in that same region.

Italian processed tomato production has undergone a following trend in the past few years: from 2002-2004 there was a considerable increase in volume (+55 per cent); then from 2004-2006 the volume fell (by about -31 per cent).

This was the scenario in which the CMO reform in fruit and vegetables was introduced in 2007. In particular, the introduction of decoupled payments for processed tomato –integrating the Fischler CAP Reform –brought much uncertainty to the market concerning the likelihood of a considerable production decrease and the consequent difficulty of commodities supplies. This concern largely explained the more than 60 per cent increase in one year in prices stipulated in 2008 to guarantee the achievement of a production target of 4.6 million tons established as part of the interbranch agreement.

Since 1 January 2008 the CMO reform in fruit and vegetables introduced by Regulation (EC) No 1182/2007 has integrated this sector into a simple payment scheme, allowing Member States to choose, during a transition period, the adoption of partially decoupled payments for processed production.

For processed tomatoes Italy chose to maintain the transitional coupled payments at 50 per cent of the national ceiling until the end of 2010. More specifically, during the three-year transition period (2008-2011) a proportion of the subsidy in the amount of EUR1,300 per hectare in 2008 is in coupled form while the other 50 per cent of national ceiling (EUR91,984 million) moves to the single farm payment scheme. The latter amount is distributed to farmers who received historical payments in the reference period of 2004-2006, while the coupled amount of the payments is subjected to the condition that farmers be members of a producer organization and have a contract for processing.

Furthermore, the CMO reform amended Article 51 of Regulation (EC) No 1782/2003 to remove the ban on eligible hectares for permanent fruit and vegetable crops, ware potatoes and nurseries, but allowed Member States to opt for implementation of the rules until 2011. Italy decided that no land would be eligible for ware potatoes, nurseries or permanent fruit other than citrus fruit.

The fruit and vegetable sector in the CAP Health Check

The 2003 Fischler Reform of the CAP is what the European Parliament defined in a recent Working Document 'the most thorough one to which the CAP has so far been subjected', though it was originally considered a mid-term review of the existing agricultural subsidy mechanisms (European Parliament 2008). The 2003 CAP Reform included a number of review clauses aimed at comparing the adopted measures with the market scenarios and the priorities will be outlined in the near future. For this reason the decision of the European Commission on the Health Check of the CAP (HC) represents an intermediate transition rather than a radical reform. The Commission decided to approve adjustments that aim to outline its future profile in a way that will 'promote a sustainable and market-oriented agriculture'. By doing this the EU is laying the groundwork for a deeper reform that will be implemented after 2013, along with the review of the EU budget.

The principal HC instruments contemplated by Regulation (EC) No 73/2009 refer to the simplification, modulation, revision of Article 69 and regionalisation.

The first issue regards simplification. With the progressive integration of direct payments in the single payment scheme, the Commission deems it necessary to shift the various subsidy regimes to a single Common Market Organisation. However, some exceptions and a progressive adjustment are provided.

The modulation is maintained in the HC as the principal instrument addressing financial reinforcement of the second pillar by draining resources from the first pillar. In consideration of the new issues identified in the HC as the new challenges to be faced under the rural development plan (climate change, renewable energies, water management, the protection of biodiversity, innovations to deal with these issues and the accompanying measures in the dairy sector), the Commission decided to reinforce compulsory modulation. The HC document also suggests introducing a mechanism based on a progressive increase of the modulation rate from 5 per cent to 10 per cent and an additional cut of four on payments of over EUR300,000.

Another important issue dealt with in the HC document concerns the amendment of Article 69 of the CAP regulation. Said article becomes the new Article 68 which should allow more flexibility in its use by the Member States because it broadens the scope of former Article 69.

The new Article 68 supports certain risk-management measures, including crop, animal and plant insurance, as well as mutual funds for animal and plant diseases and environmental incidents. More specifically it proposes to grant up to 65 per cent financial subsidies to insurance policies and mutual funds.

The issue of so-called regionalization is at the core of the HC. The Commission approved the regional model based on a mere flat rate of decoupled support as the main form of support. In other words, the Member States who implemented the single payment scheme based on the historical approach will be given the opportunity to switch to a regional approach and approximate the unit value of payment entitlements. By doing this European Commission is asserting that maintaining the historical model on the basis of the reference period, with its various payments, will become increasingly difficult to understand and justify in the years ahead. This issue has been the object of an extensive debate pointing to the discrimination among farmers caused by the current historical payments scheme (Anania 2008). The regional payment scheme seems to be a more equitable support to farmers since the farm payment will no longer be connected to the historical entitlement rights based on production typologies.

In the HC document regionalization is an option for Member States; nevertheless it certainly shows the future trends of CAP regarding the issue of single farm payments.

As regards the fruit and vegetable sector, the HC confirms the option introduced by its CMO reform to defer its integration into the single payment scheme (Article 51 of Regulation (EC) No 1782/2003). In other words, the parcels used for fruit and vegetable production (as well as ware potatoes and nurseries) will not be eligible until 31 December 2010. Furthermore, the HC specifies that no review of the partially coupled payments scheme is necessary in the fruit and vegetable sector because of the recent introduction of a similar scheme and only as a transitional measure. On the other hand, the fruit and vegetables CMO reform was decided and implemented after the 2003 Fischler Reform. For this reason processed fruit and vegetables can still receive, though only temporarily, a partially coupled support, as seen in the first paragraph. Nevertheless Member States should be allowed to review their decisions, provided they move towards a greater degree of decoupling.

The regional model definitively disconnects the payments from the farm's history. Farmers achieve further freedom to adapt their production decisions to market developments.

From this point of view it seems paramount to ask what kind of possible effects the HC might have on farmers' behaviours and then on the supply chains involved. The implementation of the historical single farm payment has determined considerable changes in production processes and in the profitability of farms. It is likely that the consequences deriving from the extension of the regional single farm payment will be equally important, accompanied by greater modulation and the removal of set-aside and milk quotas.

Some analyses of the effects of regionalization show that the regionalization of the payment process could considerably influence farmers' decisions and, what is more important, their economic results. These analyses agree that the higher the percentage of regionalization, the wider the payment redistribution effect will be within the same region. The redistribution will be brought in as the referred region broadens and as the historical production processes diversify (Anania 2008, Pupo D'Andrea 2008). A region characterized by a certain degree of homogeneity in the farm production processes will suffer less from the redistribution effects of the regionalised payment than other regions. These expected effects were the main reason for which the most Member States like Italy chose a historical approach in the past. This choice has undoubtedly been influenced by the aversion of those who received the historical support, but the new EU guidelines brought it up for discussion again.

Furthermore, the redistribution effects of regionalization are closely connected to the production processes on which historical payments were calculated. A recent interesting analysis (Pupo D'Andrea 2008) shows that regionalization could penalize the productions that in the past received more support (milk, olive oil, tobacco, etc. and also processed tomatoes), while it might reward the ones that received less support or received no support at all (fruit and vegetables, except processed products, wine, etc.). The result is that the gains or losses of each administrative region will depend on the historic production processes and the per-hectare support received by them in a given period in relation to the average regional value.

Thus said, we must examine how the measures provided by the HC (regional single payment scheme, modulation, removal of set-aside) for the fruit and vegetable sector can influence the competitiveness of farms, i.e. their ability to adapt their organization for the purpose of improving the economic and productive performance of their farm. The ability to respond to the changes arising from the HC becomes more difficult for a sector such as fruit and vegetables which in a short span of time went from coupled support to historical decoupled payments and now to the regional single payment scheme.

Methodology for representing HC

The evaluation of the effect of HC on the processed tomato cultivation is carried out using a mathematical programming model based on the positive mathematical programming (PMP) approach. In the original formula put forward by Paris and Arfini (2000), the methodology of the PMP is based on a three-phase procedure, the main parts of which are summarised below:

- Estimation of marginal costs for the processes implemented. The aim of this phase is to recover the relevant information regarding the specific production costs the farmer uses to formulate the farm production plan.
- Estimation of the cost function. In the second phase, the PMP estimates a quadratic cost function able to provide a better representation of the production costs, coherent with the economic theory. The method of estimation used in this phase is based on the maximum entropy (Arfini and Paris 2000).
- Calibration of the model versus the year of observation. In this phase, the economic-production situation observed is reproduced using just the information on production costs estimated during the previous phase. At this point, the model can simulate the effects of the main changes in agricultural policy.

The model created for the analysis of agricultural policies follows the procedure described integrated with specific constraints and conditions about the new support instruments introduced by the Health Check document. More specifically, an important element of innovation with respect to the traditional model for CAP evaluation is represented by specific mathematical relations finalised to reproduce the SPS scheme. The relations developed inside the PMP model allow sharing the SPS among the eligible land without distortions in the relative convenience of the crops.

Also the new modulation mechanism is represented by using mathematical relations inside the policy evaluation model. In this case, the different brackets foreseen by the HC have been reproduced so that for each level of subsidy is applied a different level of modulation. But more than this last policy mechanism reproduction, an important set of conditions introduced inside the model concerns the permanent crops. As it is known, the mathematical programming models are very useful for estimating possible effects in the short and medium run of the agricultural policies with respect to the annual agricultural activities, but it is difficult to consider inside the analysis the permanent crops because they can produce problems of marginal substitution with the annual crops. Thus permanent crops are frequently excluded from such analyses. The problem consists in avoiding that the permanent crops be completely substitutable with annual crops in the short and medium run. To this end, the present analysis considers that the variation of the land use for permanent crops implies adding costs for removing the plants or for planting new ones. Along these lines the model uses a negative gross margin component that is part of the objective function linked to the variation of the surface cultivated with permanent crops. This condition can be represented as follows:

$$\sum_{k=1}^{K} \left\{ h_{n,k} - h_{n,k} \, \gamma_k \right\} \quad \forall n$$

The difference in absolute value between the unknown variable of hectares cultivated with permanent crops k $(k=1,2,...,K)$ for each farm n $(n=1,2,...,N)$, $h_{n,k}$, and the level observed in the base situation,, $\bar{h}_{n,k}$ is multiplied by an average cost of removal or planting, γ_k. During the process of optimisation, this gross margin negative component introduces a rigidity component in the process of activation or deactivation of such activities with respect to the others. The entire model was developed using the algebraic packaging GAMS and it was solved using a specific discrete continuous solver GAMS/DICOPT.

Possible effects of Health Check

Policy scenarios

On the basis of the HC document, the scenarios developed consider the actual situation in terms of aid provided to the fruit and vegetables and the reform that will affect those products starting in 2011. As the data adopted for this study refers to 2006, the model foresees a specific scenario that reproduces the transitional period currently characterizing the fruit and vegetable sector currently characterizing the fruit and vegetable sector. Then, in compliance with the decision of the Italian Minister, a scenario was evaluated in which all the subsidies, fruit and vegetables included, are decoupled according to the historical approach. The two above-mentioned policy scenarios have been integrated with two others whose model reproduced the new measures of the HC, namely: the decoupling system based on a flat rate for each admissible hectare and a greater degree of modulation applied to all the subsidies received by farmers. The first HC scenario considers the previous new measures without market interferences to measure the net effect of the policy; in the second HC scenario variations in product market prices are introduced in order to evaluate the added perturbation of the likely market dynamics on producer decisions.

For the sake of greater clarity, the policy scenarios considered in this chapter are listed below:

- '*BASE*' scenario: this reproduces the situation observed in 2006 that is the total decoupling for the arable crops, transformed fruit and vegetables excluded.
- Transition scenarios '*S1*': in this case, the CMO reform for fruit and vegetables is applied according to the transition payments decided for the processed tomato and the other processed fruits. More specifically, for processed tomato, the decoupling is adopted only for the 50 per cent of the total subsidy, while the rest is paid to the producer in a coupled form; in the case of processed fruits, the model foresees a partial decoupling for prunes (75 per cent coupled), and a total coupled payment for pears and peaches. In order to consider the higher purchasing prices that the processing industries

agreed upon with producers just before the beginning of the first harvesting year with transition payments, the for tomatoes has increased from the 2006 base of 45 per cent.

- CMO fruit and vegetables in force '*S2*': the scenario considers the situation after the transition period expiration, when all the subsidies will be decoupled. The price for the processing tomato is the same as the one used in the previous scenario.
- Health Check scenario '*S3*': this attempts to simulate the possible regionalisation of aid,[3] by allocating payments calculated on a flat rate basis to each farmer. In addition to regionalised payments, this scenario takes into consideration the new rates of modulation (on the basis of the eligible brackets).
- Health Check scenario with variation in market prices '*S4*': like the previous one, in which the variations in prices and variable costs are added to scenario *S3* (Table 14.1).

Table 14.1   Variation of the prices of the main agricultural products (2006–2013)*

SOFT WHEAT	-10%	SUNFLOWERS	+3.2%
DURUM WHEAT	-10%	SOYA	+3.5%
CORN	-2%	RICE	+4.8%
BARLEY	-10%	TOMATO	+4%
SILAGE	-2%	SUGAR BEET	-4%
OTHER CEREALS	-5.00%		

* The agricultural product prices not included in the table are assumed constant.
Source: OECD/FAO 2008.

The agricultural price predictions are those provided by the outlook study developed by the OECD and FAO (2008) that projects the actual prices until 2013. The recent positive variation in purchasing prices for tomatoes was accompanied by a rise in production costs and more specifically in motor fuel, fertilizers and pesticides. To take into account this cost increase, scenario *S4* considers a 15 per cent increase in specific variable costs for every crop.

---

3   The regionalised SPS value was calculated taking into account the national maximum and the total UAA (utilised agricultural area).

Farm data

For the purposes of the present contribution, the data used in assessing the CAP scenarios refers to a sample collected by the Italian FADN from the Emilia-Romagna region. This region has a particular propensity for fruit and vegetable productions. As highlighted previously, Emilia-Romagna is the largest northern Italian region in term of processed tomato production. The farms were selected on the basis of the presence of fruit and vegetable productions within the individual production plan. The sample was done submitted to a double stratification: the first one for the farm specialization and the second one according to the GSP size. The farm specialization is identified by measuring the contribution of each group of product (fruit, vegetables and arable crops) to the total GSP. Products with a rate of contribution higher than 50 per cent give us the farm specialization type. For the second stratification, the sample of farms was divided according three size classes: up to EUR30,000, from EUR30,000 to 80,000 and more than EUR80,000 of GSP. It is important to underline that the simulations were made for each individual farm and not aggregated. The stratifications were useful for estimating the total variable cost function for each typology of farms according to its specialization and economic size. After estimating the unique total cost function per group and the differential marginal cost per farm, the simulations were made by single farm. Table 14.2 shows the principal data of the sample.

Table 14.2 The main characteristics of the FADN sample (2006)

Farm Types	n° of farms	Average UAA	Horticultural Crops (% of UAA)	Processed Tomato (% of UAA)	Fruits (% of UAA)	Average GSP (EUR/Ha)
Fruits	*318*	*16.3*	*4*	*1*	*61*	*5,378*
CL 1	122	5.6	2	0	53	2,873
CL 2	111	11.5	2	0	60	4,357
CL 3	85	38.1	5	1	62	6,310
Horticulture	*67*	*54.2*	*54*	*27*	*1*	*3,012*
CL 1	11	6.5	42	7	4	2,259
CL 2	22	21.0	42	14	5	2,571
CL 3	34	91.2	55	30	1	3,095
Arable crops	*100*	*132.3*	*6*	*3*	*2*	*1,404*
CL 1	32	13.6	8	0	11	1,231
CL 2	28	36.4	13	8	2	1,453
CL 3	40	294.3	5	2	2	1,406

Source: our processing on National FADN data.

Land use responses

The PMP model can provide a wealth of information about the production decisions of farms involved in the analysis, among which the output levels and land use per product. The variation in land use is an important signal on how the policy engenders reactions to the allocation of the major constrained agricultural factor (the land). In this case, the most important question the study would respond to concerns the decision about the production plan when the HC intervenes to redistribute the subsidy among territories, farm typologies and individual farms. To be concise, our analysis will focus on the changes in production decisions in relation to the farm typologies and with particular emphasis on the dynamics of processed tomatoes.

A quick glance at Table 14.3 is sufficient to understand how much impact policy decisions have had on farmer behaviours, in the sense that changes in the subsidy mechanism can be affect the marginal profit of each product with respect to each other. This is the case of tomatoes that undergoes a significant change in acreage when the transitional payment is introduced (*S1*). In *S1* the relative convenience of processed tomatoes changes because an important part of its coupled subsidy (50 per cent) is stabilized within the farm revenue. In this context, the effort made by industries to increase purchasing prices has protected the food chain against the risk of higher production losses. The horticulture farm typology presents the situation illustrated previously: a curb in the production level of fruit and vegetables in the CMO transitional phase scenario and a worse productive situation in the total decoupling scenario (*S2*). The results achieved seem to underline that bigger farms (size class 2 and 3) are more affected by total decoupling. In other words, the more professional and specialized farms can benefit from the decoupled payment more than the small farms: the decrease in production means lower costs and a consistent annuity in the medium run. This is also confirmed by the economic results.

Table 14.3 Changes in tomato land use

Farm Type	Class of Size	Base	S1	S2	S3	S4
		ha	Var. % wrt BASE			
Tomato land use						
Fruits	1	0.0				
	2	0.0				
	3	34.0	2.0	2.0	2.0	-4.2
Horticulture	1	4.7	-12.4	-12.4	-12.4	-20.4

Farm Type	Class of Size	Base	S1	S2	S3	S4
	2	65.2	-3.9	-9.1	-9.1	-24.5
	3	928.4	-18.0	-31.9	-31.9	-49.2
Arable crops	1	0.0				
	2	82.2	-8.5	-25.9	-25.9	-39.1
	3	293.2	-14.9	-30.5	-30.5	-47.0
Fruit land use						
Fruits	1	212.8	1.1	1.1	1.1	-2.7
	2	540.9	1.2	1.2	1.2	-2.5
	3	1543.7	1.9	1.9	1.9	-3.0
Horticulture	1	1.5	0.1	0.1	0.1	-2.5
	2	3.1	-0.2	1.2	1.2	-1.5
	3	4.5	0.0	0.0	0.0	-4.0
Arable crops	1	29.1	0.6	0.6	0.6	0.3
	2	19.6	-1.7	-1.7	-1.7	-3.6
	3	169.3	-0.0	-0.0	-0.0	-3.5

Source: our processing using PMP model.

The results that may seem surprising if compared with the previous situation are the results obtained from the HC scenarios. If in the previous phases the policy mechanisms played an important role in changing tomato hectares, the HC does not engender any productive changes in the other scenarios. This output is consistent with the reasons that have compelled farmers to reduce tomato growing when the transition phase is applied. If the policy changes the relative convenience of the crop, farmers react by changing the production plan; otherwise the production plan remains stable. In the case of HC, these measurements do not affect the relative convenience of products in the case of historical total decoupling. The flat rate imposed on all farmers and the entire eligible agricultural surface produces a redistributing effect in term of subsidy but leaves the relative convenience among processes stable. Consequently *S3* brings more stability in production levels than *S2* does for all types and sizes of farms.

The changes in tomato land allocation are more evident in the last scenario, where the HC is associated with the market price hypothesis and variable cost increases. The combined effect of price and cost variations significantly reduced tomato production on all the farm classes specialized in horticulture and arable crops, while the decrease was less significant on farms specialized in fruit. There is hard evidence that the price and cost shocks affected the largest and most specialized farms indicating that an increase in production costs produces negative product profits for a large portion of production. This means that the farms most specialized in tomato production are also the most intensive variable input users. A 15 per cent increase in production cost exposes such farms to the risk of not being able to cover the specific costs sustained for cultivating the crop.

The PMP model used in this framework allows us to evaluate the effects of the various farm processes, including the permanent processes. Actually, one methodological innovation of this simulation model involves the possibility of understanding the likely dynamics in long run activities, like olive oil, wine growing and fruit. The second part of Table 14.3 shows the results of fruit activities along the various scenarios presented. As you can see from the table, the CMO transitional payment for fruit and vegetables, the historical total decoupling and the flat rate payment did not significantly affect the land allocated to fruit in the medium run. In the last scenario, considering the increase in production costs, the acreage used for fruit decreases, but the absolute change remains quite small.

The rather large negative variation in processed tomato cultivation implies a new way of organizing the land for different crops. The question is which crops are occupying the land lost by tomatoes. It is interesting to note that the new production plan takes on a configuration in which grains are predominant. It is quite clear that on specialized farms, like the third size class of the horticulture typology, the loss in tomato land is compensated by a significant increase in cereal acreage. This trend is reinforced by the S4 scenario, in which the positive variation in cereal prices (about 10 per cent) leads to a considerable increase in the amount of land for growing grains. The fodder crops also contribute to replacing processed tomatoes, thereby decreasing production costs with a positive effect on the gross margin.

Economic outcomes

The variation in production activities is reflected by the changes in the major economic variables, like the levels of gross margin and subsidies. The discussion about the economic information can be divided in two steps, one regarding the first two scenarios and the second one regarding the HC results.

The modifications introduced by the CMO reform in fruit and vegetables have caused farms to adopt a production cost minimization strategy. This behaviour has led to an overall increase in gross margin for every farm typology in scenarios S1 and *S2*, especially on horticultural farms, for which the partial and total decoupling of processed tomato aid has led many farms to reduce production in order to minimize costs and keep the farm payment. The greatest increases are seen on the largest horticultural farms. In this case, the amount of decoupled payment with respect to the total revenue is much greater than in the other size classes.

Moreover, it is important to consider that the positive variation in gross margin levels partly due to the observed increase in tomato prices (+45 per cent). Hence, on farms where processed tomato production is prevalent, received the largest increases (Table 14.4).

Table 14.4 Dynamics in the main economic variables

Farm Type	Class of Size	Base	S1	S2	S3	S4
		EUR/ha	Var. % wrt BASE			
		Gross Margin				
Fruits	1	1151.8	-1.0	-1.0	15.2	-13.0
	2	1765.1	-0.8	-0.8	10.4	-18.0
	3	1728.4	1.7	1.7	11.2	-32.0
Horticulture	1	1686.3	3.9	3.9	6.2	-10.4
	2	1600.2	8.9	8.7	4.4	-17.4
	3	1228.7	32.1	33.1	3.7	-29.1
Arable crops	1	747.2	-0.5	-0.5	10.9	-10.8
	2	941.9	5.6	5.7	-6.5	-25.5
	3	878.8	0.5	0.5	-8.2	-27.8
		Subsidies				
Fruits	1	134.3	-1.0	-1.0	137.5	137.5
	2	113.8	4.6	4.6	177.2	177.2
	3	128.8	1.6	1.4	128.7	128.7
Horticulture	1	284.5	-2.2	-2.2	11.4	11.4
	2	389.0	-2.3	-3.7	-21.6	-21.6
	3	671.2	-3.0	-3.6	-57.5	-57.5
Arable crops	1	215.6	5.9	5.9	45.3	45.3
	2	406.6	1.6	1.0	-27.2	-27.2
	3	351.8	-1.2	-1.5	-23.2	-23.2

Source: our processing using PMP model.

A different sort of comments can be made regarding scenarios S3 and S4. The HC thoroughly redistributes the subsidy among farms, and horticultural farms are especially hit by a reduction in the subsidy amounts. This leads to a substantial resizing of the gross margin. One can say that the positive variations noted in S3 for the horticultural typology is only due to the positive price increase in processed tomatoes. The other farm types show an increase in gross margin due to the shift-effect produced by the flat rate. So, farms with a low level of subsidy per hectare receive more, while farm types with a high level of subsidy per hectare receive less. Only the two largest size classes, specialized in arable crops, have received smaller subsidies, leading to a decrease in their gross margins. Among the horticultural and arable crop farm types, only the smallest classes have received larger subsidies with positive effects on the gross margin. In particular, it is evident that the first class in the horticultural farm type includes many farms with horticulture crops other than processed tomatoes.

At any rate the greater impact on farm performance should be attributed to market prices and variable input markets. The 15 per cent increase in variable costs associated with all crops (see scenario S4) leads to a market decrease in the gross margin. This result is widely distributed throughout each farm typology and more specifically and the largest size classes. A response to this effect has already been suggested: the high intensive use of variable inputs by the largest and most specialized farms makes for a more fragile economic equilibrium. This explains how a 15 per cent increase in specific variable costs leads to a decrease of as much as 30 per cent in the gross margin on the largest farms (Table 14.4).

In addition to these negative performances of the largest farms, the writer levels of modulation have also decreased the amount of aid received by the farms. In this respect, the modulation mechanism and the flat rate system represent two CAP instruments that could guarantee solidarity among farms, allowing a major transfer of resources from the big historical beneficiaries to new farmers and farmers with scant support for their agricultural activities.

Conclusions

The HC is open to a series of questions concerning the future of the CAP interventions and how the new measures could impact European agriculture in terms of production dynamics and economic results. It is evident that the objective of HC is to trace a connection between the 2003 CAP and the post-2013 CAP. The reintroduction of regionalisation and greater modulation are two good examples of what might become the future agricultural intervention in Europe. How farmers will react to the HC is the issue to analyse.

In this changing framework, the fruit and vegetable sector is still characterized by a partial decoupling system that will become totally decoupled between 2011 and 2012. In this context, the tomato sector is submitted to a process of adaptation directly depends on the amount of subsidies coupled to the production. The high level of coupled subsidies can lead farmers to reduce production and keep a large portion of the revenue originated by the product. At the beginning of the transitional period industries compensated for the decrease in coupled payments by increasing production market prices to counteract this risk.

The evaluation proposed in this study adopts a model based on the PMP methodology, in which the new HC mechanisms are reproduced and the permanent crops are considered part of the farm production plan.

The analysis carried out on a FADN sample of farms located in Emilia-Romagna shows how the HC new measures affect the economic performance of farms but not the input allocation choice. The flat rate does not disrupt the relative convenience of the crops because it maintains the degree of substitution among activities. Only when the CAP mechanism moves from a coupling scenario to a total decoupling scenario, and in the case of a variation in price levels, the modifications in the production plan become evident.

Farms specialized in horticultural productions, particularly processed tomatoes, are subjected to a generalized resizing due to the progressive achievement of total decoupling for the sector. But regionalisation has no effect on the total decoupling situation as they neutral component in changing cross-crop conveniences. As regards economic results, regionalization changes the economic results of farms. The spread-effect of the flat rate helps to transfer financial resources from the greater historical beneficiaries of subsidies to the marginal farms and the lesser historical beneficiaries of subsidies, such as fruit or fresh vegetable producers. Regionalisation could reinforce marginal agriculture and contribute to a better distribution of funds within a given region. This is a solidarity effect of the regionalization mechanism that enables the distribution of payments according to objective and actual elements (i.e. eligible land).

Variations in prices and variable input costs lead to significant changes in land allocation and economic results. This is in line with the decision to decrease production in cases of decoupling. The assumed increase in prices in the last scenario precludes the ability to meet the greater variable costs. Output and input market prices are the main variables in defining farm allocation decisions and the role of the CAP will merely become a positive component of farm revenue without interfering with crop profitability.

Even in this CAP evolution context, the traditional raw material area of origin could be preserved only if the market allows producers to have favourable expectations in term of long-run profitability. The food-chain will continue to be affected by the CAP measures until the transitional period ends and total decoupling is applied; after the transitional period, the intervention mechanism (historical or regionalised) will essentially have a neutral effect on productive decisions, leaving the role of production orientation to the market.

References

Arfini, F., Donati, M. and Zuppiroli, M. 2005. Un'analisi degli effetti della riforma Fischler della PAC sull'agricoltura italiana utilizzando il modello Agrisp, in *La riforma delle politiche agricole dell'UE ed il negoziato WTO* edited by *A*nania G. Milano: Franco Angeli.

Arfini, F. and Paris, Q. 2000. Funzioni di Costo di Frontiera, Auto-selezione, Rischio di Prezzo, PMP e Agenda. *Rivista di Economia Agraria*, 3(2), 211-242.

De Filippis, F. 2008. *L'Health Check della PAC: una valutazione delle prime proposte della Commissione.* Quaderni Gruppo 2013, Edizioni Tellus.

European Commission. 2008. *Commission Staff Working Document accompanying the 'Health Check' Proposals.* Available at http://ec.europa.eu/agriculture/healthcheck/ full impact_en.pdf.

European Commission. 2008. *CAP Health Check – Impact Assessment Note n. 8*, Brussels, 20 May 2008.

European Parliament. 2008. *Working Document on the proposal for a Council regulation establishing common rules for direct support schemes for farmers under the common agricultural policy and establishing certain support schemes for farmers*, Rapporteur Luis Manuel Capoulas Santos, 1 July 2008.

Paris, Q. and Howitt, R.E. 1998. An Analysis of Ill-Posed Production Problems Using Maximum Entropy. *American Journal of Agricultural Economics*, 80, 124-138.

Pupo D'Andrea, M.R. 2008. *Un'analisi degli effetti dell'applicazione in Italia di ipotesi alternative di pagamento unico regionalizzato:* Paper to the *XLV Convegno di Studi SIDEA: Politiche per i sistemi agricoli di fronte ai cambiamenti: obiettivi, strumenti, istituzioni.* Portici, Italy, 25-27 September 2008.

Chapter 15

Effects of CAP Reform on Peri-urban Agricultural Area in Umbrian Valley (Central Italy)

Biancamaria Torquati, Giulia Giacché, Chiara Taglioni and Francesco Musotti

Introduction

In peri-urban areas, agriculture has strategic value in the balance and quality of the urban environment. Recognition of this is contained in the statement of the European Economic and Social Committee on the question of 'peri-urban agriculture' (EESC, Brussels 2004). For the first time in an official European Union document, peri-urban areas are described as complex territories playing important economic, environmental and social roles, especially in view of their spatial proximity to and mutual dependence on nearby cities. In the EESC document peri-urban areas may include less favoured areas (Art. 20, Council Regulation (EC) 1257/1999) on support for rural development, i.e., areas with natural or environmental problems. In this case, local farmers would be entitled to an additional indemnity, proportionate to the natural or environmental limitations in which they are forced to operate.

The later EEC Regulation 1698/2005, to support rural development in the period 2007-2013, indicates careful reflection about the characteristics of peri-urban areas, from rural activities to socio-economic inter-relationships. The aim is an action strategy to create conditions for development linked to the principles of sustainability, with particular attention to needs and demands through local participation. These concepts are repeated in the Community Strategic Guidelines for Rural Development (Council Decision 2006/144/EC) and National Strategic Plan for Rural Development (MiPAF 2007). Many regional Rural Development Programs (RDP) were set up for the period 2007-2013, including peri-urban areas suitable for investment and development. One example is that of the Umbria Region, providing for 'improvement and management of peri-urban areas' under measures for 'environmental protection in relation to agriculture, forestry, conservation of natural resources and animal welfare'.

The complexity and value of relationships in peri-urban areas is also highlighted in some specific initiatives which have given rise to "networks of exchange of methodologies'.[1]

The definition of peri-urban rural areas is not a simple exercise in classification but an essential feature in defining intervention strategies against potential threats and opportunities for agricultural activities (Branduini and Sangiorgi 2004). Fleury and Donadieu (1997) argue that peri-urban agriculture, in a strictly etymological sense, is agriculture located on the outskirts of a town, whatever the nature of its production systems, and it can only have either a relationship due to spatial proximity, or some functional mutual relationship with the town.

If these relationships vary, they give rise to changes in the type of peri-urban agriculture and spatial planning characteristics as a whole. In the light of intervention policies, study of such relationships and their dynamics must be accompanied by analysis of the characteristics of peri-urban agriculture and city-dwellers' perception of the role of agriculture in their daily lives, both positively and negatively. Therefore, faced with the new objectives of the CAP, we felt it was important to use a case study to illustrate the effects of the Fischler reform and how to target future interventions according to specific and territorially local situations. The future of peri-urban areas will undoubtedly be heavily influenced by the new CAP and local planning measures. It is therefore necessary to understand how the phenomenon arose and the mid- and long-term effects in these areas of adequate governance and planning of process under way (Stolfi 2004).

In particular, the Fischler reform included three new principles that will certainly affect the future of peri-urban agricultural areas: decoupling, compulsory modulation (transfer of resources from first to second pillar) and cross-compliance. The impact will depend on the economic and social structure of the territory, as inherited from the recent past, and on the interaction of several factors, including market prices, alternative farming, job opportunities in other sectors, and level of city-dwellers' sensitivity regarding the value of land and soil as limited resources.

This chapter has two aims: to assess the effects of the Fischler reform of 2003 (and 2004) on farms in the peri-urban area between the city of Assisi and the two towns of Santa Maria degli Angeli and Bastia Umbra, highlighting their economic and productive dimensions as well as their social and cultural roles; and to analyse the community demand for landscaping and environmental services produced by

---

1   Among these are: the Resource Center for Urban Agriculture and Forestry (RUAF), at world level; Peri-Urban European Regions Platform (PURPLE) and European Federation of Metropolitan and Peri-urban Natural and Rural Spaces (Federnatur), at European level; and *'Terres en Villes'* in France and the Institute for the Protection and Enhancement of Peri-Urban Agriculture (ISTVAP) in Italy, at national level; the Triangle Vert des Villes Maraîchères du Hurepoix and the Rural Park south of Milan, at local level.

local agriculture, by estimation of willingness to pay for positive externalities such as landscape conservation, maintenance of biodiversity and recreation services. The results will indicate the extent to which peri-urban agriculture can carry out and ensure, over time, the natural, environmental and landscape needs expressed by the population through CAP interventions.

The chapter is organized as follows. First, the methodology used to estimate reform effects on farms, and city-dwellers' willingness to pay is introduced. Results are then discussed from the empirical application of the Fischler reform assessment and the contingent evaluation to estimate willingness to pay. Some final considerations are then made.

Materials and methods

The peri-urban rural area analysed here lies in the Umbria Region of central Italy, on the plain of the Umbrian valley, between the city of Assisi, the only one located on a hill, Santa Maria degli Angeli, the most populous urban agglomeration of the municipality of Assisi, and Bastia Umbra, one of the Umbrian municipalities with the highest building density. The resident populations of Assisi and Bastia are almost the same, but the density is very different: the 25,300 residents of Assisi live in a large area (18,679 ha, 134 inhabitants/km²) and more than 7,000 live in the urban agglomeration of Santa Maria (estimated density: 600 inhabitants/km²). However, the 21,400 residents of Bastia occupy an area of 2,764 ha (751 inhabitants/km²).

In the territory of Assisi the countryside located far from the centre of Santa Maria degli Angeli shows a diversified scenario with an outsourcing of processes and functions, once associated with household agricultural production. Currently composite agricultural activities are now also connected with tourism, with many enterprises marketing various grades of rural life 'experiences'. Their quality is proportional to the level of conservation of the traditional country territory and eco-compatible forms of agriculture.

Conversely, the countryside round Santa Maria and Bastia shows a homogenization of farmland and land that used to be agricultural, respectively due to the use of monocultures with abundant use of chemicals and urban sprawl. The magnificent characteristic views of and from the city of Assisi, together with accessibility to urban services, have also attracted the attention of the wealthier members of the population, who have moved there, leading to sharp price increases for country homes and surrounding land.

The aim of this study was to identify various types of farms in the peri-urban area and, for each of them, to analyse the effects of the Fischler reform in terms of cultural adaptation and entrepreneurs' attention to environmental and landscape problems.

Farms types were identified according to the National Agricultural Information System (SIAN) and National Livestock Register for crop year 2005-2006.

The former source yielded information on farmers (age, gender, place of residence) and farms (land tenure, legal status, Total Agricultural Area (TAA), Utilised Agricultural Area (UAA), land use, single payments, modulation, number of licenses required under the coupled payment regime, and statements required by conditionality rules). The second database showed that there are no sheep, pig or cattle farms but only ones for horses for recreational use.

The economic size of each farm was determined. For each crop, the economic data examined were: Standard Value of Production, Standard Value of Proportional Specific Costs, Standard Gross Margin (SGM)[2] and Annual Work Unit[3] required (AWU). The estimated figures are expressed in Economic Size Units (ESU), each equivalent to EUR1, 200

According to the SGM, farms were grouped into five classes: i) less than 8 ESU; ii) 8.1-16 ESU; iii) 16.1-40 ESU; iv) 40.1-100 ESU; v) more than 100 ESU. Following Sotte (2006), the first two limits of ESU 8 and ESU 16 were chosen, since they are considered critical and representative of comparable incomes; the first match of EUR9,600 (SGM) per year is less than a pension income (EUR12,039 per year in 2001); the second match of EUR19,200 (SGM) per year is equivalent to the gross yearly income of employees (Sotte 2006). In this way, farms below 8 ESU were considered 'non-enterprises', because they cannot generate enough income for one person, and those with more than 16 ESU 'enterprises', because they can generate a reasonable income for one or more persons (i.e., medium farms from 16.1- 40 ESU, large farms from 40.1-100; very large farms with more than 100 ESU). Farms between 8.1 and 16 ESU were classed as 'small (or potential) enterprises'.

The effects of the Fischler reform in terms of income, cultural adaptation and farm strategies were analysed by examining the productive system in the season 2007/2008 and by structured interviews to obtain information on: family employment; level of diversification (direct sales, recreation and tourism, educational activities, landscape management, production of alternative energies); business strategy; the influence of the CAP reform and fluctuations in the prices of materials and products; the relationship between city and rural areas; disadvantages of working near the city centre with respect to urban planning; and positive externalities attributed to peri-urban agriculture by farmers.

Referring to the main externalities of peri-urban agriculture as defined by Pascucci (2008) and to specific knowledge of peri-urban agriculture in the study area, a survey was planned about town-dwellers' demand for environmental and landscape services of local agriculture and their willingness to pay for positive externalities linked to it. In particular, respondents were asked to state the importance of the multifunctional role of agriculture, scores ranging from 1 (not important) to 5 (very important), to positive environmental externalities (e.g.,

---

2   The Gross Standard Margin was calculated from the average data of the three-year period 2003-2005 for the Umbria Region.

3   One work unit is estimated at 275 days per year.

maintaining open spaces), positive socio-economic externalities (e.g., preserving farmer traditions), to negative environmental externalities (e.g., ground water salinization). The technique used was contingent valuation (CV) with dichotomous choice format, and was applied to verify to what extent people are prepared to fund farmers, together with the regional administration, so that they can continue to practise agriculture to guarantee the maintenance of the landscape and rural environment.

Contingent valuation presumed a hypothetical market in which, in certain conditions, town-dwellers state their ability to sustain the cost of a tax increase, necessary to achieve the proposed event. In this case, the market was defined by the following elements: the good is the landscape and agricultural spaces of the peri-urban area; the actors are, on one hand, farmers supplying services which preserve and enhance the value of the rural landscape and, on the other, the Umbria Region, as the institution which manages landscape and environmental policies. The payment vehicle is a domestic tax and welfare change is linked to the possibility of preventing deterioration of the rural peri-urban environment, for which more taxes are required. The kind of procedure of WTP elicitation, called dichotomous choice, includes the following values: EUR20, 30, 40 and 50.

The bid amounts were set through a preliminary survey, since they greatly influence the research results, as the literature shows (Cooper 1993, Kannien 1993). Analysis of the answers in this study follows the model of Hanemman (1984, 1989), which formulates a function of answers that can be taken back to the concept of utility according to neoclassical consumer theory, presuming that consumer utility depends both on the environmental good in question and the consumer's own income. Since the bid listed by the interviewer was modest, compared with individual incomes, a simply solved, linear model income was applied (Hanemann 1984). Median WTP values were estimated by a univariate model with only one variable: the bid proposed. Following a parametric approach, a logit model was used to estimate WTP. Then an estimation of parameters was carried out with a logit model.

The sample population of 204 residents were interviewed by questionnaire from June to August 2008. Interviews were carried out near the study area and, before they began, all respondents were asked if they were residents of Assisi, Santa Maria or Bastia and if they were familiar with the area. These questions were regarded as fundamental for continuing the interviews; they meant a higher level of respondent involvement, and also avoided the use of photographs.

Results

Effects of CAP reform

The area's crop system for agricultural season 2005/2006 was the reference situation with regard to the effects of CAP reform since the reform became effective in 2005 for arable crops (cereals, oil crops), proteic crops, linseed and hemp, leguminous crops, (chick-peas, lentils, vetches), and in 2006 for olive oil.

In 2006, the study area contained 111 farms, covering a TAA of 1,168 hectares, of which 1,034 hectares were (UAA. As regards land tenure, 57 per cent of the cultivated surface area was owned by farmers, 41 per cent was rented. Almost all farms were directly managed (92 per cent), 63 per cent managed by men and 37 per cent by women. Only 24 employers were under the age of 55 (Giacchè and Torquati 2010).

Regarding land use, 47 per cent of the UAA was under cereals, with a prevalence of barley (14 per cent), wheat (12 per cent), corn (10 per cent) and durum wheat (9 per cent); after cereals, the most important crops were sunflower (9 per cent), forage (7 per cent), olives (seven per cent), proteic crops (three per cent) and wine grapes (2 per cent). The uncultivated surface area was 9 per cent of the UAA. The above productions involved 9,440 days of work per year, or about 8.5 days per year per hectare of UAA, or 38 annual work units (without considering seasonality of work).

For the whole area, the standard value of production was estimated at EUR854,000, according to prices for marketing year 2006-2007. The value of single payments amounted to EUR316,000, or 27 per cent of the estimated standard total gross margin. These data reflect a mainly rural situation, characterized by extensive agriculture and little labour. Specialized vineyards and olive groves are also typical of the plains. The standard production value was substantially lower, around EUR764 per ha of UAA, although this is irrigable land. To the value of agricultural production must be added the tourism turnover of the six holiday farms (agritourism) and, where they are marketed wines (Giacchè and Torquati 2010).

Analysing the farms by economic size, it is clear that 'non-enterprises' were 86.5 per cent of the total, occupying 38.5 per cent of the UAA and producing 29.1 per cent of the SGM; 'small enterprises' totalled 5.4 per cent, occupied 9.1 per cent of the UAA, and produced 7.6 per cent of SGM; 'medium, large and very large enterprises' were 8.1 per cent of the total, but occupied 52.4 per cent of the UAA and produced 63.3 per cent of the SGM.

Comparing the situation in 2005/2006 with that of 2007/2008, after two years of entry into force of the reform, the production system had clearly not undergone substantial changes, while the number of farms had declined by only 2 units.

To analyse crop system changes, farms were divided into two groups: 1) farms which did not change their surface area by selling/purchasing land or increasing/decreasing the surface area in rent ('in' farms); 2) farms which changed their surface area, extending outside the peri-urban area ('out' farms).

In group 1 (92 per cent) the decrease in UAA was on the whole not great (-5 per cent), even after reducing set-aside (-50 per cent).[4] Cultivation of olives and vines decreased (-25 per cent), also fruit trees (+25 per cent), sunflower and proteic pea

---

4   In 2008, the requirement of set-aside for surfaces areas due to retirement titles was not in force.

(+75 per cent), whereas alfalfa and maize (-50 per cent), soft wheat, broad beans and little broad beans (+75 per cent) and durum wheat (+100 per cent) increased (Figure 15.1).

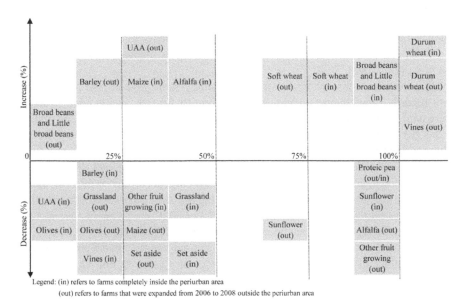

Figure 15.1   Increase and decrease of share of arable land for several crops in the season 2007-08 with respect to season 2005-06.

These changes show that the presumed abandon of cultivated land did not occur. There was also a progressive production simplification to cereals and alfalfa; as regards leguminous crops, peas were replaced by broad beans. There was a drastic reduction in sunflower, and a limited decrease in olives, vines and fruit trees, due both to explant of old vineyards, and to the abandon of small olive groves and fruit orchards.

Group 2 (8 per cent), comprising farms of substantial economic size, decreased the set-aside surface area (-50 per cent), olives (-25 per cent) and other fruit trees (+75 per cent); at the same time, it increased the surface areas of vineyards specialising in quality wines. Group 2 also considerably decreased cultivation of sunflower and proteic pea (+75 per cent); unlike the case of group 1, the reductions in maize (-50 per cent) and alfalfa (-75 per cent) are worth mentioning. The increases in soft and durum wheat were more than decoupled. These changes first show the existence of some dynamic farms which can invest in land capital by extending their own properties outside the study area. Secondly, the following elements were identified: sharp increases in soft and durum wheat, a decrease in sunflower specialisation in professional vineyards, and abandon of fruit growing and domestic olive growing.

The direct survey showed that changes in crop systems were caused more by market price trends than by changes in the support structure. Referring to the new payment structure, farmers perceive decoupled payment as an element of financial security by means of which they can react better to market changes, as the European Commission hoped in its proposal for the maintenance of decoupled payment (Borchardt 2008).

In particular, considering the value of a single farm payment as 100, 14 per cent regards supplementary payments according to Art. 69: 11 per cent of these were paid to farmers who had used certified seeds for durum, soft wheat and maize and 3 per cent of farmers who had adopted at least a two-year rotation system. The payment was perceived as financial security, because it covers 57 per cent of production costs equal to the expense for seeds, pesticides and tillage made by third parties, as indicated by the balance estimated according to average local prices from July to September 2006. Considering the same balance sheet estimated according to average prices from July to September 2008, the payment covers only 38 per cent, because of a steady increase in procurement costs of raw materials.

In addition, estimations show that single farm payments range from EUR300 to EUR370/ha of UAA in medium, large and very large farms, although the highest values (EUR680/ha) were found in some 'non-enterprises' entirely under olive groves and the lowest (EUR70/ha) in some 'non-enterprises' under forage. Instead, referring to days of work in medium, large and very large farms, single farm payments were EUR25-45 per day, with the highest value (EUR150 per day in some 'non-enterprises' under cereals) and the lowest (EUR3 per day in some 'non-enterprises' entirely under olive groves).

Market price trends from January 2006 to September 2008 showed steep increases in the selling prices of cereals which, under the single payment system, greatly influenced farmers' choices toward them. Quotations were very high from September 2007 to July 2008, the highest values being between January and March 2008, when durum wheat was on average about EUR487/ton, soft wheat EUR277/ton and maize EUR260/ton. These values are respectively 96 per cent, 123 per cent and 238 per cent higher than those for June and July 2006. In this period, farmers also faced steep price increases in productive factors; e.g., seed and tillage rose by 80 per cent and chemical products such as pesticides by 50 per cent.

Willingness to pay for peri-urban agriculture

The main aim of the survey was to assess whether peri-urban agriculture played the multifunctional role explicitly required by its population of town-dwellers: a role comprising some social functions such as maintaining the landscape, sustainable management of resources, preservation of biodiversity, leisure-time usability of the territory, ability to maintain good living conditions, and the security and healthiness of the area and its urban system.

The questionnaire was administered directly to residents and was composed of four sections: 1) topic scope of survey, and study area; 2) respondents' habits regarding use of the area for leisure activities, since it contains a greenway; 3) multifunctional role of agriculture and resident' willingness to pay for it; 4) respondents' socio-economic characteristics. The core of the interview was the contingent market, in which the concept of peri-urban agriculture was illustrated, with its specific characteristics as a territory unlike either urban centres or the countryside. Factors which may threaten ongoing agriculture and its externalities were then discussed. Financial aspects linked to maintaining agriculture in the area were introduced, stressing reduced financial resources for farmers, due to the expansion of the EU, and their replacement with regional funds. Respondents were asked if they would contribute to the maintenance of agriculture in the peri-urban area by paying a household tax ('payment vehicle'), involving sums of EUR20, 30, 40 and 50. A total of 204 residents were contacted: 41 per cent from Santa Maria, 32 per cent from Bastia and 27 per cent from Assisi. Most of the interviewees were accustomed to using the greenway, although to varying degrees. Indeed, answers referring to individual and family habits showed that it was used frequently, about 47 per cent of the sample stating 'very often' and 'often'. This result may overestimate the actual average level of use of the greenway by the town, probably due to the location of the interviews. It also suggests a high level of interviewees' involvement in the topic, as confirmed by their stated opinions regarding the beauty of the peri-urban landscape and their perception of changes to it over the past few years. Opinions were on the whole very positive, 31 per cent of respondents defining the landscape as 'very beautiful' and 46 per cent as 'beautiful'. Regarding perception of change, 36 per cent mentioned changes in it over the past few years; 15 per cent identified the main factor of change in the spread of urbanization, and 10 per cent in the rebuilding which started after the 1997 earthquake. The most frequent changes due directly to agriculture were the disappearance of sunflowers (10 per cent), decreases in cultivated land (6 per cent), new implants of vineyards and olive groves and, to a lesser degree, simplification of crop systems.

The willingness to pay (WTP) of families to support agriculture in the study area was obtained by estimating a Binomial Logit Model. Two models were estimated: dichotomous and Multivariate Log-linear. The results of both are shown in Table 15.1; Table 15.2 lists estimates of marginal effects in the second model. The average WTP was EUR40.6 and the median WTP EUR45. The multivariate model results showed that the WTP was significantly affected by household income and respondent's perception of positive externalities of agricultural production. The signs of these coefficients are positive, as expected.

Multiplying the average WTP value by the number of residents in the urbanized zones close to the peri-urban areas (about 7,000 families) gives the annual social benefits from peri-urban agriculture, which was approximately EUR284,000.

Table 15.1   Estimated Dichotomous and Multivariate Logit-linear Model

Variables	Dichotomous Model			Multivariate Model		
	Coefficient	Standard error	Level of significance	Coefficient	Standard error	Level of significance
Constant	1.625	0.37153	0.002	-5.930	1.533	0.000
Bid	-0.04	0.14	0.005	-0.064	0.017	0.000
Household income	-	-	-	0.069	0.015	0.000
Positive externalities	-	-	-	0.101	0.224	0.000
N° of. observations	204	-	-	-	-	-
Log Likelihood	-272.924	-	-	-114.884	-	-
Prob. > Chi²	0.004	-	-	0.000	-	-
Pseudo R-square	-	-	-	0.1829	-	-
correctly previewed cases	-	-	-	71,08%	-	-
WTP: mean value (EUR)	40.6	-	-	-	-	-
WTP: median value (EUR)	45	-	-	-	-	-

Table 15.2   Estimates of marginal effects in Multivariate Logit-linear Model(*)

Variables	Coefficient	Standard error	Level of significance
Bid	-0.016	0.004	0
Household income	0.017	0.004	0.000
Positive externalities	0.025	0.006	0

(*) calculated with STATA 10.0 on the mean value of variable

Discussion

The agriculture of the study area, composed of 91 'non-enterprises' and 20 'enterprises', was not greatly affected by the CAP reform, and single farm payments did not influence farmers' choices regarding the ways of distributing factors of production and their business strategies. Farmers perceived decoupled payments as an element of financial security by means of which they could react better to market changes. In addition, single farm payments represent an instrument to avoid abandoning agriculture because, although they are not high, they do ensure a certain amount of income.

Changes in production reveal both, that the presumed abandon of cultivated lands did not occur, and that the area was already characterized by simple agriculture, leading to further progressive productive simplification.

The combined effects of CAP reform and variations in market prices, of both products and inputs, have led to variations in farmers' incomes in different ways, depending on farm size and production decisions.

During direct interviews, the owners of 'non-enterprise' expressed their recent difficulties in carrying on work. Farmers who have not differentiated or diversified argue that, over the next few years, they will probably stop cultivating those few hectares of cereals, olives and vines, as the work is tiring and the economic returns are not sufficient. Increased prices of materials and price fluctuations of outputs have become unsustainable, and the single payment system often constitutes an impediment to abandonment because, although moderate, it only guarantees a minimum income. Most 'non-enterprises' and small farms are managed by non-farmers emotionally linked to the land, who spend their free time there. Many medium and large farms have diversified their activities, either into the commercial sector (e.g., opening farm shops, the sectors of mechanical (commission manufacturers) or recreational services, or industry (one mill)). In one case, a farm is managed by an industrialist who wanted to invest part of his income in the wine-producing sector.

The picture is of a type of agriculture that emerges which as a reflection of the city, but not from an economic point of view. Apart from holiday farms and investments in the wine sector, there are no other elements differentiation such as educational services, short chain development, organic products, etc. Most farms are managed by industralists 'loaned to' agriculture and by hobbyist farmers while only a small number of farms are managed by professional farmers. It is agriculture which lives as a reflection of the history and art treasures of Assisi, partly because of the strong restrictions exerted by local institutions to preserve the beauty of an area inscribed on the World Heritage list.

The farmers interviewed did acknowledge some advantages of working in a peri-urban area, which are mainly related to services offered by nearby towns including social (educational and recreational), and economic (presence of markets for sale of products and supplies of raw materials) (Table 15.3).

Table 15.3   **Positive externalities attributed to peri-urban agriculture by town-dwellers and farmers. Negative externalities and disadvantages attributed by town-dwellers and farmers respectively.**

Positive externalities – environmental	Citizens	Farmers	Negative externalities – environmental	Citizens
Landscape preservation	4.70	4.65	Seepage of pesticides, fertilizers	4.84
Maintaining open spaces	4.50	4.30	Excessive water consumption	4.50
Ground water protection	4.41	4.00	Ground water salinization	4.47
Soil conservation	4.34	4.80	Toxic gas emissions	1.94
Maintaining biodiversity	4.10	2.00	Production of bad odours	1.73

Positive externalities – socio-economic	Citizens	Farmers	Disadvantages for farms in peri-urban area	Farmers
Agroturism farms	4.50	4.80	Master plan bonds	4.50
Leisure-time services	4.40	3.20	Low recognition of agriculture role by other economic actors	4.50
Continuation of olive-growing	4.30	4.50	High traffic/ low quiet	4.30
Continuation of vine growing	4.30	4.30	Excessive control by municipality	4.00

Positive externalities – environmental	Citizens	Farmers	Negative externalities – environmental	Citizens
Maintaining rural buildings	4.10	2.80	Precariousness	3.45
Preserving farm traditions	3.50	4.00		
Production of cereals for the local industries	3.40	4.50		
Production of protein crops for local industries	3.40	4.40		
Improved access to foods	3.00	2.50		

None reported the benefits deriving from direct sales to consumers, as this would involve a completely different type of organization and would be incompatible with the reduction of labour which is gradually taking place on farms. Owners of holiday farms emphasize the advantages of being in an area of very high tourist interest: those who have purchased land in the last ten years stress the enhanced value of their farms. As for the disadvantages, the majority complained about the constraints imposed by the municipal land use plan and the poor recognition of the role played by agriculture in protecting the landscape from other economic actors, especially hotel-keepers. Other disadvantage is the very high land value faced with a moderate production value. Although the value of bare land amounts to more than EUR45 million (40,000 per hectare of UAA), this figure may reach EUR56 million when considering market values for fractions of a hectare (50,000 per hectare of UAA). This means that the value of agricultural production does not exceed 2.1 per cent of the value of the land used to produce it. However, at the same time, this area has improved the quality of life of the inhabitants of neighbouring towns and, every year, brings one million tourists, who pass through it to visit the city of Assisi.

Society's demand for the landscape and environmental assets of local agriculture clearly shows that the most greatly appreciated functions are those of conservation of the landscape and maintenance of open spaces. Conversely, less appreciated assets are improved access to foods, production for local industry, and conservation of country traditions. The most negatively perceived externalities are percolation into the soil of synthetic chemicals, excessive water consumption, and ground water salination (Table 15.3).

Study concerning the recognized functions of peri-urban agriculture shows that the function of 'the landscape' is the most significant. Also of great importance is the recreative function – the possibility of leisure activities, partly associated with agritourism – and productive functions, mainly production of cereals and protein crops for local industry. With respect to city-dwellers, farmers assign less importance to maintenance of rural buildings and conservation of biodiversity.

City-dwellers therefore seem to be more attentive towards environmental problems, highlighting the risks for types of agriculture which use too many chemicals.

Conclusions

Our analysis showed that, compared with a gross production of EUR854,000, about EUR316,000 were paid out as single payments while estimated social benefits arising from maintenance of the agricultural landscape and environment amounted to EUR284,000.

These results suggest linking single payments to the production of the landscape/environment; seeking actions suitable for creating conditions for development based on the principles of sustainability, and focusing on needs and demands through local participation.

Peri-urban areas may thus be assimilated to disadvantaged areas according to EESC proposals. However, this would not be with the aim of securing additional allowances in proportion to the natural or environmental limitations in which farmers operate, but to enhance the functions of landscape and of maintenance of open spaces recognized by the population. Agriculture can play a strategic role in improving the welfare of the urban community (residents and tourists) by renewing a "dialogue" which has been interrupted between urban and peri-urban areas, built-up and open spaces, daily life and leisure.

Lastly, we stress that the new Regional Rural Development Plan is a great opportunity to set up initiatives involving private citizens and associations, enterprises and business associations, together with public entities like institutions, regions, municipalities and public-private joint aggregations based on objective protocol agreements.

References

Borchardt, K.D. 2008. Speech on Health Check, in Future Challenges for agriculture: A day of scientific dialogue: Paper to the XXII EAAE Congress: People, Food and Environments: Global Trends and European Strategies, Brusselles session, 28 August 2008.

Branduini, P. and Sangiorgi, F. 2004. Verso la progettazione integrata delle aree agricole periurbane. International Conference: Il sistema rurale. Una sfida per la progettazione tra salvaguardia, sostenibilità e governo delle trasformazioni, Milan, 13 and 14 October.

Cooper, J. C. 1993. Optimal bid selection for dichotomous choice-contingent valuation surveys. Journal of Environmental Economics and Management, 24, 25-40.

European Economic and Social Committee (EESC), N.1209/2004, OPINION of the European Economic and Social Committee on Agriculture in peri-urban areas, Bruxelles 16 September 2004.

Fleury, A. and Donadieu, P. 1997. De l'agriculture périurbaine à l'agriculture urbaine. Le Courrier de l'environnement de l'INRNE, 31.

Giacchè, G. and Torquati, B. 2010. The Role of the Agriculture in the Plain of Assisi, Italy. Acta Horticulturae 881: 2nd International Conference on Landscape and Urban Horticulture. Prosdocimi Giaquinto, G. and Orsini, F. 2010 (eds), Bologna, Italy295-300

Hanemman, W. M. 1984. Welfare evaluations in contingent valuation experiments with discrete responses. American Journal of Agricultural Economics, 66, 332-341.

Hanemman, W. M. 1989. Welfare evaluations in contingent valuation experiments with discrete responses, Reply. American Journal of Agricultural Economics, 71, 1057-1061.

Kannien, B. J. 1993. Optimal experimental designs for double-bounded dichotomous choice contingent valuation. Land Economics, 69, 138-146.

Ministero per le Politiche Agricole e Forestali (MiPAF). 2007. PSN 2007-2013 Piano strategico nazionale per lo sviluppo rurale.

Pascucci, S. 2008. Agricoltura periurbana e strategie di sviluppo rurale: una riflessione. QA_Rivista dell'Associazione Rossi-Doria, 2.

Sotte, F. 2006. Quante sono le imprese agricole in Italia?. Agriregionieuropa, 5.

Stolfi, N. 2004. L'agricoltura negli spazi periurbani: caratteristiche e tendenze: Jornadas Europeas de Agricultura Periurbana. Viladecans Barcelona, 12 – 14 may 2004.

PART 5
The Environmental Impact of CAP Reform

Chapter 16

Impact of the 2003 CAP Reform on Organic Farming in Germany

Jürn Sanders, Hiltrud Nieberg and Frank Offermann

Introduction

The financial performance of organic and conventional farming is highly influenced by the direct payment policy of the EU Common Agricultural Policy (Häring and Offermann 2005). While organic farms receive considerable support from agri-environmental programmes, the design of the first pillar put organic farming at a disadvantage in the past. The 2003 CAP reform has changed this situation by decoupling direct payments and reducing price support. In Germany, a transitional decoupling scheme has been implemented in 2005 which will be transferred stepwise into a regional Single Farm Payment (SFP) model by the year 2013. In recent years, a number of studies estimated that the CAP reform could reduce some of the unfavourable elements of the CAP depending on the national implementation of the reform (Offermann 2002, Offermann and Nieberg 2006, Schmid and Sinabell 2007). Against this background, this chapter aims to identify and assess quantitatively the present impact of the reform on the relative profitability and production structure of organic farms in Germany using book-keeping records from the German Farm Accountancy Data Network (FADN) as well as results from an extensive farm survey as empirical base.

CAP reform in Germany

In 2003, the EU decided to fundamentally reform the CAP. Key elements of the reform were the decoupling of direct payments via a Single Farm Payment, the linkage of this payment to agricultural and environmental standards and the revisions of the market policy of the CAP. The reform provided a variety of options for the national implementation, especially with respect to the design of the single payment scheme (SPS) and the degree of decoupling. In Germany, all payments were fully decoupled in 2005, with only a few exceptions (tobacco and starch potatoes). For the single payment scheme, a dynamic hybrid model was implemented in 2005. The initial values of the farm individual payment entitlements are partly based on historic receipts and partly based on regional flat rates for arable land and grassland. In Germany, 13 different regions were established for the SPS. Table 16.1 provides an overview of the allocation of the formerly coupled first pillar payments to the level of these entitlements. The values of these entitlements increased between 2005 and 2007 and will further increase in the coming years as a result of the inclusion of payments for the sugar market reform, the incremental payments for the third step of the dairy market reform, and the further decoupling of payments for tobacco. From 2010 to 2013, the values of the entitlements will then dynamically be transformed into pure regional flat rates. It is expected that by 2013 the regional flat rate payments for all eligible area will range from 258 to 359 EUR/ha, depending on the region.

Table 16.1 Decoupling scheme in Germany

	Level of		Decoupled payments included in the	
	decoupling	coupling	farm individual amount	regional amount
	in percent			
Crop payments				
Arable are premiums	100	0		X
Seed premium	100	0		X
Hops premium	75 [1)]	0		X
Legume crop premium	100	0		X
Tobacco premium	40	60	X	
Potato starch premium	40	60	(25 %)	(75 %)
Dried fodder premium	52	48	X	
Premiums for edible nuts	0	100		
Energy crops	0	100		
Protein crop premium	0	100		
Livestock payments				
Slaughter premium for beef	100	0		X
Slaughter premium for calves		0	X	
Special premium for beef	100	0	X	
Suckler cow premium	100	0	X	

Extensification premium for beef	100	0	X	(50 %)	X	(50 %)
Milk premium	100	0	X			
Ewe premium	100	0	X			

25 per cent of the premium volume is given directly to hops producer associations.

Source: Own presentation based on BMVEL (2005)

The 2003 CAP reform included also a time schedule for a review of the agreed policy changes aiming to streamline and modernise the CAP. Based on the first experiences the EU Commission has proposed modifications of the direct aid system, market instruments and rural development policies. More specifically, the so-called Health Check (HC) of the CAP comprise of a simplification of the existing single payment schemes, further reduction of market intervention, phasing out of the milk quota regime and additional policy measures to address new challenges such as management of production risks, fighting against climate change, more efficient management of water and the use of bioenergy as well as the preservation of biodiversity.

Methodology and data

In order to assess whether the relative financial performance of organic farms compared to conventional farming systems (i.e. the economic incentive to conversion) has already changed as a result of the implemented hybrid decoupling model, FADN data of 224 organic farms from the year 2003/04 and 2006/07 were analysed and compared with the corresponding data of conventional farms that were similarly structured. The selection of the conventional farms was carried out in accordance with a differentiated, internationally harmonised method (Nieberg et al. 2007) using various natural and geographic factors, resource endowment (ha UAA, milk quotas) and general farm type as selection criteria. Family Farm Income plus wages per annual work unit (FFI+W/AWU) were used as criteria to measure the profitability of organic and comparable conventional farms before and after the CAP reform. FFI+W/AWU is an indicator for the return for labour that allows a comparison of the incomes of farms with different legal forms. This is of relevance especially in the Eastern part of Germany, where a relatively high share of farms is managed as Limited or Joint stock companies.

To get a deeper understanding of the impact of the undertaken policy changes on farm profitability, 3,000 organic farmers were surveyed. The farm sample was based on a random selection. In total, 915 returns were included in the analysis which corresponds to approximately 5 per cent of all organic farms in Germany. Besides a financial assessment the survey asked also for possible changes on the farm in response to the CAP reform. A combination of closed questions (partly multiple choice and scale type questions) and open questions was used. In addition, respondent had the possibility to add comments and provide additional information.

Results

Impact of decoupling

As indicated in Table 16.2, first pillar direct payments increased on organic and comparable conventional farms between 2003/04 and 2006/07. This increase is mainly a result of the inclusion of the milk premium into the SFP as well as due to the larger farm size and the fact that more area is now eligible for direct payments. While organic farms and comparable conventional farms obtain similar receipts from the single payment scheme, the latter one are much more affected by the "reduction" of coupled direct payments. This results in a relative plus of EUR1,000 for organic farms.

An increase of first pillar direct payments can also be observed for different farm types (see Figure 16.1). Not surprisingly, the increase is particularly high for specialist dairying farms due to the introduction of dairy premiums and the eligibility of grassland for payment entitlements, while arable farms receive only slightly more direct payments. In the year 2003/04, differences in first pillar direct payments between organic and comparable conventional farm types varied between 2 per cent (arable farms) and 35 per cent (dairy farms). Two years after the introduction of the SFP, the difference decreased on organic arable, dairy and mixed farms. Organic mixed farms received on average even higher direct payments compared to their conventional counterparts, which is mainly due to the fact that these farms have substantially more clover leys in their rotation (that are now eligible under the single payment scheme). In absolute terms, the greatest differences exist between organic and comparable conventional dairy farms; i.e. for those farms that have an especially high share of the SFP determined on the basis of farm individual historical reference premiums. Under the "old CAP" this difference was mainly due to the fact that (comparable) conventional dairy farms received more headage premiums (larger number of beef cattle) and crop premiums (more crops that were eligible for direct payments) which is still mirrored in the current level of the SFP. The existing difference will however decrease after the introduction of the regional flat rate. Since the main part of organically managed farms in Germany belong to this farm type, this fact is particularly relevant for the average profitability of organic farming. One may assume that, compared to (comparable) conventional farms, relatively more organic farms will experience an increase in value of their payment entitlements after the introduction of the regional flat rate.

Table 16.2 Comparison of farm accountancy data from organic and comparable conventional farms before and after the implementation of the CAP reform 2003.

		Organic farms			Comparable conv. farms		
		2003/04	2006/07	Difference	2003/04	2006/07	Difference
Number of farms	N	224	244	-	224	224	-
Revenues							
Output from agr. production	EUR	88,024	109,011	20,987	89,866	101,925	12,059
Total direct payments	EUR	42,234	46,003	3,768	29,288	31,527	2,238
Organic area payments	EUR	10,016	12,883	2,867	0	0	0
Single Farm Payments (SFP)	EUR	0	22,803	22,803	0	23,170	23,170
Coupled payments (CP)	EUR	17,741	297	-17,444	19,333	509	-18,824
Net balance (SFP+CP)	*EUR*	-	-	*5,359*	-	-	*4,346*
Prices							
Cereals	EUR/kg	305	328	23	112	106	-6
Potatoes	EUR/kg	365	448	83	152	179	27
Milk	EUR/kg	0.34	0.36	0.02	0.30	0.30	0.00
Beef	EUR/Head	1018	1140	121	886	1033	147
Fattening pigs	EUR/Head	207	240	34	124	134	11
Production costs							
Total variable costs	EUR	45,451	55,396	9,945	55,163	62,066	6,903

Feeding costs	EUR	9,153	10,231	1,077	16,465	19,995	3,530
Personnel costs	EUR	14,868	17,594	2,726	9,212	9,683	471
Income							
FFI	EUR	29,760	43,701	13,941	20,802	31,328	10,526
FFI+W/AWU	EUR/AWU	20,318	27,905	7,587	15,295	21,393	6,098

Source: Own calculations based on German FADN

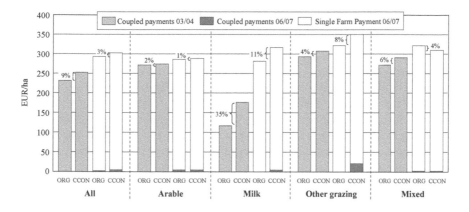

Figure 16.1 Changes in first pillar direct payments for different farm types[1]

Source: Own calculation based on German FADN data

As indicated in Table 16.1, the Family Farm Income plus wages per agricultural work unit increased on average by approximately EUR7,500 on organic farms and EUR6,000 on comparable conventional farms. As described above, the decoupling of direct payments contributed to this development – but it is not the only reason for higher average profits on organic farms. Absolute profits increased between 2003/04 and 2006/07 also due to higher producer prices (potatoes, cereals, milk, and pork meat) and higher organic area payments. The relative profitability of organic farms improved besides decoupling as a result of higher price premiums (potatoes, milk, and pork meat), higher organic area payments and a lower increase of fodder costs.

Furthermore, the calculation points out those neither organic nor comparable conventional farms have substantially changed their farm structure as a result of the decoupling of direct payments. The farm size increased on average by 2 ha (organic farms) and 1.8 ha (comparable conventional farms), respectively. Minor changes can also be observed with respect to the number of livestock kept on the farms and the labour input per farm. The only small changes are probably due to the fact that the CAP reform was implemented just two years ago.

Farmers' perspective

While the comparison of FADN data from 2003/04 and 2006/07 revealed that decoupling had in general a positive impact on the profitability of organic farms in Germany, results of the farm survey suggest a different view. A direct comparison

---

1   Sample size: All (n=448), Arable (n=122), Milk (n=194), Other Grazing (n=34), Mixed (n=72) – each with 50 per cent organic and 50 per cent comparable conventional farms. Farm groups with less than 15 farms were not included in the figure.

of the FADN data and the survey data is however difficult due to structural differences in both samples.[2] According to the survey results only 11 per cent of the organic farmers think that decoupling has had a positive impact on their farm profits (see Table 16.3). Every second farmer stated that profits have not changed as a result of the single payment scheme. The share of farmers with constant profits is particularly high among organic arable (61 per cent) and pig & poultry farmers (62 per cent). According to the survey results, dairy farmers have experienced most frequently a positive effect (23 per cent) of decoupling. This could be due to the fact that compensation payments for lower intervention prices for dairy commodities were included in the SFP as well as that dairy farms have a very high share of their SFP determined on the basis of historical receipts.

Table 16.3   **Farmers' assessment of the impact of decoupling on farm profits**

	All	Arable	Dairy	Other Grass	Granivores	Mixed	Others
Numer of farms	**850**	195	169	284	34	98	70
Percentage of farms							
Profit increased	**11**	8	23	9	6	8	4
Profit decreased	**27**	23	17	43	15	22	10
Profit did not change	**50**	61	47	42	62	56	40
Don't know	**12**	8	12	6	18	13	46

Source: vTI survey 2008

Though a concrete example was given the question on the financial impact of decoupling was certainly not easy to answer. In order to assess properly the decoupling effects, detailed economic farm data are needed. Since the number of part-time farmers is relatively high in the survey sample, this was probably not always the case. For this reason, one may assume that a number of farmers were not able to distinguish the decoupling effects from other policy or market changes.

As indicated in Table 16.4, 65 per cent of the organic farmers surveyed stated that they have not yet changed their farm in response to the introduction of the

---

2   For example, 49 per cent of the farmers who participated in the survey stated that they are part-time farmers. By contrast, the FADN sample comprise of only 9 per cent part-time farmers.

single payment scheme. The share is particularly high for organic dairy farms (77 per cent). Not surprisingly, the survey results suggest that particularly farms with more than 200 ha UAA changed their farm structure or farm organisation (57 per cent), while the opposite is the case for farms with less than 10 ha UAA (25 per cent). Consequently, it can be assumed that the adaptation pressure is particular high for large farms. Another explanation would be that farm managers of large farms have dealt more with the subject than others. The opposite is probably true for small holdings that are mainly managed as part-time farm. Furthermore, only 20 per cent of all farms intend to adapt their farm in the coming years. Thus, the majority of farms (54 per cent) do not see any need for changes as a consequence of CAP reform. On the other hand, one may critically question whether the majority farmers have already sufficiently identified the new possibilities given through decoupling. On the other hand, technical constraints could also be a reason why the greater part of farmers is currently not intending to response to the changes in the direct payment system.

Table 16.4   Farmers' response whether changes were made on the farm following the decoupling of direct payments

	All	Arable	Dairy	Other Grass	Granivores	Mixed	Others
Number of farms	**883**	201	171	289	35	98	89
				Percentage of farms			
Changes made	**35**	34	23	40	43	40	31
Changes not made	**65**	66	77	60	57	60	69

Source: vTI survey 2008

Those farmers who stated that they adapted their farm in response to decoupling changed in most cases the level of individual production activities. Furthermore, a number of farmers stated that they increased their off-farm activities or are now more involved in direct marketing. In general, it is not possible to identify a general adaptation pattern. Instead, partly contrasting answers were given. For example, 17 per cent of the farmers stated that they decreased the number of suckler cows while 12 per cent did the opposite. A similar contrasting result concerns changes of cereal production. According to the survey results, 17 per cent of the respondents increased and 11 per cent of the respondents decreased their cereal production. Again, one may assume that a number of farmers stated what kind of changes they made in the last two years which however were not necessarily related to the decoupling of direct payment.

The survey also contained also a number of questions that addressed issues of the current Health Check proposal. The EU Commission proposed for example a modification of the existing modulation scheme. According to the results of the survey the majority of the farmers surveyed would agree to a capping of single payments for large farms and a transfer of funds from the first to the second pillar. However, to sound a critical note, most of the farmers who agreed would probably not be affected by the proposal. In this connection, it is not a surprise when the lowest acceptance for this proposal can be observed among large farmers (21 per cent, respectively), i.e. among those farmers who would mainly be affected.

Conclusion

The 2003 CAP reform changed substantially the policy environment for organic and conventional farms. Farm accountancy data of the German FADN suggests that the decoupling of direct payments has generally been favourable for organic farms. As a result of the hybrid decoupling scheme, absolute and relative profitability of organic farms increased compared to comparable conventional farms. It is expected that the benefits for organic farms will even further increase when the existing hybrid decoupling model will be changed to a pure regional flat rate. The question whether organic farms are more profitable than (comparable) conventional farms will however also strongly depend on other factors such as changes in farm-gate prices. In this connection, it would be interesting to monitor the effects of the current rise in prices for food and fibre on the relative profitability of organic farming in Germany.

Results from the farm survey suggest that farmers have probably not yet fully internalised the changes resulting from the reform and therefore are not always aware of the specific economic consequences. For this reason, one may assume that adjustments will be lagged and decided on within the coming years. Thus, the outcomes of this investigation suggest that organic farmers still require more specific information and advice in order to use the new possibilities given through decoupling. In doing so, it would be necessary to consider the technical constraints of organic farming systems and the large number of part-time farmers that have presumably less capital and time to adapt their farm.

References

Häring A.M. and Offermann, F. 2005. *Impact of the EU Common Agricultural Policy on organic in comparison to conventional farms*. Paper to the XIth EAAE Congress on The Future of Rural Europe in the Global Agri-Food System, Copenhagen, Denmark, 24-27 August 2005.

Nieberg, H., Offermann, F. and Zander, K. 2007. *Organic farms in a changing policy environment: impacts of support payments, EU-enlargement and Luxembourg reform*. Hohenheim: Institut für landwirtschaftliche Betriebslehre.

Offermann, F. 2002. *The influence of the Common Agricultural Policy of the EU on the competitiveness of organic farming*. Paper to the OECD Workshop on Organic Agriculture, Washington D.C, USA, 23-26 September 2002.

Offermann, F. and Nieberg, H. 2006. *Impacts of the 2003 CAP reform on structure and competitiveness of organic farming systems*. Paper to the 93rd EAAE Seminar on 'Impacts of Decoupling and Cross Compliance on Agriculture in the Enlarged EU', Prague, Czech Republic, 22-23 September 2006.

Schmid, E. and Sinabell, F. 2007. *Impacts of Alternative Implementations of the Single Farm Payment on Organically and Conventionally Producing Farms in Austria*. Paper to the 9th. Wissenschaftstagung Ökologischer Landbau, Stuttgart-Hohenheim, Germany, 20-23 March 2007.

Chapter 17

Pressure Factors Affecting the Lombardy Agricultural System: The Environmental Consequences of the Fischler Reform

Claudia Bulgheroni and Guido Sali

Introduction

This chapter expounds the application of a mathematical programming model to evaluate the implementation of adjustment measures in the agricultural sector of Italy's irrigated Lombardy lowland, in the presence of new environmental pressures resulting from an EU move towards environmental protection and the implementation of the change in market dynamics brought on by the implementation of the Fischler reform. The elimination of direct payments as result of the latter is in fact linked to the serious upheaval in the European agricultural products market during recent years.

With regards to the Lombard lowland, the vast availability of water has always been a fundamental feature of this area and this environmental characteristic has, over the centuries, led to a flourishing agriculture based on dairy cattle breeding, cereals and rice crops. Since the beginning of the decade, and especially since 2003, water crises have become recurrent. This has reflected on the availability of water for irrigation, especially during the summer when farms in Lombardy have been forced to deal with conditions of water shortage that in some cases causing declines in production. The repetition of periods of water shortage has moved regional authorities to take political initiatives in the management of the water resources in order to contain consumption.

The planned and implemented actions include an improvement in the efficiency of the distribution network, the conversion of the irrigation systems to sprinkle irrigation, the adjustment of the water concession fees that also involves the introduction of *pricing* policies, and the reduction in the number of water concessions to irrigation consortia.

In particular, provision through irrigation consortia is closely linked to recent regional water management practices to safeguard the integrity of the waterways, as demonstrated by the will to maintain the so-called *Low Flow Limit* to preserve the aquatic ecosystems. The possibility that the availability of agricultural water supplied through irrigation consortia could be reduced makes it necessary to study what adjustments will be necessary in the agricultural sector, to achieve the market objectives, despite the difficulties due to the cuts in water supplies for irrigation. The objective of this chapter is to relate the dynamics of agricultural prices with the possible restrictions on the availability of agricultural water, and within this framework, verify whether the current, Fischler reform-induced, production structure of the Lombard irrigated plain is compatible with the scenarios of reductions in water availability for agriculture.

Theoretical Framework

Positive Mathematical Programming (PMP) has shown considerable versatility in recent years, being used in both sectorial and regional analyses to evaluate the effects on agriculture of both agricultural and environmental policies (Röhm and Dabbert 2003, Schmidt et al. 2006).

The reasons why PMP has proved popular are the advantages derived from its positive approach. As such, the first and main contribution that PMP brings to policy modelling, and in general to issues in agricultural economics, is the capability of maximizing information from agricultural data banks such as for example FADN, and REGIO (all at a European-wide level, Arfini et al. 2003, Paris and Howitt 1998). This is because PMP requires a lower bulk of information compared to other mathematical programming techniques, providing useful analyses to policy makers even in presence of a limited set of information, as it generally happens when European agricultural databases are adopted. A second advantage is that through PMP it is possible to exactly represent the situation observed, in this case farmers' production plans: the methodology's total variable cost estimation makes possible the reproduction of the observed farm allocation plan and the decision variables (total specific variable costs) that drive farmers in selecting such production plan (Howitt 1995). A third important advantage is that PMP allows results to continuously change, subject to farmers' reaction, as a consequence of changing exogenous variables (Buysse, Van Huylenbroeck and Lauwers 2006). Hence, PMP can respond with flexibility to a large spectrum of policy issues, typically concerning land-use change, production dynamics, and variation in gross margin and in other economic variables (costs, subsidies, etc.).

Among all the PMP models developed to forecast, at a country's regional level, changes in farmers' behaviour as a consequence of CAP reforms, it is worth mentioning the following models: CAPRI (Heckelei and Britz 1999, 2001, 2005, Britz et al. 2003); the INRA of Nancy model (Barkaoui et al. 1999, Barkaoui and Butault 2000); the University of Madrid model, PROMAPA, (Júdez et al. 2001, 2002, 2008); FARMIS (Offermann et al. 2003, 2005); AGRISP (Arfini et al. 2005). All of them share the assessment of the CAP impact through the forecast of the changes in the productive system due to the new conditions imposed by the policy.

At the same time, in recent years especially after the approval of the so-called Water Frame Directive (2000/60/CE), many PMP models have also been implemented, particularly at regional level, in order to assess the impact on irrigated agriculture of new principles introduced by the normative (e.g. Bartolini et al. 2007, Bazzani and Scardigno 2008, Cortigiani and Severini 2008). Main WFD novelties refer to the Full Cost Recovery, the Polluter Pays Principle (PPP) and the introduction of water pricing. Since all of them aim to reduce water use and water pollution, it is reasonable to think that these instruments will deeply influence irrigated agriculture, one of the largest water-consuming sectors.

Materials and Methods

The Model

The model we propose is based on the traditional phases of Positive Mathematical Programming (Howitt 1995). Specifically, these consist of:

- Definition of a linear programming model where the land allocated to each production process is the only constraint adopted. The marginal cost values of the soil factor in each activated production process are obtained from the dual structure coming from the profit maximization.
- Use of marginal costs of the soil factor, estimated in the first phase, to calibrate the parameters of the non-linear cost function. This curve is hypothesised as a quadratic function with respect to the quantities produced.
- The construction of a non-linear model that has as its optimal solution the same distribution of land among the various production processes established in the first phase.
- Use of a non-linear model, accordingly constrained on the basis of the availability of resources and the characteristics of the system, to construct scenarios of production choices and consequent land use.

In the model, the production units assumed include *agrarian regions*[1] rather than farms. This choice was based on several considerations.

Firstly, the structural conditions that influence production costs depend to a great extent on the conditions of the farmland: the pedo-climatic conditions, the quality of the soil, water availability and methods of distribution, and so on. This fact leads to note that in agriculture the contextual conditions are important at least as much as the conditions of the organization of the farm. Furthermore, taking into account that in optimization-based models fixed costs, which are mostly due

---

1 The agrarian region represents the unit of the *sub-regional* partition established by the Italian National Institute of Statistics to split the Italian land into homogeneous areas according to territorial and agricultural features. Albeit having slightly different meaning, in this paper "agrarian region" and "territorial unit" are used interchangeably.

to structural factors, are ignored, contextual conditions become the prevailing conditions.

Secondly, the assumption of a homogeneous land that includes all the relative farm holding factors considerably reduces the distortions in the model due to the specificity of the analysed farm samples and the choices made by the farmers. In fact, a territorial analysis like ours carries a strong risk that the sample of farm holdings is not sufficiently representative of the production trends prevalent in the area of study, especially in the case where the sample is already set, as in the case of FADN (Farm Accountancy Data Network). By contrast, the assumption of a real use of agricultural land highlights exactly which crops are more suitable or simply possible on that land on the basis of the contextual conditions. Finally, it should be remembered that in assuming the agrarian region, the rigidity of determining the production alternatives in each farm is greatly diminished. The main rigidities are due on the one hand to the need for rotation, which imposes certain sequences in the choice of crops, and on the other hand the feeding needs of the livestock, that restrict the allocation of part of the farm land. In both cases, the assumption of the agricultural region considerably reduces the aforementioned rigidity.

Anyway, the FADN sample farms located in the *agrarian regions* are fundamental for the acquisition of data on a number of economic variables. This is true for production costs and for sales prices. The former are calculated by crop, and compared with the cultivated area within the *agrarian region*. The latter are calculated as the average prices taken at the farm level.

Fourth, the N+1*th* production unit is introduced in the model and represents the entire area of study. With reference to phase 4 above, given a non-linear cost function $C(\mathbf{x}) = \mathbf{x}' \, \mathbf{Q} \, \mathbf{x} \, / \, 2$, the problem consists on the search for the unknown values of vector $\mathbf{x}$ in order to optimise Z given the constraints assigned to the system. In this case, the function to maximise is the gross income of each *agrarian region*, expressed as follows,

$$(1) \quad Z = \mathbf{p}' \, \mathbf{x} - \mathbf{x}' \mathbf{Q} \, \mathbf{x} \, / \, 2$$
$$\text{s.t. } \mathbf{Ax} \geq \mathbf{b}$$
$$\mathbf{x} \geq 0$$

where $\mathbf{p}$ is the vector of product prices, $\mathbf{x}$ is the production vector, $\mathbf{Q}$ is the coefficient matrix. $\mathbf{Ax} \leq \mathbf{b}$ is a set of linear inequalities representing the equations of the constraints, and $\mathbf{x} \geq 0$ is the non-negative constraint on the variables.

In this model a number of constraints have been introduced on vector $\mathbf{x}$ to make allowances for both the structural and land characteristics of the production system, as well as the analysis of the effects of a possible reduction of agricultural water availability should the regional sector authorities introduce concession restrictions to the irrigation consortia.

The model's profit-maximization objective takes into account a series of specific constraints important to the production conditions of the area, as described below:

- The area of study is specialised in dairy production. The diet of the animals is strictly based on locally produced forage and this must be guaranteed even in the case of economic instability that may affect production of other crops. Furthermore, adjustments to livestock numbers are difficult to implement, even in the face of strong fluctuations in milk and meat prices: on the one hand, the current system of milk quotas does not allow adjustments for increasing livestock numbers; and on the other hand, adjustments are likely to occur only in the long run, when price changes in the market are structural, and do not reflect short term fluctuations. Therefore, we introduced a constraint that controls for the production quota (QL) present in every agrarian region, and for the associated forage requirements necessary to feed the present livestock. The restriction for forage corn is as follows:

(2) $$\left| \frac{QL_n(m_v + m_r)}{\gamma} - x_{nj_M} \right| \leq r \cdot q_{nj_M}$$

where m_v and m_r are the average annual feeding requirements of cows and other livestock respectively, γ is the average annual milk production per cow, is the current production of forage corn and r is the percentage that indicates the allowed deficit or surplus level of production activated with respect to the feeding requirements. In this chapter the value of r is set at 0.20.

- Similarly to corn, alfalfa represent feed that is usually used in the feeding rations of milk cows. For this reason, the production constraint is as follows:

(3) $$\left| \frac{QL_n e_v}{\gamma} - x_{nj_E} \right| \leq r \cdot q_{nj_E}$$

where e_v is the average annual per head requirement of alfalfa in production, q_{nj_E} is the production of alfalfa activated by the model in agricultural region n and q_{nj_E} is the current production of alfalfa. Equation (3) also takes r = 0.20.

- A further constraint is introduced with regards to grassland. In this case it is necessary to consider the characteristics of the agricultural land of many areas of the Lombard lowland, where the fertile layer is extremely fine and the gravel subsoil does not allow the cultivation of arable land. Hence, permanent grassland becomes the only possible use. A constraint has been therefore introduced for grasslands where the surface assigned to its cultivation, , is not lower than quota r of surface h_{njp}, currently used (in this case r=0.80): SUP_{njp}

(4) $$SUP_{nj_P} r \leq h_{nj_P}$$

Finally, a restriction relating to water has been introduced for each agrarian region *n*, imposing in some scenarios that the water requirement simulated could not exceed 80% of the current water availability, as follows:

(5) $$\sum_j F_{jn} q_{jn} \leq G_n \cdot r$$

where F_j is the water requirement of the J-*th* crop and G_n the current availability of water in the n-*th agrarian region.*

Data Sources and Model Input

The input data for the territorial economic model proposed are drawn from two different data sources, the *Farm Accountancy Data Network* (FADN) and the *Agricultural Information System of the Lombardy Region* (SIARL), which includes the land use statements presented by farmers to obtain subsidies. The land use information is aggregated at *agrarian region* level, resulting in a list of all the agricultural land uses of each *agrarian region*. At the same time, farms in the FADN sample are aggregated in order to obtain average prices and costs.

Since the territorial connotation plays a key role in agricultural policy analysis, and considering that the *agrarian region* represents the minimum homogeneous unit, the need for the reproduction of the true agricultural land allocation among the different production processes (and the associated techno-economic parameters) becomes evident.

In addition, water requirements for all the main crops have been introduced in the model as an input variable. The data comes from a hydrologic model, which calculated the water requirements as the average of all irrigation seasons of a reference period (1993-2005) developed by the Department of Agricultural Engineering of the University of Milan (Facchi et al. 2004, 2007). Table 17.1 shows the complete list of the model input variables with the corresponding source.

Table 17.1   Model input

Source	Input variables	Unit	Year
	Land uses	ha	2008
SIARL	Animal heads	n	2007
	Milk quotas	t	2007
	Variable costs for sold and re-employed goods	EUR/ha	2007
FADN	Sold goods prices	EUR/t	2007
	Sold and re-employed production	t	2007
	Subsidies and payments	EUR/ha	2007

Study Area

The study of CAP effects was carried out in a territorial-based analysis using the watered lowland of of Lombardy, a region situated in the North of Italy, as the main focus.

The study area is bordered by administrative regional limits on the East and West, while the northern and southern boundaries follow the limits of Irrigation Districts, which are the territorial-based unit of the irrigation water management system of the region.

The study area covers an overall area of 888,243 ha and contains 814 municipalities that belong to 45 *agrarian regions*. Each of them includes 10-20 municipalities and an average regional UAA of around 15,000 ha.

According to the Agricultural Information System of the Lombardy Region (SIARL) (2008) the Utilized Agricultural Area (UAA) covers approximately 87 per cent of the study area. Farming is mainly dedicated to cereals: specifically 283,938 ha of Maize, 93,255 ha of Rice, 76,831 ha of Soft Wheat, 61,303 ha of Silage Maize, 55,244 ha of Grassland, 46,800 ha of Alfalfa, 27,259 ha of Barley, 21,386 ha of Hard Wheat, 11,473 ha of Soya, 6,657 ha of Sugar Beet and 5,692 ha of Tomato. Furthermore, 6,995 ha are allocated to the Set Aside scheme.

Along with crops, dairy cow breeding is also extremely common and very much representative of the area: here the 36 per cent of the Italian production of milk is obtained (3.7 millions of tons) by 500.000 dairy cows, which represent respectively the 26 per cent of the total amount of bovines and the 29 per cent of dairy cows of the Country. Furthermore, 5 millions of swine (55 per cent of the total swine livestock of the Country) are bred in the area, in the most intensive Italian farms.

The most common farming types are: *Specialist dairying* (TF41) and *Specialist cereals*, *oilseed* and *protein crops* (TF13). Also *General field cropping* (TF14), *Specialist granivores* (TF50) and *Field crops-grazing livestock combined* (TF 81) play an important role in the area.

Results and Discussion

Simulated Scenarios and Results Analysis

As previously stated, the objective of this chapter is to combine in a unique PMP model the assessment of both CAP issues and the potential reduction in irrigation water supplies due to the WFD policy, by linking potential water supply changes with information on the dynamics of agricultural prices.

Accordingly, 11 scenarios have been simulated, by considering, on the one hand, possible increases or decreases in the price of milk and cereals, and, on the other, the possible reduction of the water availability for agricultural use.

The plan of the examined scenarios and the related results is given in Table 17.2.

In the absence of irrigation constraints, the analysis of price variations of agricultural products shows different effects.

Variations in the price of milk have an effect on the value of forage transformation (maize grain, dried alfalfa and permanent grassland), without however strongly influencing final production. This reflects the stability of the animal husbandry production sector, which is characterised by costly fixed investments that limit its capacity to adjust in the short to medium run (SIM_01). Moreover, with a decrease in the price of milk (SIM_03), the effects are mainly evident in those agrarian regions where forage production exceeds local requirements, which translates into a reduction of stocks for sale (alfalfa and grassland). In general, the production of silage maize in the study area almost entirely covers the feeding requirements of farms; therefore, even in a scenario of loss of competitiveness of the zootechnical production, the surface necessary to guarantee supply self-sufficiency is preserved.

Table 17.2   Simulated scenarios and variation (%) of the land allocated to the main crops of the area, in respect to the *baseline*

					Scenario hypothesis						
	SIM_01	SIM_02	SIM_03	SIM_04	SIM_05	SIM_06	SIM_07	SIM_08	SIM_09	SIM_10	SIM_11
Price Variation Product	+20% milk	+20% milk	-20% milk	+20% cereals	+20% cereals	-20% cereals	-20% cereals	+20% milk cereals	+20% milk cereals	-20% milk cereals	-20% milk cereal
Water Variation	--	-20%	--	--	-20%	--	-20%	--	-20%	--	-20%
Variation in land allocation (%)											
Land use	SIM_01	SIM_02	SIM_03	SIM_04	SIM_05	SIM_06	SIM_07	SIM_08	SIM_09	SIM_10	SIM_11
Beet	0.2	9.9	4.6	8.8	--	17.6	113.7	--	--	30.6	97.1
Alfalfa	8.8	8.8	-20.1	-28.0	-19.5	25.4	8.8	-16.9	-19.5	-11.3	14.4
Durum wheat	-4.0	5.0	10.0	18.9	32.6	-8.2	81.9	-21.1	32.6	36.7	63.7
Soft wheat	-6.2	4.8	9.2	-19.7	74.1	-4.0	31.4	38.5	74.1	28.7	29.0
Maize corn	-4.2	-6.0	3.4	20.9	6.8	-2.2	-27.4	21.8	6.8	-9.0	-47.6
Silage maize	7.9	7.9	7.9	-2.1	7.9	4.4	7.9	-1.9	7.9	-2.1	3.0
Barley	-7.0	5.3	10.0	-1.9	13.2	-25.1	89.4	-54.8	13.2	39.4	71.9
Tomatoes	-0.6	-0.4	1.5	-58.4	-71.0	69.0	85.0	-70.4	-71.0	21.1	70.4
Grassland	35.2	23.3	-20.0	-20.0	-20.0	-10.9	18.7	-20.0	-20.0	-20.0	35.9
Set aside	--	--	-97.1	--	--	57.4	--	--	--	83.8	539.0
Rice	-1.6	-3.3	-0.1	5.8	-45.8	-13.5	-55.9	-24.9	-45.8	-27.1	-35.1
Soya	-4.9	-6.9	4.4	--	--	61.1	344.6	--	--	103.3	315.8

Source: own estimates

On the other hand, variations in the price of cereals (wheat, barley, maize grain, rice) seem to cause the greatest changes in the production objective of arable land surfaces in the study area. In the case of price increases (SIM_04), maize (+21 per cent) was the item most affected, followed by durum wheat (+19 per cent);[2] whilst the surface destined for rice also increased. Other cereals lost competitiveness due to a lower increase in gross income with the increase in price: this is the case for barley and soft wheat.

On the other hand, with the decrease in the price of cereals (SIM_01), we can see a diversified withdrawal of the cereal surfaces, greater for barley and durum wheat, less for maize and soft wheat.

In the case of the joint increase of the price of milk and cereals (SIM_08), there is a concentration of production of corn and soft wheat; whilst a decrease (SIM_10) reflects an expansion of wheat and barley, characterised by a modest contribution of capital and work.

The introduction of the constraint of agricultural water availability highlights the trend to shift towards crops with lower water requirements. So, comparing scenarios with the water constraint to those without constraints (SIM_02 vs SIM_01, SIM_05 vs SIM_04, SIM_07 vs SIM_06, SIM_09 vs SIM_08 e SIM_11 vs SIM_10), the maize grain and rice crops show the greatest decrease to the advantage of autumn-winter cereals and grassland.

It is worth noting that in some cases the agricultural water constraint assumes significance for the purpose of sowing equal to or greater than the price effect. This means, for example, that the increase in the surface allocated to maize grain in case of a price increase (SIM_04) tends to cancel when the water constraint is introduced (SIM_05). Similarly, the modest decrease in surface due to the reduction in price (-2.2 per cent in SIM_06), becomes considerably greater if we introduce the water constraint (-27.4 per cent in SIM_07). Both rice and corn lose approximately 50 per cent of surface, especially in the presence of both the price decrease and the water constraint. Since both crops are widely distributed over the study area, a loss of approximately 170,000 ha is implied. Such a reduction can deeply change the production profile of the Lombardy's watered lowland.

In fact, scenarios SIM_07 and SIM_11 show a production situation that if realized, would have a significant impact on the economic balance of farm holdings. Elsewhere, the strong increase of the set-aside in the last two scenarios (SIM_10 and SIM_11) is also significant: in both cases, the set-aside land exceeds the surface it covered during the 2005-2006 periods when the guarantee of EU aid and the difficulties of the market re-orientation process (driven both by the CAP reform) triggered the resort to the set-aside.

---

2   The surface allocated to durum wheat has increased in the Lombard lowland since the introduction of the European aid regime based on the single payment.

Conclusions and Recommendations

In this work, we analysed the impact of water and other European level policies at a regional, with a case study of the Lombardy region in the North of Italy. This was done using PMP programming, simulating the effect of price and quantity dynamics on farmers' production decision and input allocation.

In summary, the predicted scenarios due to price variations show some stability with regard to the zootechnical matrix of the production structure. The presence of animal husbandry makes farmers less willing to make sudden changes to the kind of crops they cultivate, especially those crops used for forage. Cereal and industrial crops are more sensitive: the transfer from direct payments to a single payment makes the role of EU aid on the choice production non-influential and that price variations may determine significant shifts in production choices.

It is worth remarking, however, that the process of adjustment to price variations foreseen by the model has in reality certain forces that limit it and delay its occurrence. For instance, price variations need to assume a structural character and not simply reflect a seasonal or economic shift, at least as perceived by those in the sector. The time periods between the selection of the seeds and the sale of the product often exceed one year, hence even significant –but not structural – price dynamics, such as those recorded between 2007 and 2008, did not cause equally significant variations in the distribution of arable land in the area of study.

By contrast, the results of the model relative to the adjustments caused by lower agricultural water availability could be considered to have a higher degree of reliability. In fact, the supply of water from irrigation consortia is quantified with certainty and is made available in the same quantities in the years to come, with reductions in level occurring over a relatively long period of time and with small levels of uncertainty.

From the simulation results the possibility of lower profitability of the agricultural areas emerges, as a result of the increased reliance on winter cereals and grasslands, and the re-emergence of the set-aside. In short, lower water availability causes a shift towards more agricultural production.

From the methodological point of view, PMP models like the one undertaken serve to shed light on both benefits and drawback of policies that affect the economic variables of the agricultural sector, but also the impact on agricultural production systems of environmental policies, which are bound to interact and influence those systems.

References

Arfini, F., Donati, M. and Paris, Q. 2003. *A national PMP model for policy evaluation in agriculture using micro data and administrative information*. Paper to the International Conference: Agricultural policy reform and the WTO: Where are we heading?, Capri, June.

Arfini, F., Donati, M. and Zuppiroli, M. 2005. Agrisp: a regional model for the assesment of the effects of agricultural policies changes in Italy, in *The EU agricultural policy reform and WTO negotiations* (in Italian) edited by G. Anania. Milano: Franco Angeli, pp. 81-128.

Barkaoui, A., Butault, J. P. and Rousselle, J. M. 1999. Mathematical Programming and oilseeds supply within EU under Agenda 2000, in *Proceedings of Eurotools Seminar*, edited by F. Arfini. University of Parma, Parma, Italy.

Barkaoui, A. and Butault, J.P. 2000. Cereals and Oilseeds Supply within the EU, under AGENDA 2000: A Positive Mathematical Programming Application, *Agricultural Economics Review*, 01(2), 1-12.

Bartolini, F., Bazzani, G.M., Gallerani, V., Raggi, M., and Viaggi, D. 2007. The impact of water and agriculture policy scenarios on irrigated farming systems in Italy: An analysis based on farm level multi-attribute linear programming models. *Agricultural Systems*, 93, 90-114.

Bazzani, G. and Scardigno, A. 2008. *A simulation model for regional economic analysis of water use and CAP reform: a methodological proposal, and the first applications to agriculture in Puglia.* Paper to the XLV SIDEA Conference, Portici, 25-27 September.

Britz, W., Pérez, I. and Wieck, C. 2003. Mid-Term Review Proposal Impact Analysis with the CAPRI Modelling System, in *European Commission, Directorate-General for Agriculture Mid-Term Review of the Common Agricultural Policy.* Brussels: pp.111-140.

Buysse, A., Van Huylenbroeck, G. and Lauwers, L. 2006. Normative, positive and econometric mathematical programming as tools for incorporation of multifunctionality in agricultural policy modeling. *Agriculture, Ecosystems and Environment,* 120, 70–81.

Cortigiani, R. and Severini, S. 2008. *Water Frame Directive n°60/2000 and the increase of agricolture water cost: an assessment through PMP models.* Paper to the XLV SIDEA Conference, Portici, 25-27 September.

Facchi, A., Bernardini, M., Gandolfi, C. and Ortuani, B. 2007. Estimation of water requiremets at territorial level: time series and scenario analysis, in *Agriculture and Water: models for sustainable management: the case of reorganization of Treviso area irrigation system*, edited by C., Giupponi and A., Fassio. Bologna: Il Mulino, pp. 191-228.

Facchi, A., Gandolfi, C., Ortuani, B. and Maggi, D. 2005. Simulation supported scenario analysis for water resources planning: a case study in Northern Italy. *Water Science and Technology*, 51(3), 11-18.

Heckelei, T. and Britz, W. 1999. Maximum Entropy specification of Pmp in CAPRI, *CAPRI Working Papers,* University of Bonn, Bonn.

Heckelei, T. and Britz, W. 2005. Model based on positive mathematical programming: state of the art and further extensions, in *Modelling agricultural policies: state of the art and new challenges* edited by F. Arfini. Parma:, MUP, pp. 48-74.

Heckelei, T. and Britz, W. 2001. Concept and Explorative Application of an EU-wide Regional Agricultural Sector Model (CAPRI project). in *Agricultural Sector Modelling and Policy Information Systems* edited by T. Heckelei, H.P. Witzke and W. Henrichsmeyer. Bonn: Vau Verlag Kiel, pp. 281-290.

Howitt, R.E. 1995. Positive mathematical programming. *American Journal of Agricultural Economics*. 77, 329–342.

Júdez, L., Chaya, C., Martínez, S. and González, A.A. 2001. Effects of the measures envisaged in "Agenda 2000" on arable crop producers and beef and veal producers: an application of Positive Mathematical Programming to representative farms of a Spanish region. *Agricultural Systems*, 67, 121-138.

Júdez, L., De Miguel, J.M., Mas, J. and Bru, R. 2002. Modeling crop regional production using Positive Mathematical Programming. *Mathematical and computer modeling*, 35 (1-2), 77-86.

Júdez, L., De Andres, R., Ibanez, M., De Miguel, J.M., Miguel, J.L. and Urzainqui, E. 2008. *Impact of the Cap Reform on the Spanish Agricultural Sector*. Paper to the 109th EAAE Seminar: The CAP after Fischler reform. Viterbo, 20-21 November.

Offermann, F., Kleinhanss, W. and Bertelsmeier, M. 2003. Impacts of the decisions of the mid-term review of the common agricultural policy on German agriculture, *Landbauforschung Volkenrode*, 53 (4), 279-288.

Offermann, F., Kleinhanss, W., Huettel, S. and Kuepker, R. 2005. Assessing the 2003 CAP Reform Impacts on German Agriculture, Using the Farm Group Model FARMIS, in *Modelling Agriculture Policies: State of the Art and New Challenges*, edited by F. Arfini. Parma: MUP, pp. 546-564.

Paris, Q. and Howitt, R.E. 1998. An Analysis of Ill-Posed Production Problems Using Maximum Entropy. *American Journal of Agricultural Economics*, 80, 124-138.

Röhm, O. and Dabbert, S. 2003. Integrating agri-environmental programs into regional production models: an extension of positive mathematical programming. *American Journal of Agricultural Economics,* 85, 254–265.

Schmidt, E., Sinabell, F. and Hofreither, M.F. 2006. Phasing out of environmental harmful subsidies: consequences of the 2003 CAP reform. *Ecological Economics*, 60, 596-604.

Chapter 18

Investigating the Economic and Water Quality Effects of the 2003 CAP Reform on Arable Cropping Systems: A Scottish Case Study

Ioanna Mouratiadou, Graham Russell, Cairistiona Topp, Kamel Louhichi and Dominic Moran

Introduction

The 2003 Common Agricultural Policy (CAP) Reform aimed to increase the prominence given to the sustainability of agricultural systems in both socio-economic and environmental terms. An important question is whether the Reform has indeed effectively encouraged farmers to achieve broad economic and environmental goals. The effects of the Reform on economic decision making and associated viability of farms can be explored by analysing data on current farmers' decisions and related economic indicators. However, the comparison of such figures before and after the implementation of a policy is complicated because they represent the combined effects of all the changes that took place during that period including the non-policy changes in prices of inputs and outputs, structural changes, etc. As it is unlikely that the individual effects of each of these factors can be identified, Mathematical Programming Modelling provides an attractive alternative for policy assessment.

Mathematical Programming Models (MPMs) have been widely used for agricultural economics policy analysis. An optimisation-based MPM selects the optimal allocation of farm resources to a large number of alternative agricultural activities, through the optimisation of an objective function subject to technical, agronomic, economic and policy constraints. For each of the policy scenarios modelled, the parameters or constraints representing the scenario are altered, invoking changes in land use and the economic and environmental outcomes of the optimisation. The comparison of those outcomes with a base scenario facilitates the ex-ante impact assessment of policies and consequently their design. Even though MPMs are predominantly used for ex-ante policy assessment, their use for ex-post assessment is particularly useful as the impact of different factors affecting agricultural production can be studied separately. Using MPMs for ex-post analysis and comparing the results with the actual effects of policies can be also fruitful for testing the reliability of models, an aspect that is of increasing importance for the quality assurance of models used for ex-ante assessment of future policies.

The environmental effects of the Reform are not easy to predict. This is because first, they are the result of the interaction of changes in farmers' production decisions with biophysical factors such as soil type and climate and second, they are subject to significant time lags between the cause and effect of the environmental problems. Water is a major environmental asset that is directly impacted by agricultural production. The effects of the Reform on farmers' production and management decisions through the decoupling of payments, the imposition of cross-compliance measures and the potential agri-environmental measures of Rural Development Programs can directly impact on water resources. An investigation of the effects of the Reform on water resources is essential, if there is to be a reconciliation of the economic and environmental objectives of the CAP. The effects of farmers' decisions on water resources can be estimated with biophysical agronomic simulation models (BSMs). BSMs deal with the effects of weather, soil types, inputs, management practices and their interactions on agricultural productivity and yields, while also providing information on specific environmental attributes of different agricultural activities. Effectively these models consist of a set of non-linear mathematical equations describing the complex biophysical processes that take place within the agricultural system. If constructed appropriately, they provide a reliable way to estimate production and pollution functions, overcoming the scarcity of consistent data and allowing the combined or separate assessment of varying levels, timing, type and application methods of fertilisers and irrigation water, crop rotations and alternative tillage techniques.

The aims of the chapter are to use and evaluate these methodologies to explore the agricultural and environmental effects of the 2003 CAP Reform on arable cropping systems in Scotland, using as a case study area the Lunan Water catchment, a representative catchment of Eastern Scotland. The effects of the CAP Reform will be first assessed by analysing land use for the farms of the catchment. Secondly, the results of a bio-economic modelling exercise, integrating the outcomes of a BSM into a MPM, will be presented and discussed. The Mathematical Programming component of the Farm Systems Simulator (FSSIM-MP) (van Ittersum et al. 2008, Louhichi et al. 2007) has been used for modelling farmers' decision making. The BSM NDICEA (van der Burgt 2004, van der Burgt et al. 2006) has been used for the estimation of nitrate leaching associated with the agricultural activities. Finally, conclusions will be drawn on the appropriateness of these data analysis and modelling methodologies for assisting decision-making for the establishment of future agricultural and water policies.

CAP Policies in Scotland

The aim of the CAP Reform is to promote sustainable, market-focused agricultural systems throughout Europe. Under Agenda 2000, the payments to farmers were coupled to production. The compensation rate per hectare was estimated by multiplying the regional yield by the compensation rate for each crop category. In Scotland, for areas out with the Less Favoured Areas, the payment was £264.71/ha for protein crops and £230.02/ha for cereals, linseeds, flax, hemp, oilseeds and set-aside. Producers were obliged to set-aside 10 per cent of the total claimable area in order to receive the payments. The policy was subject to criticisms of distorting the markets and directing farmers towards a subsidy rather than a market oriented behaviour. The response to these criticisms was the 2003 CAP Reform.

In Scotland, the Reform was brought into effect in 2005. The model chosen was the historic Single Payment Scheme under which each farmer was granted entitlements per hectare relating to the reference amounts and the reference areas that gave rise to the direct payments in the reference period 2000-2003. There were standard entitlements that corresponded to arable and grassland, and set-aside entitlements. The value of the entitlements was equal to the reference amount divided by the reference area. The reference amount was calculated on the basis of average claims made during the reference period. The total number of entitlements equated to the average reference area, adjusted for the overshoot of the base area and the National Reserve. The overshoot corresponded to 3.13 per cent reduction of payments on average over the three years (Scottish Executive 2005). The National Reserve, which aimed to help producers that would be seriously disadvantaged by the Reform, was equal to three per cent of all entitlement allocations.

For an entitlement to be activated it has to be matched with an eligible hectare of agricultural land. The only payments that remain coupled are the protein crop premium (55.57 EUR/ha) and the energy crops premium (45 EUR/ha). In Scotland, both compulsory and voluntary modulation are being used to fund Pillar II payments. The rates in 2008 were 8 per cent for the voluntary modulation and 5 per cent for the compulsory one. To receive full payment, farmers must conform to Statutory Management Requirements and achieve minimum standards of Good Agricultural and Environmental Conditions, as defined by the Member State. One Statutory Management Requirement is the Protection of Water in Nitrate Vulnerable Zones (NVZs). Farmers with land in NVZs must follow the rules of the Action Programme for Nitrate Vulnerable Zones (Scotland) Regulations 2003, as set out in the Guidelines for Farmers in Nitrate Vulnerable Zones (2003). The measures can be broadly classified as a) restrictions on the quantity of N applied; b) restrictions on the timing of N applications; c) manure storage requirements; d) record-keeping requirements; and e) other restrictions on N application.

Methodology

The 134 km² Lunan Water catchment was chosen as a case study as it is representative of intensive arable cropping in Scotland (SEPA 2007). The area includes three rivers (Lunan Water, Gighty Water, Viny Water) divided into five water bodies and is partly groundwater fed. It is one of the two priority catchments monitored under the Diffuse Agricultural Pollution Action Plan of the Scottish Environment Protection Agency (SEPA), as it is vulnerable to failing to meet the environmental objectives of the Water Framework Directive (SEPA 2007). The whole catchment lies within a designated river nutrient sensitive area and a nitrate vulnerable zone.

The June Census Data (JCD) (Scottish Executive) were analysed to quantify the changes in land use after the Reform. The data set consists of information on cropping areas of different crops for the individual farms in the area, for the years 2000-2007.[1] The JCD use the UK Farm Classification System, to classify the individual farms by type (http://statistics.defra.gov.uk/esg/pdf/farmclass.pdf.). This typology was also used in our analysis. The JCD were used for the estimation of land use per crop during 2000-2007 for 1) the whole case study area, 2) the average general cropping farm and 3) the average cereal farm. To compare figures before and after the CAP Reform, two reference periods were chosen: 1) average values of 2001, 2002 and 2003 representing the Agenda 2000 period and 2) average of 2006 and 2007, representing the 2003 CAP Reform. The intermediate years have not been used for the comparison, as they constitute the transition period from the one policy to the other.

These land use data were then multiplied by nitrogen input and nitrogen leaching coefficients, "to explore the environmental effects of land use changes. Two different levels of fertilisation per crop, medium and intensive, were considered. As no relevant fertiliser input data were available, it was assumed, based on discussion with experts, that "medium" fertilisation was equivalent to the RB209 (MAFF 2000) recommendations for the relevant soil types and that "intensive" fertilisation represented an increase of 20 per cent. The values were checked against those of the Farm Management Handbook (FMH) (Chadwick 2002) and the British Survey of Fertiliser Practice (DEFRA 2004). The NDICEA (van der Burgt et al. 2006), a process-based simulation model, was used to estimate the nitrogen leaching coefficients as it only requires relatively easily obtainable data on initial states, parameters and driving variables. It simulates on a weekly time-step, soil water, and nitrogen dynamics in relation to weather conditions and crop demand.

NDICEA was run for the main crops of the catchment, for the two main soil types and both fertilisation levels. The spatial distribution and characteristics of the soil series within the area came from the Scottish Soils Knowledge and

---

1   The catchment is situated within an area of 12 agricultural parishes which extend beyond the boundaries of the catchment. As no information on the spatial distribution of the farms is available, and the areas outside the catchment are similar to those within, the JCD of all the farms within the 12 parishes have been analysed.

Information Base of the Macaulay Institute. After consultation with soil experts, the soil series were allocated to light or medium soil categories. As heavy soil types represented a very small part of the catchment they have not been included in our analysis. Weather data for the period 1984-1998 were obtained for the meteorological station at Mylnefield, which although outside the catchment is exposed to similar weather (as shown in a preliminary analysis). The weather was averaged over the 15 years for the simulations and rotational effects were ignored. Each of the crop scenarios consisted of the simulation of two crops at a time, with the first crop always being spring barley. Sowing, harvest and fertilisation dates were obtained from experts. The FMH yield estimates were used. For most crops, this provides three levels of yields, low, medium and high. It was assumed that the medium soils gave medium yields and that the light soils gave the average of the lowest and medium yields. Yield estimates were increased by 10 per cent for intensive fertilisation. All these data were validated by experts. In this chapter, only the results of the intensive fertilisation scenario will be presented, using the average coefficients of the two soil types.

Finally, the average general cropping and average cereal farms were modelled using bio-economic modelling. Bio-economic modelling is a specific type of mathematical programming modelling that links BSMs to MPMs to integrate socio-economic and agro-ecological information. While the MPM describes farmers' production and management decisions, the BSM describes the relevant production and environmental processes. It is thus used to establish agronomic and environmental pollution relationships, which serve as an input to the MPM. The bio-economic MPM that was used for modelling farmers' decision making is FSSIM-MP, developed under the EU FP6 Project SEAMLESS. The model is based on profit maximisation and risk aversion and includes a detailed specification of the agricultural activities in terms of rotations, soil types and management techniques. The non-linear objective function represents expected income and risk aversion towards price and yield variations (Louhichi et al. 2007, van Ittersum et al. 2007):

Max U = Z– φσ

Where: U: Utility; φ: the risk aversion coefficient; σ: the standard deviation of income according to states of nature and market defined under two different sources of instability: yield (due to climatic conditions) and price; Z: expected income.

$$Z = \sum_{c,prd} Price_{c,prd} Sales_{c,prd} - \sum_{r,s,t,p,sys} Costs_{r,s,t,p,sys} \frac{X_{r,s,t,sys}}{N_r}$$

Where: c+ $\sum_{r,s,t,sys,c} Prme_c \frac{X_{r,s,t,sys}}{N_r} - PMPterm - \sum twage.Tlabour$ on
techniques;   :ts,

Sales$_{c,prd}$: total sales of each crop, Costs$_{r,s,t,p.sys}$: variable cost per crop within agricultural activity, X$_{r,s,t,sys}$: level of selected activity, N$_r$: number of years of each crop rotation, Prme$_c$: compensation payment for each crop, PMPterm: the Positive Mathematical Programming term, Twage: labour cost, Tlabour: average number of hours rented labour.

The model is calibrated using the risk approach, and subsequently complemented by an extension of the Positive Mathematical Programming approach (Howitt 1995). Positive Mathematical Programming is a methodology that adds quadratic cost terms to the objective function, ensuring that the model outcomes in the base run calibrate exactly to the observed production levels (Janssen and van Ittersum 2007).

FSSIM-MP follows a joint production approach using discrete production/ pollution functions for the incorporation of yield and environmental information, as opposed to the incorporation of continuous production and pollution functions or of cost functions as a proxy for environmental damages. Agricultural activities are defined as vectors of technical and environmental coefficients that describe their inputs, agricultural outputs and environmental effects. The model has a high technical specification and allows a multi-dimensional specification of the agricultural activities as discrete and independent options, whether they refer to different crop or livestock activities, to different technologies for the same activity, or to variations of the same technology.

Agricultural activities are defined as combinations of rotation, crop, soil type and technique. The two different soil categories and fertilisation scenarios that were used for NDICEA were also used for the bio-economic modelling. Forty-nine rotations were composed based on advice given by experts. This resulted in input-output matrices of around 1,600 rows, which required data modelling modules to feed the information into FSSIM-MP, using MS Access and the MDB2GMS utility. The fertilisation and yield data that were used for the NDICEA simulations also served as an input to FSSIM-MP. Variable costs were taken from the FMH, after subtracting the FMH fertiliser cost estimates and adding the quotient of fertiliser input by fertiliser price. The FMH was also used for labour requirements per crop category. The nitrate leaching coefficients were the result of the NDICEA simulations. To calculate family labour availability, the JCD items relating to the work of the occupier or spouse were multiplied by their hourly equivalent, assuming full time labour to be 1,900 hours per year. The percentage distribution of each soil category within the area was calculated and then attributed to the average cereal and general cropping farm, where their average size was calculated using the JCD. Although this is a rather crude assumption, lack of additional information on the spatial distribution of farms within the parishes, offered no alternative. Finally, the JCD were used to calculate the average land use pattern of each of the two farm types that was used for model calibration.

The scenarios of JCD Analysis and FSSIM-MP modelling are shown in Table 18.1.

Table 18.1   Scenarios for Modelling and JCD Analysis

	Baseyear Agenda 2000	JCD Average 2006-2007	Scenario 1 CAP Reform	Scenario 2 Price Changes	Scenario 3 Price Changes 2
Source of Scenario	JCD used also for FSSIM-MP calibration	JCD analysis	FSSIM-MP modelling		
Exogenous Assumption	2001-2003 prices		2001-2003 prices	2006-2007 prices	
EU CAP	Agenda 2000		2003 CAP Reform		
Measures			Cross-compliance NVZ: 60% cut of premiums if average N application >170kg/ha		

The prices (£/ton) used for the period 2001-2003 are: barley: 68; wheat: 75, winter oilseed rape: 148; oats: 70; main crop potatoes 97; seed potatoes: 140; beans: 72; peas: 230; carrots: 220. The prices for 2006-2007 are: barley: 70; wheat: 80; winter oilseed rape: 160; oats: 65; main crop potatoes: 140; seed potatoes: 130; beans: 79; peas: 230; carrots: 240.

Results

An analysis of the JCD regarding the average percentage of the number of farms and agricultural area occupied for each farm type for the periods 2001-2003 and 2006-2007, showed that there has been a slight decrease in the number and areas of general cropping farms and a minor increase in the area of mixed farms. This was accompanied by a small increase in the number of cattle and sheep and mixed farms. There is also a slight increase in the area of cereal farms which is associated with a decrease in their number.

The main land use changes in the period 2000-2007 were a decrease in the area of barley (6.12 per cent) and seed potatoes (1.04 per cent) and an increase in the area of wheat (2.95 per cent), main crop potatoes (1.52 per cent) and vegetables (1.12 per cent). Oats increased by 0.74 per cent, while set-aside decreased by 0.61 per cent, temporary grass by 0.20 per cent and winter oilseed rape by 0.01 per cent. Very similar changes were observed for the average general cropping farm due to the large number of such farms in the sample. Regarding the average cereal farm, fluctuations in the levels of barley and temporary grass accompanied by fluctuations in the opposite direction of wheat, oilseed rape and potatoes, appeared more pronounced after 2003. The most significant change was, as for the general cropping farms, a decrease in the area of barley followed by an increase in the area of wheat. The areas of winter oilseed rape and oats also rose, while set-aside and temporary grass declined.

The nitrogen inputs and nitrogen leaching coefficients per soil type and on average which have been used in our analysis are shown in Table 18.2. The nitrate leaching differs by less than 5 per cent between the two soil types, because the fertilisation levels take the soil type into account. Even though spring crops have much lower inputs than the equivalent winter crops, the average leaching (kg/ha/year) is higher, since land is left bare for longer periods of time.

Table 18.2   Nitrogen Inputs and Nitrate Leaching Coefficients (kg/ha/year)

	Nitrogen Input			Nitrate Leaching		
	Medium	Light	Aver.	Medium	Light	Aver.
S. Barley	144	120	132	84	82	83
W. Barley	216	192	204	85	84	84.5
W. Wheat	240	192	216	77	72	74.5
W. OS Rape	198	220	209	66	76	71
S. Oats	144	120	132	81	79	80
W. Oats	156	144	150	64	66	65
Main Potato	216	216	216	79	78	78.5
Seed Potato	114	114	114	87	85	86
Beans	0	0	0	56.5	57	56.75
Peas	0	0	0	69	67	68
Carrots	72	132	102	73	89	81

A combination of these coefficients with the land use of the average cereal and average general cropping farms for the years 2000-2007, provides the average nitrogen use and nitrate leaching per hectare for each of the two farm types for the period 2000-2007. The cereal farm has higher inputs compared to the general cropping farm. These range from 157 kg/ha/year in 2001 to 175 kg/ha/year in 2006, while for the general cropping farm they range from 156 kg/ha/year in 2001 to 164 in 2007. On the other hand, the average nitrate leaching estimates per ha per year are very similar between the two farms and between years, despite some land use changes and slight increases in the fertilisation levels particularly for the cereal farm. This is due to nitrogen uptake by nitrogen intensive crops being higher.

Figure 18.1 shows the percentages of total land occupied per crop under each of the scenarios, for the average general cropping and cereal farms. The Baseyear (Agenda 2000) and JCD (Average 2006-2007) scenarios represent the actual percentages of crop levels as an average of the years 2001-2003 and 2006-2007 respectively, estimated through the JCD. The former has also been used as the Baseyear for model calibration. Scenario 1 (CAP Reform), Scenario 2 (Price changes) and Scenario 3 (Price changes 2) correspond to scenarios modelled with FSSIM-MP. The results of Scenario 2 and Scenario 3 were identical for the general cropping farm as the cross-compliance limit was inactive. Under Scenario 1, the areas of barley and winter oilseed rape in the general cropping farm decrease, while the areas of seed potatoes and wheat increase. Minor increases were also observed in the areas of oats, maincrop potatoes and vegetables. Scenario 2 indicates decreases in the areas of barley, seed potatoes and vegetables and increases in the areas of wheat, main crop potatoes, winter oilseed rape and oats.

For the average cereal farm, under Scenario 1 there are some slight area increases for all crops due to replacement of set-aside land.[2] Scenario 2 indicated considerable reductions in the levels of barley, which was mainly replaced by winter wheat and winter oilseed rape. Maincrop potatoes also slightly increased, and oats and seed potatoes decreased slightly. When the nitrogen cross-compliance limit is inactive (Scenario 3), the changes in relation to barley, wheat and winter oilseed rape are more pronounced.

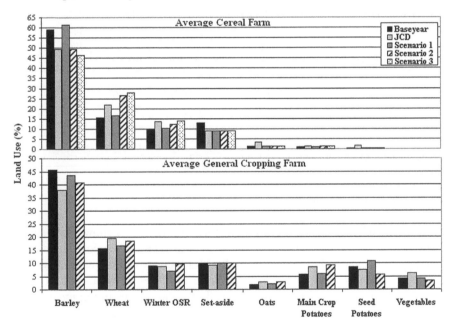

Figure 18.1 Land Use – Modelling Results and Current Levels

The main economic and environmental results of the modelling scenarios are summarised in Table 18.3. The income of the general cropping farm is slightly reduced under Scenario 1. However, when changes in the prices are taken into account (Scenario 2) it actually increases considerably and the premiums as a share of income decrease. The average nitrogen use is below the nitrogen cross-compliance limit for all scenarios. Neither nitrogen use nor leaching

2   Set-aside land was kept to the actual level using a model constraint.

differ significantly between the scenarios. The cereal farm appears to generate less income per hectare than the general cropping farm, while the premiums as a share of total income are much higher. Income does not decline significantly under Scenario 1, while the highest income is achieved under Scenario 3. Under this scenario, however, the nitrogen use surpasses the quota, even though nitrate leaching does not increase.

Table 18.3   Economic and Environmental Modelling Results

	Ag. 2000	Sc. 1	Sc. 2	Sc. 3
Average Cereal Farm				
Utility (£)	15405	15010	17164	17364
Income (£/ha)	463	462	514	527
Premium (% of income)	41	39	35	34
Nitrogen use (kg/ha)	162	170	170	180
N leaching (kg/ha)	70	73	72	71
Average General Cropping Farm				
Utility (£)	66677	66136	84530	
Income (£/ha)	635	632	792	
Premium (% of income)	27	26	20	
Nitrogen use (kg/ha)	152	151	159	
N leaching (kg/ha)	72	72	71	

Discussion and Conclusions

The analysis of both the JCD analysis and modelling results shows only small changes in the cropping pattern of the two farm types. The most significant changes are decreases in the area of barley and increases in the area of wheat. This is explained by the higher relative price increases of winter wheat, and its higher yield compared to spring barley. The same applies to the increase of maincrop potatoes in both farm types in all modelling scenarios. The decrease of seed potatoes in the case of the general cropping farm could be attributed to the reduced price, as also demonstrated by the modelling Scenario 2. The price change of oilseed rape also showed an increase to its production in both Scenarios 2 and 3 and in the data analysis of the cereal farm. Even though vegetables increase, this is only captured by Scenario 1 and not Scenario 2, as the price changes were not sufficient to invoke upward effects. The effects of Scenario 1 for the cereal farm showed an increase in the area of all crops after a reduction of set-aside. While this is consistent with the data analysis for most crops, it is not the case for barley. This effect is reversed with the introduction of the new prices, as predicted results largely match the actual changes shown by the JCD analysis. In the case of the general cropping farm, Scenario 1 captures the direction of changes for most crops. However, these are further augmented and much closer to the changes shown in the data analysis after the introduction of the new prices. Overall, a shift towards higher yielding and therefore more profitable crops can be observed, partly as a result of the CAP Reform, but mainly due to changes in crop prices.

Farm incomes do not decline significantly after the introduction of the Reform. Indeed, significantly higher incomes are achieved after the introduction of new crop prices. As would be expected, the premiums as a share of income decline under all scenarios. The cereal farm appears to have lower income per hectare and to be more dependant on premiums than the general cropping farm. It also seems to be more reactive to price changes, as demonstrated by higher crop fluctuations of the yearly JCD analysis and by stronger model predictions in relation to the main crops.

AOverall nitrogen use and nitrate leaching do not differ considerably between the scenarios for each of the two farm types. Although the average nitrogen use only exceeds the nitrogen quota in Scenario 3 for the cereal farm, it is very close to the 170 kg/ha limit. Thus, such a measure might constrain farmers' flexibility in relation to high input crops. This is more the case for cereal farmers, since some of the most profitable cereals require high fertiliser inputs. Scenario 3 provides the highest income for the cereal farm type. While Scenario 2 is associated with lower profits and lower nitrogen use invoked by the cross-compliance limit, it does not result in lower nitrate leaching levels due to higher nitrogen uptake by nitrogen intensive crops. This is also why nitrogen leaching estimates remain unaltered through 2000-2007, despite observed land use changes. This suggests that the assumption that increased nitrogen fertilisation levels inevitably lead to increased leaching might not always be valid, and that measures of more restrictive input quotas might yield no major improvements. On the other hand, targeting fertilisation in relation to soils and crops might be more efficient.

The methodological framework of combining analyses of actual land use data with bio-economic modelling has been shown to be a useful technique for overcoming the problem of specifying the driving forces of land use changes, which could be the result of many interacting factors such as policy and/or price changes. A key conclusion is that in spite of a lack of detailed input data, bio-economic modelling can help in explaining the drivers of changes demonstrated by modelling results and actual data analysis. The structure of the models lends itself to further sensitivity analysis and an exploration of the boundaries of resilience of farming systems. Although the comparison of model predictions with actual data constitutes a form of model testing and increases confidence in the model outcomes, not all potential modelling scenarios have been tested and it does not imply blind acceptance of model predictions. Rather, the outcomes should be considered as hypotheses that become the input to further discussions with experts and policy makers.

The present research represents a pilot study for testing the methodology which is being further developed. More work is needed on the farm classification to take into account farm size and more differentiated land endowments. Fixed costs, rotational effects, and weather variability need to be taken into account and a comparison of simulated and actual nitrate losses is needed. Further research will also include the examination of more measures to counter water pollution, such as target fertilisation, and the exploration of the effects of the Reform on livestock and mixed farming systems.

Acknowledgements

The authors wish to acknowledge the funding from RERAD. Part of the work presented benefits from the SEAMLESS integrated project, EU 6th Framework Programme. We gratefully acknowledge Guillermo Flichman for valuable advice on the work with FSSIM-MP. We also thank Andy Vinten, Mike Rivington, Allan Lilly and Malcolm Coull from the Macaulay Institute for help with the provision of land use and soil data, SCRI and the Meteorological Office who supplied the weather data, and Steve Hoad and John Elcock from SAC for advice on agronomic data.

References

Chadwick, L. 2002. *The Farm Management Handbook 2002/2003*. Edinburgh: SAC.

DEFRA 2004. *The British Survey of Fertiliser Practice*. London: DEFRA.

Howitt, R. 1995. Positive mathematical programming. *American Journal of Agricultural Economics*, 77, 329–342.

Janssen, S. and van Ittersum, M.K. 2007. Assessing farm innovations and responses to policies: A review of bio-economic farm models. *Agricultural Systems*, 94, 622-636.

Louhichi, K., Flichman and G. Blanco, M. 2007. *A generic Template for FSSIM for all Farming Systems*, PD3.3.11, SEAMLESS integrated project, EU 6th Framework Programme, contract no. 010036-2.

Ministry of Agriculture, Fisheries and Food 2000. *Fertiliser Recommendations for Agricultural and Horticultural Crops (RB209)*. London: The Stationary Office.

Scottish Executive 2005. *Single Farm Payment Scheme – Information Leaflet 9*: Understanding your statement of provisional payment entitlements. Edinburgh: Scottish Executive.

SEPA 2007. Press Release 2007: Rural water quality project is up and running *3 Apr 2007 – EXT01-B01*. [Online]. Available at: http://www.sepa.org.uk/news/releases/view.asp?id=513&y=2007 [accessed: 25 September 2008].

Van der Burgt, G.J.H.M. 2004. Use of the NDICEA model in analyzing nitrogen efficiency, in *Controlling Nitrogen Flows and Losses: Proc 12ᵗʰ Nitrogen*

Workshop, edited by D.J. Hatch et al. Wageningen: Wageningen Academic Publishers, 242-243.

Van der Burgt, G.J.H.M., Oomen, G.J.M., Habets, A.S.J. and Rossing, W.A.H. 2006. The NDICEA model, a tool to improve nitrogen use efficiency in cropping systems. *Nutrient Cycling in Agroecosystems*, 74, 275-294.

Van Ittersum, M.K., Ewert, F., Heckelei, T., Wery, J., Olsson, J.A., Andersen, E., Bezlepkina, I., Brouwer, F., Donatelli, M., Flichman, G., Olsson, L., Rizzoli, A.E., van der Wal, T., Wien, J.E. and Wolf, J. 2008. Integrated assessment of agricultural systems – A component-based framework for the European Union (SEAMLESS). *Agricultural Systems*, 96, 150-165.

PART 6
Cross-Compliance and Agro-Environmental Measures

Chapter 19

Agri-Environmental Measures Adoption: New Evidence from Lombardy Region

Danilo Bertoni, Daniele Cavicchioli, Roberto Pretolani and Alessandro Olper

Introduction

Recent reforms of the Common Agricultural Policy (CAP) increasingly recognize the central role the European agricultural sector plays in environmental management. This acknowledgement has led legislators to partly modify CAP objectives and, consequently, to set-up new policy instruments. Indeed, agri-environmental measures (AEMs) have progressively gained centrality in CAP.

AEMs, since their introduction in 1992, have been widely implemented in all the European Union Member States, becoming a familiar instrument for farmers. Given their high adoption rate, the question that arises is whether AEMs are really effective policy instruments or, differently, do they simply represent a form of disguised agriculture protection (Anderson 2000, Swinbank 2001). This question calls for a rigorous and reliable analysis into this policy's instrument implementation, starting from the determinants influencing farmers' enrolment in agri-environmental schemes. Indeed, it is to be noted that the patterns of farmer participation strongly affect the policy's objective attainments (e.g., Wilson 1997).

Many studies have investigated the determinants of farmer participation in rural development schemes, starting from the assumption that such participation is mainly the outcome of a farmer utility maximization process (e.g., Vanslembrouck et al. 2002, Defrancesco et al. 2007, Barreiro-Hurlé et al. 2008). The relevant literature also considers factors like social commitment and the environmental attitude of farmers as drivers in their participation (e.g., Damianos and Giannakopoulos 2002, Wossink and Wenum 2003). However, with the notable exceptions of Vandermeulen et al. (2006) and Hackl et al. (2007), factors related to the policy decision-making environment have been largely neglected, despite the central role played by regional and local political bodies in the design and implementation of AEMs.

This chapter contributes to this literature by assessing the determinants of farmer AEMs implementations in an important North Italian region, that of Lombardy. We add to the previous literature in two main directions. First, by focusing on the political institution constrains that potentially affect the level of AEMs uptake at the local level. Secondly by working, not through a farm survey as does all the previous relevant literature, but on the total population of 62,454 regional farms receiving CAP payments.

More specifically, in this chapter we test whether AEMs adoption depends both on farms and farmers' characteristics, and also on the political institutional framework. Indeed the relevant decisions concerning AEMs design and implementation could be affected by transaction costs embodied in the bargaining process among farmers, other interest groups and regional and sub-regional governments. Thus, disentangling genuine farmer incentive and attitude towards AEMs from the role played by the local political-institutional environment appears a crucial step toward a better understanding of agri-environmental schemes (Bertoni and Olper 2008).

To deal with this kind of issue we exploit the sample of *all* eligible farmers in the Lombardy region agri-environmental programme, taking advantage of the Regional Agricultural Information System (SIARL) database. Such a database includes all the farms that received payments on both Pillar I and Pillar II of the CAP.

With respect to the methodology, we applied a parametric approach (Probit model) to explain the probability of farmer AEMs adoption conditioned upon three broad categories of determinants: farm and farmer characteristics, territorial and geographical context and, finally, political and institutional environment.

Conceptual framework

The conceptual model to analyse factors affecting the adoption of AEMs follows the micro-economic modelling framework developed in Vanslembrouck et al. (2002) and Dupraz et al. (2003), and recently applied by Barreiro-Hurlé and Espinosa-Goded (2007) and Barreiro-Hurlé et al. (2008). According to this literature, the determinants of AEMs farmers' adoption can be divided into *extrinsic* and *intrinsic* factors (Vanslembrouck et al. 2002). The former rely on programme characteristics, like the nature of the specific agro-environmental scheme, and market conditions (supply and demand) for both food and environmental goods. Differently, the latter rely on *farm* characteristics like size, location, type of farming, and *farmer* attributes, such as age, education, and composition of the family farm. More recently, a further factor has been included in the analysis of AEMs adoption: the 'governance structure' often called 'social capital' (Jongeneel et al. 2008, Barreiro-Hurlé et al. 2008, Mathijs 2002), which emerges from the interaction between the extrinsic and intrinsic factors with political and institutional context.

Following Barreiro-Hurlé et al. (2008), the farmers' choice to uptake AEMs is based on the assumption that they derive utility from four key components: economic benefit (m), provision of agri-environmental goods (v), farmer individual characteristics (Z^U), and farmer's social capital (Z^{SC}).

The farmer problem can be expressed as follows:

$$\underset{m,v}{Max} U\left(m, v, Z^{U}, Z^{SC}\right) \text{ s.t.} \qquad (1)$$

$$m \le \overbrace{\pi^{R}\left(p, v, Z^{\pi}\right)}^{a} \overbrace{+ pv}^{b} - \overbrace{TC\left(Z^{U}, Z^{\pi}, Z^{C}, Z^{SC}\right)}^{c} \quad (2)$$

$$v > 0 \quad (3)$$

Thus, farmers maximize their utility given by equation (1), subject to restrictions (2) and (3). Restriction (2) implies that farmers' economic benefit is derived from farm activity (a) and their participation in AEMs (b), minus the transaction costs (c) due to their participation in the AEMs. More specifically, the economic benefit from farming (π^{R}) is a function of relevant prices (p), the area devoted to AEMs (v) and farm technical characteristics (Z^{π}). The benefit from AEMs participation depends on the premium (ρ) multiplied by the intensity level v. The transaction costs component (TC) is a function of farmer (Z^{U}), contract (Z^{C}) and farm characteristics (Z^{π}), as well as the farmer's social capital (Z^{SC}). Finally, restriction (3) simply recalls the logic of the AEMs, namely that in order to obtain a subsidy the level of agri-environmental goods production should be greater than a minimum level, currently defined by the good farming practice.

The modelling framework above suggests the following basic relationships. First, an increase in income derived from farming (a), as an effect of, let us say, food price increase, should reduce the environmental goods provision, and thus the income coming from the AEMs (b) because it increases the opportunity costs of AEMs adoption. Differently, an increase in the AEMs premium ρ, or in the marginal utility of the environmental provision should increase the surface devoted to AEMs. Moreover, a reduction in the transaction costs component necessary to implement AEMs (c), should increase the provision of environmental goods (v).

It should be noted that all similar previous papers have been mainly based on sample farm surveys, collected from the population of AEMs eligible farms. From this point of view, the main contribution of our chapter is to work on data gathered from the entire farming population rather than from a restricted sample. This massive data availability offers both advantages and some drawbacks, however. The key advantage of working on the entire population is to overcome the problem of sample selection bias that always takes place in this kind of analysis. However, this happens at the cost of some over-simplification in terms of our ability to control for 'all' the relevant factors affecting AEMs adoption. The second main contribution of the analysis is to put particular emphasis on the identification of factors affecting transaction costs, particularly referring to several district level political-institutional variables. More specifically, our basic assumption is that interaction within the specific local institutional environment affects the bargaining process between farmers and local institutions and should, in turn, influence the farmers' sign-up decision.

Variables definition

We use data extracted from the agricultural information system of the Lombardy Region (SIARL). SIARL is the instrument by which the regional administration collects, and processes, farmers' applications for public funding (and consequently for RDP funds). The SIARL dataset contains information concerning farm and farmers' characteristics and CAP administrative proceedings. These data have been integrated with territorial, institutional and political information in order to control for potential determinants of AEMs participation.

With respect to a survey approach, the exploitation of the SIARL dataset allows us to work with a sample representing almost the entire universe of Lombardy farms. Thus problems of sample representativeness have been totally overcome. On the other hand we lack some information which is only directly available by survey (for example farmers' attitudes toward the environment). We partially reduce this problem by replacing missing sample information with proxy variables measured at a very detailed territorial level.

In financial terms, AEMs represent the main policy instrument within the Rural Development Programme (RDP) 2000-2006 of the Lombardy region. During the programme implementation the AEMs absorbed almost EUR328 million, 165 of which was derived from the EAGGF contribution (36 per cent of the total public expenditure in Lombardy RDP, 45 per cent if we consider only the EAGGF allocation). Over 215,000 hectares were under agri-environmental commitments, corresponding to the 20 per cent of the regional utilized agricultural area (UAA).

Within the AEMs framework farmers could choose among four different categories of schemes:

- farming input reduction and integrated production (AEM1);
- organic farming (AEM2);
- management of meadows and pastures (AEM3);
- landscape conservation, restoration and creation (AEM4).

2005 has been chosen as a reference year for our analysis. In 2005 there were 10,793 farms participating in at least one agri-environmental scheme; the most adopted schemes were AEM1 and AEM3, while AEM2 and AEM4 represented a small share of the total AEMs expenditure (Table 19.1).

Table 19.1   **Farmers participation and public funds expenditure in Lombardy AEMs (2005)**

Scheme	AEM_TOT (At least 1 scheme)	AEM1	AEM2	AEM3	AEM4
Sample	62,454	62,454	62,454	62,454	62,454
Eligible farms	58,766	37,396	40,409	43,412	58,766
Participants	10,483	3,555	443	5,801	2,324
Expenditure in euros (2005)	45,922,813	25,394,003	3,106,996	14,493,273	2,928,540
% of expenditure (2005)	100.0%	55.3%	6.8%	31.6%	6.4%

The size of the selected sample is 62,454 farms, among which 10,483 adopted at least one AEMs scheme. As not all the farms contained in the dataset met the AEMs eligibility criteria, non-eligible farms were excluded from the analysis. It should be noted that the number of eligible farms can change, depending on the considered scheme (Table 19.1). Furthermore, sample size is influenced by a lack of observations for some variables.

Dependent variables

The dependent variable, *AEM_all*, is a dichotomous variable indicating the participation (= 1) or non-participation (= 0) of eligible farms in at least one agri-environmental scheme. However, because agri-environmental schemes differ in terms of their asset specificity (Barreiro-Hurlè et al. 2008), we expect that factors affecting participation will vary across the different instruments.

Thus, to deal with, and to test this hypothesis, we also considered participation choices within each scheme, by creating four different dichotomous dependent variables, namely *AEM_1*, *AEM_2*, *AEM_3*, and *AEM_4*, respectively referred to the participation of eligible farms in each single scheme.

Independent variables

Farm and farmer characteristics.

Current literature on farmers' willingness to participate in AEMs indicates that variables related to farms and farmers' characteristics are the main explanatory factors for the sign-up decision. Given the lack of reliable information about the family and non-family agricultural labour force, our key variables aimed at describing the farm level context are mainly represented by farm characteristics (rather than those of the farmers). Data available for the attributes of the farmer are: farm heads age (*age*), which is also a proxy of the education level; the percentage of property land (*landown*); and the average farmer income of the farm district (*farmer_income*).

For farm characteristics, we include farm economic size (*esu*); type of farming, distinguishing among field crops (*field_crop*), permanent crops (*permanent_crop*) and dairy production *(dairy)*; farming intensity, expressed by the number of livestock units per hectare (*lsu_ha*) and the number of horsepower per hectare (*hp_ha*); and, finally, the share of grasslands and pastures area (*pasture*) indicating the inverse degree of farming intensity as well. A summary description of the explanatory variables discussed above is reported in Table 19.2.

Table19.2   Variables definition

Variable	Description and measurement
Farm characteristics	
Esu	Number economic size units per farm.
lsu_ha	Number of livestock standard units per hectare.
hp_ha	Number of horsepower per hectare.
field_crop	Dummy variable indicating field crops type of farming.
permanent_crop	Dummy variable indicating permanent crops type of farming.

Variable	Description and measurement
Dairy	Dummy variable indicating dairy specialized type of farming.
Pasture	Share of pasture and grasslands on the agricultural utilized area.

Farmer characteristics

Age	Age of the farm holder.
Landown	Share of the property land on the total agricultural area.
farmer_income	Average farmer income of the 'farm district' (EUR).

'Social capital'

Investment	Dummy variable indicating farms participating in the 'investment in agricultural holdings' RDP measure.
lfa_payment	Dummy variable indicating farms taking the LFA payment.
Income	Average income of the 'farm district' (EUR)
Education	Share of the population having an education level ISCED 3 or upper.
Participation	Share of participation in the regional Lombardy elections (2005) at the municipal level.
Greens	Share of votes obtained by the Green Party in the regional Lombardy elections (2005) at the municipal level

Variable	Description and measurement
Left	Share of votes obtained by the left-oriented parties in the regional elections (2005) at the municipal level.
Euroskeptic	Share of votes obtained by the euroskeptic parties in the regional Lombardy elections (2005) at the municipal level.
Beds	Number of beds per inhabitant in accommodation establishments at the 'tourism district' level
Agtourism	Number of rural tourism establishments per inhabitant at the 'tourism district' level
Sisa	Dummy variable indicating if farmer is enrolled to Sisa farmers group
Copagri	Dummy variable indicating if farmer is enrolled to Copagri farmers group
Cia	Dummy variable indicating if farmer is enrolled to Cia farmers group
Confagri	Dummy variable indicating if farmer is enrolled to Confagricoltura farmers group
Coldiretti	Dummy variable indicating if farmer is enrolled to Coldiretti farmers group

Location and other determinants

pillar_1	Dummy variable indicating farms receiving the CAP single payment
Plain	Dummy variable indicating if farm is located in a lowland area
Mountain	Dummy variable indicating if farm is located in a mountain area
Periurban	Dummy variable indicating if farm is located in a periurban area

Variable	Description and measurement
Park	Dummy variable indicating if farm is located in a municipality included in a natural park
Nzv	Dummy variable indicating if farm is located in a 'nitrate vulnerable zone' according to the Directive (91/676EC)

Social capital characteristics.

In our framework 'social capital' describes the complex relationships between farmers and socio-economic and institutional environment. In fact, farmers are part of a complex social network in which different categories of stakeholders interact (farmer groups, commodities and public goods consumers, taxpayers, institutions, etc.). The nature and size of the relationships within this social network and, generally speaking, the social, economic and institutional context in which farmers operate, should be considered an important determinant of their choices. In this work we extend such an assumption to the AEMs implementation.

As proxy variables for social capital we include in the analysis the average per capita income of the 'farm district' (*income*), the educational attainment of the population (*education*), the tourism intensity (*beds, agritourism*), the shares of specific political parties and 'ideology' orientations at the municipal level (*greens, euroskeptic, left*).[1] These variables should proxy for social needs and demands, farmer's attitude toward agri-environmental issues, and the district political orientation.

Furthermore, farmer participation in Rural Development measures other than AEMs, like farm investments (*investment*) and less-favoured areas payments (*lfa_ payment*), should reveal the farmer familiarity with EU policies, thus reducing transaction costs involved in their activation.

Finally, farmers' affiliation to a specific farmer organization distinguishing among five different existing associations (*sisa, copagri, cia, confagri, coldiretti*) has been considered. Membership in a farmers organization has been proxied utilizing the proceedings of single farm payments; for that reason it is only available for a sub-sample.

Geographical location and other covariates.

For different reasons, also farm geographical location should represent a relevant factor affecting farmers' involvement in voluntary schemes. This is even more true if we refer to a policy intervention strongly related to rural areas management like

---

1   Classification of left-oriented and euroskeptic parties follows Kemmerling and Bodenstein (2006), including in these categories parties and political movements enrolled in specific EU Parliament political groups.

AEMs. The relevant territorial levels are related to altimetry, with two dummies for *mountain* and *plain*, respectively (reference, omitted dummy, is *hill*); a dummy for periurban location (*periurban*); dummies for farms located in a natural park (*park*) and in a 'nitrate vulnerable zone' (*nzv*). Moreover it is important to emphasise that the few regional priorities on AEMs implementation are, in fact, related to mountain areas, and natural and rural parks.[2] Thus controlling for location should represent an ex-post evaluation of the Regional priorities accomplishment. On this ground also *nzv* checks whether the AEMs targeting in environmentally sensitive areas has been reached.

Finally the *pillar_1* variable was included in the model to represent both farmers income integration through CAP Pillar I and, more generally, to verify the issues of overlapping between Pillar I and Pillar II payments.

Econometric model and results

Econometric model

Our dependent variable is a dichotomous choice variable taking the value (1) when a farmer participates in at least one agri-environmental scheme, and (0) if he does not. Thus, the econometric model is based on a binary response model, where we are interested in the so-called response probability, namely the probability that a farmer uptake an agri-environmental scheme conditioned to a set of endogenous variables.

Following the previous literature, we model this probability as the latent variable, y^*, in a Probit model. This latent variable represents the conditional participation in the AEMs, and can be interpreted as the result of the farmer utility maximization process, discussed in Section 2. Formally, we have

$$y^* = \alpha + \beta x_i + u_i \quad (3) \qquad Y_i = \begin{cases} 0 & if \ y^* \leq 0 \\ 1 & if \ y^* \geq 0 \end{cases} \quad (4)$$

where y^* is the latent variable reflecting the marginal utility from AEMs adoption; Y_i is a binary variable reflecting what we really observe, namely whether the farmer adopts AEMs or not (Y_i takes the value 1 when the latent variable is positive and 0 when is negative); x_i is a vector of covariates affecting the farmer's participation choice and is related to farm, farmer and other determinants of the adoption choice; α and β are the estimated model constant and coefficients parameters, respectively.

---

2   The main characteristic of Lombardy regional parks is that they include many agricultural areas, some of which are exclusively or mostly dedicated to the preservation of agricultural landscape.

Denoting with $\Phi(\bullet)$ the cumulative normal distribution function and σ the standard error, the probability of up-taking an agri-environmental scheme is then defined by $P(y^*>0)= \Phi(x'\beta k/\sigma)$. The parameters β/σ have been estimated via the maximum likelihood estimator (MLE), correcting the standard errors for unknown correlation of the residual within each district (clustered standard error).

Results

Table 19.3 reports the MLE results of five different models, the first related to the adoption of all types of AEMs, while the others refer to each single scheme. In selecting the final specification we adopt the following strategy (Jongeneel et al. 2008). In a first step we model a specification that considers the effect of several potential determinants of AEMs adoption (Table 19.2). The criterion adopted for the final specification is to include a variable only if it turns out to be significantly different from zero in at least one model.

Figures in Table 19.3 report the marginal effect (dF/dx) calculated at the sample mean, that is, the change in predicted probability associated with changes in explanatory variables (e.g., Greene 2003), as well as their respective *p-values*. All five models have a significant χ^2, meaning that all the regressors are jointly significantly different from zero, thus the set of our explanatory variables plays a role as a whole in explaining the probability of the farmer's enrolment in AEMs. Indeed, the fraction of correct predictions is quite high, ranging from 83.8 per cent for the overall model to 99 per cent for the organic farming scheme.

However, the goodness of fit, measured by McFadden (Pseudo) R^2, is quite low but in line with similar studies (e.g., Vanslembrouck et al. 2002, Barreiro-Hurlé et al. 2008, Jongeneel et al. 2008). For the overall model (*AEM_all*) the Pseudo R^2 is equal to 0.2, and ranges between 0.11 (*AEM_2* – organic farming), and 0.26 (*AEM_3* – management of meadows and pasture). Thus, several other unknown factors are at work in explaining AEMs adoption, other than those considered here. However, in evaluating this general conclusion, it should be remembered that the sample employed for this analysis was of a huge dimension, counting more than 50,000 farmers.

In what follows we discuss the results by grouping the set of explanatory variables into the above mentioned categories of determinants.

Table 19.3 Estimation results for the general model and the single schemes

PARAMETER	MODEL AEM_All		AEM_1		AEM_2		AEM_3		AEM_4	
	dF/dX	p value	dF/dX	p value	dF/dX	p value	dF/dX	p_value	dF/dX	p value
Farm characteristics										
ESU	0.0222	0.000	0.0063	0.001	0.0003	0,499	0,0114	0,000	0,0060	0,000
LSU_HA	-0.0023	0.000	-0.0019	0.017	-0.0005	0,003	-0,0005	0,005	-0,0002	0,018
HP_HA	-0.0004	0.000	-0.0003	0.005	-0.0001	0,000	-0,0002	0,020	-0,0001	0,130
FIELD_CROP	0.0456	0.049	0.0789	0.000	0.0017	0,136	-0,0538	0,000	0,0231	0,000
PERMANENT_CROP	0.0295	0.191	0.1474	0.000	0.0046	0,007	-0,0838	0,000	0,0059	0,148
DAIRY	0.1568	0.000	0.0129	0.467	-0.0025	0,023	0,1298	0,000	0,0061	0,213
Farmer characteristics										
AGE	-0.0019	0.000	-0.0010	0.000	-0.0002	0,000	-0,0003	0,070	-0,0007	0,000
LANDOWN	-0.0353	0.000	-0.0087	0.027	-0.0005	0,601	-0,0161	0,004	-0,0006	0,780
FARMER_INCOME	-0.0059	0.491	-0.0052	0.309	-0.0002	0,497	-0,0005	0,947	-0,0022	0,099
Social capital										
INVESTMENT	0.1192	0.000	0.0721	0.000	0.0069	0,000	0,0127	0,045	0,0390	0,000
LFA_PAYMENT	0.3572	0.000	0.1447	0.000	0.0068	0,000	0,1307	0,000	-0,0023	0,847
INCOME	0.0116	0.012	0.0016	0.612	0.0002	0,359	0,0058	0,280	0,0027	0,002
EDUCATION	0.0014	0.317	0.0009	0.295	-0.0002	0,069	0,0023	0,019	0,0002	0,495
GREENS	-0.0065	0.285	-0.0143	0.000	0.0005	0,190	0,0092	0,068	-0,0041	0,006
EUROSKEPTIC	-0.0052	0.001	-0.0046	0.000	-0.0001	0,109	0,0007	0,543	-0,0025	0,000
LEFT	-0.0006	0.552	0.0009	0.040	0.0001	0,028	-0,0015	0,158	0,0000	0,782
Location and others										
PILLAR_1	0.0888	0.000	0.0318	0.000	-0.0001	0,897	0,0302	0,002	0,0260	0,000

PLAIN	-0.0958	*0.004*	-0.0405	*0.064*	-0.0080	*0.000*	-0,0398	*0,228*	0,0026	*0,656*
MOUNTAIN	0.0083	*0.850*	-0.0018	*0.956*	0.0001	*0.970*	0,1831	*0,000*	-0,0046	*0,829*
PARK	0.0129	*0.324*	-0.0239	*0.001*	0.0008	*0.257*	0,0364	*0,005*	0,0101	*0,010*
NZV	-0.0029	*0.893*	-0.0190	*0.005*	-0.0001	*0.928*	0,0245	*0,250*	-0,0007	*0,862*
No.of observations	54,177		37,142		40,101		43,346		54,177	
Chi square (*p_value*)	0.000		0.000		0.000		0.000		0.000	
Pseudo R2 (McFadden)	0.203		0.2593		0.1111		0.2653		0.1469	
% of correct predictions	83.8%		91.5%		99.1%		88.7%		95.7%	

Coefficients X 100: Farmer_income (1000), Income, Landown, ESU

Farm characteristics

Farm characteristics seem to strongly affect participation in AEMs as a whole and in single schemes as well. Farm economic dimension (*esu*) increases the probability of AEMs adoption, except for organic farming scheme that is insignificant. Previous evidence is quite contrasting on this point (Defrancesco et al. 2007, Mann 2005). However, the positive relation could be explained through reference to the transaction costs related to participation (that are mainly fixed costs). In fact, smaller farmers could have been discouraged from up taking AEMs schemes as they are not able to spread the fixed transaction costs over a reasonably large financial base (Falconer 2000). Moreover small farms, many of which are part-time, probably lack adequate entrepreneurship and sufficient information about these voluntary policy instruments.

As economic size is not necessarily correlated to farming and capital intensity, we represent this issue by using *lsu_ha* and *hp_ha* variables. In this case the signs of the coefficients are always significantly negative, confirming the well-known adverse selection effect in AEMs implementation (Hart and Latacz-Lohmann 2005, Latacz-Lohman 2004). Notably, more intensive farms are less likely to participate in AEMs as they usually incur higher opportunity costs in complying with schemes commitments. This consideration appears particularly true if we refer to the higher negative marginal effect in the *AEM_1* equation (input reduction), which is the scheme involving more farm management changes than others (Barreiro-Hurlè et al. 2008).

With regard to farming type, the probability that a dairy farm will participate in AEMs is 15 per cent higher than other specializations, a result in line with evidence reported by Jongeneel et al. (2008). Differently, *AEM_1* adoption is more likely for permanent crops type of farming. Finally, field crops specialization also somewhat affects the probability of AEMs up-take, in *AEM_all*, *AEM_1* and *AEM_4*.

Farmer characteristics.

As expected, *age* negatively affects the probability of entering AEMs, in line with the large part of the previous evidence (e.g., Vanslembrouk et al. 2002, Bonnieux et al. 1998). Thus, older farmers show a low propensity towards measures involving strong change, with respect to the usual farming practices. However, in contrast with Barreiro-Hurlè et al. (2008), also *AEM_3*, comparable with their "traditional farm management" scheme, shows a negative age coefficient, even if the marginal effect is smaller than in other schemes. *Farmer_income* is generally insignificant, except for *AEM_4*. Nevertheless this variable was calculated as a mean of the 'farm district', thus, as previously highlighted, variables related to farming intensity should better explain the role of opportunity costs in discouraging participation.

Finally we note that the share of property land (*landown*) negatively affects the farmer's willingness to participate in AEMs, indicating that landlords are less concerned about public goods production than tenants.

Social capital

With respect to 'social capital' variables, evidence has been found that farmers participating in other RDP measures (*investment* and *lfa_payment*) are more likely to sign-up for AEMs. This effect appears quite plausible if we think that some transaction costs related to participation could be spread among different measures. In any case this finding would indicate that a greater familiarity with RDP measures increases implementation probability.

The per-capita income (*income*) at the district level increases the probability of farmer enrolment in agri-environmental schemes. Income level is a good indicator of social demand for amenities and public goods and, more generally, of environmental sensitivity. Moreover it is important to know that the level of development goes hand in hand with the quality of institutions. *Education* seems, at least partially, to confirm this assumption, but its estimated effect on the probability of implementation is significantly positive only for *AEM_3*.

Interestingly, also ideological orientation influences the probability of participation in AEMs, confirming the Vandermeulen et al. (2006) and Hackl et al. (2007) interpretations of the influence of institutions and local policies on the uptake of agri-environmental and multifunctional-oriented commitments. Indeed, left parties share (*left*) positively affects the probability that a farmer will join both *AEM_1* and *AEM_2*. Such evidence appears in line with the notion that left-oriented political movements take more care of environmental issues. Nevertheless, the results of the *greens* variable seem to be at odds with the last statement. This apparent contradiction could be partly explained by the fact that, in Lombardy, the Green Party electorate tends to have little political power and lacks strong territorial variability. Consequently we ascribe environmental concerns to the entire left coalition to which the Green Party belongs.

Moreover, farmers are less likely to participate in AEMs where the share of euroskeptical parties (*euroskept*) is higher. At first glance this could indicate the rejection, or limited knowledge, of EU policy instruments. At a deeper level, it must be remembered that the euroskeptics of the Lombardy region are mainly represented by the Lega Nord party, whose members have often reaffirmed the strengthening of the productive role of agriculture *vs* the environmental/multifunctional one, giving stronger emphasis to farm competitiveness priority.[3]

To complete the discussion on social capital, the role played by farmers' associations needs to be clarified. To do this we resort to a second model applied to a smaller sample from which we have information on farmers' group affiliation (Table 19.4). Notably, the effect of other variables does not change with respect to the 'general' model, confirming the robustness of our specification. In the sample, all five existing organizations are represented. First of all *Coldiretti*

---

3   For example, during the recent CAP 'Health Check' discussion the Italian Minister of Agriculture (Lega Nord) claimed a reduction in the proposed rate of modulation from Pillar I to Pillar II.

(catholic-oriented), representing 61 per cent of farmers in the sample, followed by *Confagricoltura* (traditionally representing landlords, right-wing oriented), *CIA* (left-wing oriented), and two other minor organizations, *SISA* (moderate left-wing oriented) and *Copagri* (recently founded on an agreement between agronomists and agricultural contractors' associations).

The outcomes highlight the fact that farmers enrolled in *sisa* and *copagri* – compared with *coldiretti*, the omitted reference group – are more likely to participate in AEMs, by 7 per cent and 8.5 per cent, respectively. No particular effects were highlighted with respect to the other three main organizations, except for *CIA* in *AEM_2*, confirming the traditional positive attitude toward organic farming of the left-wing orientation. An interpretation of this outcome suggests that in small organizations transaction costs are lower because, among other things, of the deeper level of technical assistance (for example, think of the agronomists' involvement in *copagri*). This interpretation appears quite convincing if we consider AEMs implementation as needing long-term planning and substantial changes in farm management.

Table 19.4 Estimation results for models including farmers' organizations

PARAMETER	MODEL									
	AEM All		AEM 1		AEM 2		AEM 3		AEM 4	
	dF/dX	p value	dF/dX	p value	dF/dX	p value	dF/dX	p value	dF/dX	p value
Farm characteristics										
ESU	0.0232	0.000	0.0035	0.019	0.0001	0.705	0.0135	0.000	0.0078	0.000
LSU_HA	-0.0024	0.000	-0.0012	0.028	-0.0003	0.010	-0.0005	0.007	-0.0003	0.014
HP_HA	-0.0008	0.000	-0.0005	0.000	-0.0001	0.002	-0.0002	0.019	-0.0003	0.001
FIELD_CROP	0.0667	0.005	0.0591	0.000	0.0011	0.195	-0.0560	0.000	0.0315	0.000
PERMANENT_CROP	0.1302	0.000	0.1370	0.000	0.0057	0.002	-0.0744	0.000	0.0261	0.001
DAIRY	0.1878	0.000	0.0166	0.208	-0.0019	0.019	0.1513	0.000	0.0110	0.107
Farmer characteristics										
AGE	-0.0023	0.000	-0.0009	0.000	-0.0001	0.000	-0.0006	0.002	-0.0009	0.000
LANDOWN	-0.0317	0.000	-0.0024	0.449	-0.0004	0.640	-0.0249	0.000	0.0005	0.847
FARMER_INCOME	-0.0094	0.382	-0.0058	0.185	0.0000	0.843	-0.0015	0.874	-0.0034	0.076
Social capital										
INVESTMENT	0.1075	0.000	0.0525	0.000	0.0028	0.000	0.0114	0.134	0.0498	0.000
LFA PAYMENT	0.2439	0.000	0.0240	0.388	0.0074	0.000	0.1249	0.000	-0.0168	0.331
INCOME	0.0175	0.005	0.0032	0.270	0.0002	0.222	0.0084	0.201	0.0044	0.000
EDUCATION	0.0038	0.013	0.0009	0.133	0.0000	0.951	0.0030	0.010	0.0007	0.174
GREENS	-0.0126	0.114	-0.0124	0.000	-0.0002	0.355	0.0087	0.141	-0.0076	0.001
EUROSKEPTIC	-0.0054	0.006	-0.0046	0.000	-0.0001	0.035	0.0010	0.500	-0.0034	0.000
LEFT	-0.0005	0.690	0.0007	0.091	0.0001	0.057	-0.0018	0.174	0.0001	0.656
SISA	0.0704	0.009	0.0530	0.000	0.0201	0.001	-0.0108	0.518	0.0251	0.027
COPAGRI	0.0858	0.001	0.0139	0.286	0.0302	0.000	0.0332	0.068	0.0121	0.063
CIA	-0.0012	0.940	-0.0095	0.106	0.0070	0.000	0.0037	0.753	-0.0113	0.090
CONFAGRI	0.0092	0.468	-0.0022	0.543	0.0027	0.003	0.0075	0.526	0.0009	0.852
Location and others										
PLAIN	-0.1005	0.025	-0.0416	0.033	-0.0052	0.000	-0.0447	0.000	0.0027	0.765
MOUNTAIN	0.0865	0.234	0.0171	0.719	0.0014	0.372	0.2179	0.000	-0.0012	0.975
PARK	0.0247	0.133	-0.0177	0.005	0.0001	0.808	0.0428	0.005	0.0138	0.016
NZV	-0.0082	0.751	-0.0183	0.004	-0.0008	0.205	0.0233	0.340	-0.0021	0.720
No.of observations	38,447		30,076		32,477		35,424		38,447	
Chi square (p value)	0.000		0.000		0.000		0.000		0.000	
Pseudo R2 (McFadden)	0.2116		0.2599		0.1579		0.2701		0.1339	
% of correct predictions	81.7%		93.2%		99.2%		87.7%		94.5%	

Coefficients X 100: Farmer_income (1000), Income, Landown, ESU

Location and other determinants

The outcome of the variables related to farm location seems to highlight the failure of the agri-environmental schemes territorial targeting (Table 19.3). Indeed, *mountain*, *park* and *nzv* are largely not significant; furthermore, and surprisingly, farm location in nitrate sensitive areas reduces the probability of activating *AEM_1*, which is the scheme most concerned with tackling water pollution problems. Once again, the opportunity cost to participate – note that in Lombardy nitrate vulnerable areas are usually in an intensive farming context – discourage farmers from adopting AEMs. An exception to the above-mentioned lack of participation is the increase in probability of *AEM_4* implementation, characterizing farms situated in *parks*, where landscape-amenity social demand is, for obvious reasons, stronger.

Finally, the *pillar_1* positive marginal effect denotes a discrete overlapping of Pillar I and agri-environmental payments, thus it would seem that the redistributive nature of AEMs are only partially confirmed.

Concluding remarks

In this chapter we study the determinants of farmers' adoption of AEMs in the Lombardy region, our aim being to disentangle farm and farmer determinants from political and institutional ones. Working with the 'universe' of farms eligible for AEMs and with four different AEMs schemes, we obtain evidence about the effect of both farm and farmer characteristics on AEMs adoption. At the same time, our results corroborated the idea that the local institutional framework, by affecting the inter-relations of farms, local stakeholders and government bodies, influences the farmers' probability of up-taking AEMs.

The main findings of our analysis highlight how intensive farming seems to discourage AEMs implementation, while farmers' participation in other RDP measures exerts a positive effect. On the 'social capital' side, we found that local institutions affect AEMs uptake in the direction suggested by *a priori* considerations. However the weight of 'social capital' variables seems to be less important than that of farm and farmer characteristics. Finally, territorial location variables, explaining regional administration priorities, do not seem to affect, to any degree, the farmers' decision-making process to join AEMs.

This evidence leads us to highlight three main, interlinked, issues. First, a confirmation of the adverse selection phenomenon, notably the fact that the farmers entering AEMs are those who easily accomplish measure commitments (i.e. extensive farms). Secondly, the failure of specific territorial targeting of AEMs tends to suggest that the selection process of farmers' applications does not properly take into account environmental local needs. Finally, our analysis seems to suggest that, due to lack of rigorous selection, AEMs implementation favours a quantity-based rather than a quality-based funding approach.

References

Anderson, K. 2000. Agriculture's 'multifunctionality' and the WTO. *The Australian Journal of Agricultural and Resource Economics,* 44(3), 475-494.

Barreiro-Hurlé, J. and Espinosa-Goded, M. 2007. *Marginal Farmers and agri-environmental schemes: evaluating policy design adequacy for the Environmental Fallow measure.* Paper to the I Mediterranean Conference of Agro-Food Social Scientists: 103th EAAE Seminar 'Adding Value to The Agro-Food Supply Chain in the Future Euromediterranean Space', Barcelona, Spain, April 23th-25th, 2007.

Barreiro-Hurlé, J., Espinosa-Goded, M. and Dupraz, P. 2008. *Does Intensity of Change Matter? Factors Affecting Adoption in Two Agri-environmental Scheme.* Paper to the 107th EAAE Seminar: 'Modelling of Agricultural and Rural Development Policies', Sevilla, Spain January 29th – February 1st.

Bertoni, D. and Olper A. 2008. *The Political Economy of EU Agri-environmental measures :An empirical Assessment at the Regional Level.* Paper prepared to the 12th EAAE Congress: 'People, Food and Environments: Global Trends and European Strategies', Ghent, Belgium, August 26-29, 2008.

Bonnieux, F., Rainelli, P. and Vermersch, D. 1998. Estimate the Supply of Environmental Benefits by Agriculture: A French Case Study. *Environmental and Resource Economics* 11(2), 135-153.

Damianos, D. and Giannakopoulos, N. 2002. Farmers' participation in agri-environmental schemes in Greece. *British Food Journal,* 104, 261-273.

Defrancesco, E., Gatto, P., Runge, F. and Trestini, S. 2007. Factors Affecting Farmers' Participation in Agri-environmental Measures: A Northern Italian Perspective. *Journal of Agricultural Economics,* 59(1), 114-131.

Dupraz, P., Vermersch, D., Henry De Frahan, B. and Delvaux, L. 2003. The Environmental Supply of Farm Households – A Flexible Willingness to Accept Model. *Environmental and Resource Economics,* 25(2), 171-189.

Falconer, K. 2000. Farm-level constraints on agri-environmental scheme participation: a transactional perspective. *Journal of Rural Studies,* 16(3), 379-394.

Greene, W.H. 2003. *Econometric Analysis.* 5th Edition. Upper Saddle River, New Jersey: Prentice Hall.

Hackl, F., Halla, M. and Pruckner, G.J. 2007. Local compensation payments for agri-environmental externalities: a panel data analysis of bargaining outcomes. *European Review of Agricultural Economics,* 34(3), 295-320.

Hart, R. and Latacz-Lohmann, U. 2005. Combating moral hazard in agri-environmental schemes: a multiple-agent approach. *European Review of Agricultural Economics,* 32(1), 75-91.

Kemmerling, A. and Bodenstein, T. 2006. Partisan Politics in Regional Redistribution: Do Parties Affect the Distribution of EU Structural Funds across Regions? *European Union Politics,* 7(3), 373-392.

Jongeneel, R.A., Polman, N.B.P. and Slangen, L.H.G. 2008. Why are Dutch farmers going multifunctional? *Land Use Policy,* 25(1), 81-94.

Latacz-Lohmann, U. 2004. *Dealing with limited information in designing and evaluating agri-environmental policy.* Paper to the 90th EAAE Seminar: 'Multifunctional agriculture, policies and markets: understanding the critical linkage', Rennes, France, October 28-29, 2004.

Mann, S 2005. Farm Size Growth and Participation in Agri-Environmental Schemes: A Configural Frequency Analysis of the Swiss Case. *Journal of Agricultural Economics,* 56(3), 373-384.

Mathijs, E. 2002. *Social Capital and Farmers' Willingness to Adopt Countryside Stewardship Schemes.* Paper to the 13th International Farm Management Congress, Wageningen, The Netherlands, July 7-12, 2002

Swinbank, A. 2001. *Multifunctionality: a European Euphemism for Protection?,* FWAG Conference: 'Multifunctional Agriculture – A European Model', National Agricultural Centre, Stoneleigh, 29 November 2001.

Vandermeulen, V., Verspecht, A., Van Huylenbroeck, G., Meert, H., Boulanger, A. and Van Hecke, E., 2006. The importance of the institutional environment on multifunctional farming systems in the peri-urban area of Brussels. *Land Use Policy,* 23(4), 486-501.

Vanslembrouck, I., Van Huylenbroeck, G. and Verbeke, W. 2002. Determinants of the Willingness of Belgian Farmers to Participate in Agri-environmental Measures. *Journal of Agricultural Economics,* 53(3), 489-511.

Wilson, G.A. 1997. Factors Influencing Farmer Participation in the Environmentally Sensitive Areas Scheme. *Journal of Environmental Management,* 50(1), 67-93.

Wossink, G.A.A. and van Wenum, J.H. 2003. Biodiversity conservation by farmers: analysis of actual and contingent participation. *European Review of Agricultural Economics,* 30(4), 461-485.

Chapter 20

Impact Assessment of 2003 CAP Reform and Nitrate Directive on Arable Farming in Midi-Pyrénées: A Multi-Scale Integrated Analysis

Kamel Louhichi, Hatem Belhouchette, Jacques Wery, Olivier Therond and Guillermo Flichman

Background and Objectives

Impact assessment of European Union's (EU) Common Agricultural Policy (CAP) has become a central issue for researchers, stakeholders and policy-makers in all European Member States. It represents the attempt to assess systematically and in detail the potential effects of policy proposals and plays an important role in the European Commission decision making process. Several governments have already introduced regulatory impact assessment procedures for their own policy-making processes, and others are being encouraged to do the same. The aim of this study is to contribute on such issue by analysing through an integrated approach the combined effects of 2003 CAP reform and Nitrate Directive on the sustainability of arable farming using a set of economic, social, and environmental indicators.

The CAP reform of June 2003 constitutes the most significant reform of the EU's Common Agricultural Policy since its inception. Driven primarily by budgetary concerns, this reform has aimed at stimulating global markets competitiveness, better environmental performance, supporting rural viability as well as better meeting consumer demands (CEC 2003B). The main measures of this reform are the adoption of decoupled direct payment, the introduction of a new modulation system, and the enforcement of agri-environment schemes. The decoupled payment consists on the replacement of all Direct Producer Payments associated with beef, sheep, and arable crops production (and planned future dairy payments) with a 'single payment per farm (SFP)' received by beginning in 2005. Such single farm payments are calculated on the basis of 'a reference amount in a reference period 2000-2002 and are paid to those holding land with a payment entitlement. This implies that the amount of the payment would not depend on what and how much the farmer actually produces but essentially on area and historical entitlement. Farmers are free to decide what they want to produce in response to demand without losing their entitlement to support. The reform, however, gives each EU Member State the possibility to choose a 'degree of decoupling' among some options, which can be applied at national or regional level (OECD 2004).

The modulation system introduced in this reform aims to finance the additional Rural Development Regulation (RDR) measures through the reduction of direct payments by five per cent from 2007 for farms with more than EUR5,000 direct payment a year. This five per cent reduction, known as "modulation", will result in additional RDR funds of EUR1.2 billion a year (CEC 2003b).

The 2003 CAP reform has been also promoting the multifunctional role of agricultural. Farmers are viewed not only as food suppliers but also as the custodians of the countryside. This role of farmers has been acknowledged in the EU Common Agricultural Policy through a number of regulations that enforce agri-environment schemes and cross-compliance. These measures have been introduced under the Agenda 2000 regulation as optional but the 2003 CAP reform made them obligatory for all farmers receiving compensation payments. The nitrate directive is one of the oldest EU environmental programs designed to protect water quality by preventing nitrates from agricultural sources polluting ground and surface waters and by promoting the use of good farming practices. Defined at regional level, this Directive stipulates that each Member State draws up at least one code of good agricultural practices in order to reduce pollution by nitrate in the vulnerable zones. In arable farming, this directive is based on the following measures: (i) better management of mineral and organic nitrogen fertilization; (ii) respect of the restricted period for applying manure or nitrogen fertilizer taking into account the type of fertilization and the land use; and (iii) maintenance of a minimum quantity of vegetation cover during (rainy) winter periods for the uptake of the nitrogen from the soil. If one of these measures is not respected a range of penalties linked to EU premiums can be applied (EEC 1991).

Within this context, the basic purpose of this study is to attempt answering the following questions: (i) How the 2003 CAP reform and Nitrate directive would impact the economic and environmental performance of the selected arable farms? (ii) What happens if all the farmers are enforced to respect the Nitrate Directive? (iii) Which policy instruments could be applied in order to stimulate/force farmers to adopt this Directive?

In Section 2 the modelling approach is presented, followed by a description of the study area, data requirement, model calibration and simulation scenarios. In section 3 the results of policy scenarios are presented and discussed. In section 5, we conclude on the relevance of this type of approach and its capacity to assess the impact of EU policies.

Materials and Methods

During the last decade, there has been a growing interest in analysing the impacts of the EU's Common Agricultural Policy on the sustainability of the agricultural sector, and the body of literature and models for this purpose is increasing (Britz and Heckelei 2008). However, most of the existing models in the EU are mono-dimensional (i.e., often focusing on only one of the three familiar dimensions of sustainability: economic, social or environment), monolithic (i.e., based on one software component), and used for specific purposes and locations. Approaches that allow flexible impact assessment for a range of issues and functions are scarce. This seems to be due, on the one hand, to the complexity of new policy schemes, and, on the other hand, to the necessity of a multi-disciplinary approach to policy decision making. With these limitations in mind, the aim of this study is to present an integrated modelling approach enables to assess the effects of a wide range of agricultural and environmental policies at different scales and under various conditions. Model use is illustrated by simulating the response of arable farming in Midi-Pyrénées to 2003 CAP reform and Nitrate Directive.

Modelling approach: CropSyst – FSSIM model chain

CropSyst is a biophysical model developed, by the Biological Systems Engineering Department of the Washington State University, to serve as an analytic tool to simulate the effect of cropping systems management on productivity and the environment (Donatelli et al. 1997). It is a multi-year, multi-crop, daily time step crop growth simulation model. The model is partially mechanistic and partially empirical. The mechanistic component is dominant in the more complex models. This means that the relation between the quantity of fertiliser used and the yields obtained is explained by a series of cause and effect mechanisms corresponding to soil, climate, characteristics of the plant and the variety.

CropSyst was used in this application to quantify, at field scale and according to agro-ecological conditions, the effects (in term of yields and environmental externalities) of the current and alternatives activities defined as a combination of crop rotation, soil type and management type.

FSSIM is a farm model developed, within the SEAMLESS project (Van Ittersum et al. 2008) to assess the economic and environmental impact of agricultural and environmental policies and technological innovations. FSSIM was designed to describe farmer's behaviour given a set of biophysical, socio-economic and policy constraints, and to predict his/her responses under EU policy changes, using data generated from the biophysical model as well as other data sources (Farm Accountancy Data Network (FADN), expert knowledge, surveys).

The general context in which FSSIM was developed and the variety of policy questions that is called to address justifies a combination of choices that makes this model unique (Louhichi et al. 2010):

- A static programming model which optimizes an objective function for one period (i.e. one year) over which decisions are taken. This implies that it does not explicitly take account of time. Nevertheless, to incorporate some temporal effects, agricultural activities are defined as 'crop rotations' and 'dressed animal' instead of individual crops and animals.

- A risk programming model, taking into account the risk according to the Mean-Standard deviation method in which expected utility is defined under expected income and risk (Hazell and Norton 1986).
- A primal-based approach, which makes technology representation explicit, and allows for switching between current production techniques as well as between current and alternative (i.e. innovate) production systems.
- A discrete-based function to makes easily the smooth integration of engineering data or results from bio-physical models needed to assess the environmental effects of production process. This discrete function can (better) capture the technological and policy constraints than a behaviour function in econometric models.
- A positive model, where the main objective is to reproduce the observed production situation as precisely as possible by making use of the observed behaviour of economic agents. As mentioned by Howitt (2005), normative models are generally of limited use in the accurate representation of farm response to policy shifts.
- A modular setup to be re-usable, adaptable and easily extendable to achieve different modelling goals (i.e., easily activate/deactivate modules according to regions and conditions).
- A generic model designed with the aim to be applied to any farming systems across Europe and elsewhere, and to assess different policies under various conditions.

The general mathematical formulation of FSSIM is as follows:

$$U = Max\left[Z - \varphi\sigma\right] \tag{1}$$

Subject to: $Ax \leq B$ (2)
$\quad\quad\quad\quad x \geq 0$ (3)

Where: **U** is the variable to be maximised (i.e. utility), **Z** is the expected income, **x** is a (n x 1) vector of agricultural activity levels, **A** is a (m x n) matrix of technical coefficients, **B** is a (m x 1) vector of levels of available resources and upper bounds to the policy constraints, φ is a scalar for the risk aversion coefficient and σ is the standard deviation of income according to states of nature defined under two different sources of variation: yield (due to climatic conditions) and prices.

The expected income (**Z**) is a non-linear profit function. Using matrix notation, this gives:

$$Z_x = \left[gm'x - d'x - \frac{x'Qx}{2} \right]$$

(4)

Where **gm** is the (n×1) vector of the gross margin of activities, **x** is the (n×1) vector of the simulated levels of the agricultural activities, **d** is the (n×1) vector of the linear part of the activities' implicit cost function and **Q** is the (n×n) matrix of the quadratic part of the activities' implicit cost function.

The gross margin **(gm)** is defined as total revenues including sales from agricultural products and compensation payments (subsidies) minus observed accounting costs of production activities. The accounting costs include costs for fertilizers, crop protection and seeds as well as plant material and cost of hired labour. **Q** and **d** are the parameters of the implicit cost function estimated using PMP approach following Kanellopoulos et al. (2010) formalism.

Arable agricultural activities are based on crop rotations. It is assumed that in each year, all crops of a rotation are grown on equal shares of the land. A model solution can include several crop rotations on one farm. The concept of crop rotations allows accounting for temporal interactions between crops even if the model is static. Regional experts were asked to specify the most frequent rotations which are currently used by farms in their region.

Each agricultural activity is defined as a crop rotation with a specific agro-management practice growing in a predefined agri-environmental zone. Let **R** denote the set of crop rotations (including mono-crop rotations), **S** the set of agri-environmental zones and **T** the set of production techniques (i.e., agro-management practice). The set of agricultural activities **i** can be defined as follows (Louhichi et al. 2010):

$$\mathbf{i} = \{\, i_1, i_2, ... \,\} = \{\, (r_1, s_1, t_1),\ (r_2, s_1, t_1),\ ... \,\} \subseteq R \times S \times T$$

(5)

Figure 20.1 gives an overview of the integrated modelling approach as a combination of a biophysical model 'CropSyst' and a farm model 'FSSIM'.

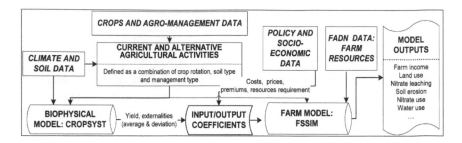

Figure 20.1   Integrated modelling approach: CropSyst-FSSIM model chain

Application of the CropSyst-FSSIM model chain

The application of this model chain to our case study is based on the following steps: (i) selection of relevant farm types representative of the arable farming in the region using the SEAMLES typology (Andersen et al. 2006) and the FADN data sources; (ii) identification of the 'average' farms (i.e. a virtual farm derived by averaging historical data from farms that are grouped in the same type) that represent adequately the whole farms that belong to the same farm type (iii) modelling each farm type separately in order to reproduce the farmer's observed behaviour (model calibration); (iv) definition and implementation of the selected scenarios and analysis of their impacts at farm scale through a set of relevant indicators and (v) aggregation of the results across selected arable farm types.

Brief description of the case study   Midi-Pyrénées is the largest region in France with a surface of 45,348 km². It is as big as Denmark and bigger than Belgium, Switzerland or Holland. Agriculture in Midi-Pyrénées is very important, with production equally divided between livestock and crops. It represents the first French region by its number of holdings (around 60,000) and the fifth by the value of its agricultural production.

The main crops cultivated in the region are cereals, protein crops and oilseeds. They represent approximately 40 per cent of the cultivated areas of the region (Agreste-annual farm statistics 2006). Five per cent of the total cultivated area of the region was lying fallow in 2006: 9 per cent of the total cultivated area is irrigated. Rainfed annual grain crops are therefore predominant in the Midi-Pyrénées region. The main soil types in the region are locally known as: calcareous clay and clay-loam.

The Midi-Pyrénées region is also known for the problem of water pollution caused by nitrate from agricultural sources. In 2002, more than 45 per cent of the water quality in term of nitrate concentration was judged as average or very bad. Only 3 per cent of the water body was considered of very good quality (Ifen 2002).

Selection of representative farm types Modelling all individual arable farms in the Midi-Pyrénées region is not feasible because of the large number and the diversity of farms. For that reason it was decided to use the farm typology developed in the SEAMLESS project. Based on FADN and Farm Structural Survey (FSS), this farm typology provides a set of farm types relatively homogeneous defined by 4 criteria: size, intensity, land use and specialisation.

From this typology we have selected three farm types to represent the arable farming system in Midi-Pyrénées. For each farm type average endowment characteristics and observed crop pattern have been computed and reported in Table 20.1.

Table 20.1 Main characteristics of the three arable farm types in Midi-Pyrénées

	Farm type 1	Farm type 2	Farm type 3
Specialisation -land use-	Cereal	Cereal/Fallow	Mixed
Farm represented (number)	2.33	990	1.74
Area by Farm (ha)	113.9	101.5	123.3
Irrigable area by Farm (%)	37	30	13
Soil Types (% of texture)	Clay (40%) Clay-loam (60%)	Clay (36%) Clay-loam (64%)	Clay (41%) Clay-loam (59%)
Available labour (hours)	2,901.6	3,260.3	3,179.0
Crop allocation (ha)			
Cereals	72.8	52.4	53.3
Oilseeds	19.5	17.7	43.3
Protein	2.9	4.3	5.9
Fallow	11.4	18.9	11.5
Perennial crops	7.3	8.2	9.3

Source: FADN database (average of 2002-2004)

Collecting required data Three types of data are required to apply the CropSyst-FSSIM model chain:

- Bio-physical data (weather information, soil data) characteristics of the agri-environmental zones used as input for the bio-physical model CropSyst.
- Farm resource data such as available farm land per soil type, irrigated land, available family labour and observed crop allocation (i.e. crop pattern). These data are collected from the FADN sources and used in the FSSIM model for the definition of constraints' RHS value and for the calibration process.
- Identification of the current and alternative activities and quantification of

theirs input output coefficients such as yield (average and variability), input use (e.g. fertiliser, water, labour), prices (average and variability), costs, premiums, etc. To collect these data in the Midi-Pyrénées region a survey has been designed and used, completed by local expert knowledge and statistical database. These data have been collected for the most frequent cropping systems in the region, taking into account climatic variation and other factors as pests and weeds. In total 65 rotations were identified, with 11 different crops. The principal types of rotations are soft wheat-sunflower, durum wheat-sunflower and maize-maize for grain. Combined to ago-management and soil types, these rotations define the so-called current agricultural activities. For each crop within agricultural activities a set of data were collected. It includes the data on amount, nature, method and temporality of management events: sowing, harvesting and tillage events, weed, pest and disease management, water management, nutrient management, labour use. Additionally, for each crop a set of economic data has been specified including producer prices (the average value and the variability), variable costs and premiums. The expected producer prices are collected from regional database and based on the 1999-2003 average. Variable costs are calculated by adding input costs for fertilizers, seeds, irrigation, biocides and the application costs associated with each event. The premiums are of the three years average around 2003 according to Agenda 2000 regulation taken as base year policy.

Model calibration   The CropSyst model was calibrated, for each crop, against observed yield during the simulated years. The values of the biomass-transpiration (KBT) and of light conversion to above ground biomass (KLB) coefficients were adjusted within a reasonable range of variation based on previous research and expert knowledge in order to have the best model estimation of the biomass accumulation observed for each crop in the calibration experiments (Donatelli et al. 1997). Adjustment ends when further modification of crop parameters would generate little or no improvement on the basis of the relative error, a statistical index is used to quantify the degree of fitness in the relationship between measured and simulated aboveground biomass (Cabelguenne et al. 1990).

The calibration of FSSIM is based on two steps: in the first step, we apply the risk approach in order to calibrate the model, as precisely as possible. The model assigns automatically a value to the risk aversion coefficient which gives the best fit between the model's predicted crop allocation and the observed values. The difference between both values is assessed statistically by using the Percent Absolute Deviation[1] (PAD). The aim of this step is to ensure that the model produces acceptable results before going to the second step. In the second step, a Positive Mathematical Programming variant (Kanellopoulos et al. 2010) is implemented in order to calibrate the model exactly to the observed situation and guarantee exact reproduction of the base year situation (Howitt 1995, Heckelei and Wolff 2003).

The base year information for which the model is calibrated stems from a three-year average around 2003. In term of policy representation the Agenda 2000 (since 2000) Regulation constitutes the base year policy.

Building baseline (reference run)   The baseline scenario is interpreted as a projection in time covering the most probable future development in term of technological, structural and market changes. It represents the reference for the interpretation and analysis of the selected policy scenarios. In our case study, the 2003 CAP reform is considered as the principal policy assumption operating in the baseline scenario. In term of technological and market change, three exogenous assumptions are adopted: (i) an assumed regional inflation rate of 1.19 per cent per year; (ii) a projection in producer prices obtained from the market model CAPRI (Britz et al. 2002) and (iii) a yield trend to reflect technical progress coming also from CAPRI database (Table 20.2). All the others parameters (including farm endowments as well farm's weight on the region) are assumed to remain unchanged up to 2013, taken as time horizon for baseline definition.

Table 20.2 Price and yield changes between base year and baseline scenarios

Crops	Price change (%)	Yield change (%)
Durum wheat	10	22
Soft wheat	4	-7
Barley	-3	15
Maize	-13	5
Sunflower	0	1
Soya	-19	-1
Rapeseed	11	21
Peas	9	-4
Oats	-8	20
Maize fodder	29.9	13.2

Source: CAPRI database

---

1      $$PAD\ (\%) = \frac{\sum_{i=1}^{n}\left|\hat{X}_i - X_i\right|}{\sum_{i=1}^{n}\hat{X}_i} \cdot 100$$ where $\hat{X}_i$ is the observed value of the variable i and X_i is the simulated value (the model prediction). The best calibration is reached when PAD is close to 0.

Layout and implementation of policy scenarios   The simulated policy scenario combines the 2003 CAP reform and the first measure of the Nitrate Directive (the other measures are not retained as they require more time in data collection and in CropSyst simulations). This measure consists to apply better management of nitrogen mineral fertilisation in order to limit nitrate lixiviation without reducing yield. It stipulates that farmers should fertilize according to the crop requirement and the soil provision of nitrogen. The implementation of this measure in the model chain CropSyst-FSSIM was achieved through the following steps (Table 20.3):

1. Generating a set of alternative activities (AA) based on current crops but with better management of nitrogen mineral fertilisation:
    –   Nitrogen from mineral fertilizers needed by AA are calculated based on the 'local advisory services' recommendations (simple nitrogen balance) using the current yield as target yield since expert observed that the yield of this type of AA are very close to the corresponding current activities (CA).
    –   Yield and yield variability of AA are generated from CropSyst.
    –   Costs of AA are calculated as the cost of the corresponding current activity minus the reduction in fertiliser costs due to reduction of N use.
    –   A five per cent transaction cost related to the collection of information on policy implementation, the participation in training sessions... was introduced for AA.
    –   Environmental externalities associated to each AA are quantified by CropSyst.
2. Application of cross-compliance restrictions related to AA: 3 per cent cut of EU premiums if AA are not applied.

Table 20.3   Definition of base year, baseline and policy scenarios

	Base year [2003]	Baseline [2013]	Policy scenario: Nitrate Directive [2013]
Exogenous assumptions		- Projection in producer prices from 2003 to 2013 - Yield trend from 2003 to 2013 - Inflation rate of 1.19% per year	
EU Common Agricultural Policy	Agenda 2000	2003 CAP reform (with an option of 25% partial coupling as arable crops area payments chosen for France and 5% modulation)	
Agricultural activities	Current activities (CA)		Current activities (CA) + Alternative activities (AA)
Measures	none		Cross-compliance restrictions:3% cut of EU premiums if AA are not applied

However before analysing the impact of the Nitrate Directive scenario, a comparison of the likely impacts of the 2003 CAP reform and the continuation of Agenda 2000 Regulations is presented and discussed. In this comparison all the exogenous assumptions adopted in the baseline scenario are deactivated in order to assess the separate impact of 2003 CAP reform.

Results and discussion

The impacts of the different scenarios are illustrated through a set of technical (crop allocation), economic (farm income and EU premiums) and environmental indicators (nitrate leaching and soil erosion). In order to make the results comparable across scenarios and farm types, the economic indicators are expressed in constant 2003 prices (i.e. deflated prices) and the environmental indicators are defined per hectare of usable farmland. First, the results for each farm type are shown. Subsequently, the aggregated results across all the simulated arable farm types are computed as the weighted sum of the results for each farm type. The weights for each farm type correspond to the share of real farms belonging to that farm group.

Impact analysis of 2003 CAP reform at farm and regional scales

Compared to the continuation of Agenda 2000 Regulations, the adoption of the 2003 CAP reform leads, as shown in Figure 20.2, to a largest change in crop allocation manifested by (i) a fall in durum wheat area explained by the fact that the supplement for durum wheat in traditional production zones was reduced and integrated in the single payment scheme; (ii) a slight increase in the land used for irrigated crops, especially for maize grain, considering that 25 per cent of the payments for these crops remain coupled and; and (iii) a rise on the area devoted to oilseeds and protein crops as these crops become more competitive under the decoupled payment. These tendencies are observed in all three farm types of the Midi-Pyrénées region, with different degrees according to farm's resource endowments.

In terms of economic impacts, the 2003 CAP reform would induce a decrease of EU support level (i.e. EU premiums) owing to modulation system and a slight amelioration of farm income, reaching 5 per cent, due to a better crop allocation. Indeed, the decoupled system stimulates farmers to choose activities according to market opportunities without losing their entitlement to support.

Regarding the environmental results, the implementation of the 2003 CAP reform would lead to a decrease of nitrate leaching from 5 to 13 per cent depending on farm types, mostly because of the drop in the level of durum wheat growing under cereal rotations in profit of soft wheat-sunflower rotation which generates less pollution levels. The impact in soil erosion is quite different across farm types. It seems positive in farm type 1, marginal in farm type 2 and negative in farm type 3. This is explained by the fact that in farm type 3 the irrigable land is low and completely used and so the substitution of durum wheat was done in favour of rained cereals that have large soil erosion coefficients.

Most of the technical, economic and environmental results obtained at the farm level remain consistent when aggregated at the regional scale: (i) a decrease of EU premiums, nitrate leaching and durum wheat area and (ii) an increase of farm income, soil erosion and oilseed and protein crop areas. These results could be explained by the large similarity between arable farms in the region but also by the fact that the flexibility inside the arable sector is very restricted (Fig. 20.2)

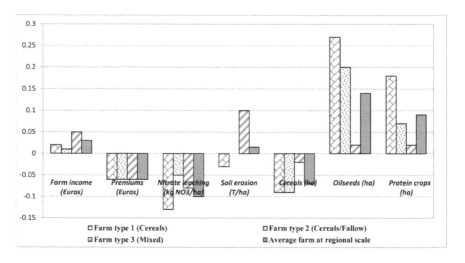

Figure 20.2   Impacts of the 2003 CAP reform at farm and regional scales (per cent change to Agenda 2000)

Impact analysis of the Nitrate Directive at farm and regional scales

Figure 20.3 exposes the technical, economic and environmental results of the policy scenario in comparison to baseline. The main result shown in this Figure is that none of the farm types has adopted entirely the first measure of Nitrate Directive. This implies that the penalty of 3 per cent is not enough to compel farmers to adopt the Nitrate Directive and to substitute entirely the current activities by the alternative ones based on better N management. In another term, it would be more profitable to accept a 3 per cent cut of premiums than to adopt fully the alternative N management since not all the alternative managements are competitive under the taken assumptions (e.g., five per cent transaction costs). This appears clearly while looking to the change in crop allocation provoked by this policy scenario. Indeed, the share of alternative activities in the total farm area is less than 36 per cent in the three farm types (i.e., 23 per cent in farm type 1; 21 per cent in farm type 2 and 36 per cent in farm type 3). The other impacts in term of crop allocation are dominated by the substitution of oilseeds by soft wheat which becomes more profitable with the adoption of better N management.

The impact on farm income is marginal either in relative or absolute terms in spite of the 3 per cent cut of premiums. This implies that the reduction of premiums was entirely compensated with the partial adoption of alternative activities which are more competitive to their corresponding current activities. However, this substitution is still marginal compared to the directive goal: a full adoption of better N management.

Regarding the environmental results, the impacts of the policy scenario seem very positive in term of soil erosion but uncertain for leached nitrate. Indeed, soil erosion decreases in the three farm types, reaching the 30 per cent in some cases. This is due mainly to the reduction of spring crops (sunflower and soya bean) and the increase of winter soft wheat, thereby reducing the bare soil area during winter. However, for nitrate leaching the impact is not regularly positive and swing between -6 per cent to +5 per cent depending on farm types. This implies that the partial adoption of better N management is not enough to ensure a reduction of leached nitrate and the solution could be through a fully implementation as proposed in the Nitrate Directive. The questions that emerges is how to stimulate/ force farmers to adopt the better N management since the 3 per cent is not enough to reach this goal and which policy instruments could be applied for that? This is the aim of the sensitivity analysis developed in the last section.

Despite some differences between farm types, the trend obtained at the farm scale was kept after aggregation at the regional scale: no change in farm income and in nitrate leaching, a slight decrease of premiums and a significant reduction of soil erosion (Fig. 20.3).

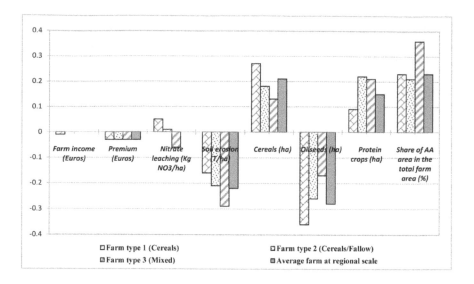

Figure 20.3   Impacts of the Nitrate Directive at farm and regional scales (per cent change to baseline scenario)

Sensitivity analysis

The aim of this sensitivity analysis is to estimate the thresholds of penalty (i.e. percentage of premium cut) to apply in each farm type in order to enforce farmers to respect the nitrate directive and also to show the sensitivity of these thresholds to the percentage of transaction costs assumed on the implementation of this directive.

As reported in Table 20.4, from 13 to 17 per cent of penalty, according to farm type, was required to force the farmer to adopt the alternative N management. These thresholds of penalty allowed a reduction of used N fertiliser and of leached nitrate in all the farm types with a slight loss of farm income (around 6 per cent).

Table 20.4   **Impacts of the compulsory application of Nitrate Directive at farm and regional scales**

| | Compulsory application of Nitrate Directive (% change to baseline scenario) | | | |
	Farm type 1	Farm type 2	Farm type 3	Average farm at regional scale
Penalty (%)	17	13	13	17
Farm income (euros)	-6	-5	-6	-6
N fertiliser used (Kg/ha)	-28	-26	-29	-28
Nitrate leaching (Kg NO3/ha)	-5	-6	-14	-9

Source: model results

Figure 20.4 summarises the sensitivity of these thresholds to transaction costs. To perform this sensitivity analysis, we shift the initial value of the transaction cost to more less 100 per cent (the use of same percentage allows assessing the degree of symmetry in the sensitivity) and then we run the model several times in order to establish the new penalty threshold for each farm type from which nitrate directive would be applied. As expected, the penalty threshold seems very sensitive and positively correlated to the transaction costs as the change in penalty threshold is important and would affect hardly the economic and environmental results of the selected farms. For this reason it would be appropriate to establish a consistent method for estimating these costs in a realistic manner.

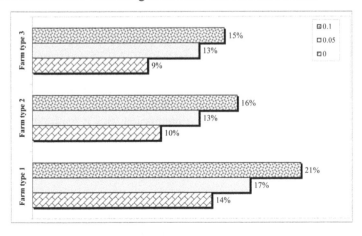

Figure 20.4   Sensitivity of penalty threshold to transaction costs

Conclusion

This chapter has presented the results of the application of the CropSyst-FSSIM model chain to assess the combined effects of 2003 CAP reform and Nitrate Directive on the sustainability of selected arable farming in Midi-Pyrénées region. The main conclusions coming up from this study in terms of policy impacts are: (i) the implementation of the 2003 CAP reform affects positively but moderately farmer's income due to a better crop allocation induced by decoupling system; (ii) the 3 per cent cut premium is not enough to compel farmers to adopt the Nitrate Directive and to substitute entirely the current activities by the alternative ones based on better N management; (iii) the impact of nitrate directive, as currently implemented, on nitrate leaching is not always positive and depends on farm types, implying that the partial adoption of better N management is not enough to ensure a reduction of leached nitrate (iv) the sensitivity analysis shows that a threshold of 17 per cent in premium cut is required to enforce all arable farmers in the region to adopt the nitrate directive. However, this threshold remains very sensitive to the transaction costs connected with the implementation of this directive.

This application highlights the relevance and the power of this type of approach for making finer and integrated analysis at disaggregated scales and for analysing complex policy scenarios integrating technical, economic and environmental aspects. It provides insights in some key methodological aspects to be considered and improved in further research. The main aspects are: i) the need for several interactions with local experts and further methodological development for a better models calibration and validation at field and farm scales, ii) the need for better consideration of transaction costs connected with nitrate directive in order to bring the analysis even closer to reality.

Acknowledgements

This publication has been funded under the SEAMLESS integrated project, EU 6th Framework Programme for Research, Technological Development and Demonstration, Priority 1.1.6.3. Global Change and Ecosystems (European Commission, DG Research, contract no. 010036-2. Its content does not represent the official position of the European Commission and is entirely under the responsibility of the authors.

References

Andersen, E., Elbersen, B., Godeschalk, F. and Verhoog, D. 2006. Farm management indicators and farm typologies as a basis for assessment in a changing policy environment. *Journal of Environmental Management 82*, 352-362

Britz W., Henrichsmeyer W., Wieck C. and Perez I. 2002. *Impact analysis of the European Commission's proposal under the mid-term review of the Common Agricultural Policy (using the CAPRI model)*. Final Report, 75p, University of Bonn.

Britz, W. and Heckelei, T. 2008. *Recent Developments in EU Policies – Challenges for Partial Equilibrium Models*. Invited paper prepared for presentation at the 107th EAAE Seminar "Modeling of Agricultural and Rural Development Policies". Seville, Spain, January 29th-February 1st, 2008.

Cabelguenne, M., Jones, C.A., Marty, J.R., Dyke, P.T. and Williams, J.R. 1990. Calibration and validation of EPIC for crop rotations in southern France. *Agricultural Systems,* 33, 153-171.

CEC. 2003b. Council Regulations relative to CAP changes, Interinstitutional Files 2003/0006, 2003/0007, 2003/0008, 2003/0009, 2003/0010, 2003/0011, 2003/0012, September, Brussels.

Donatelli, M. 2002. *Simulations with CropSyst of cropping systems in Southern France. Common Agricultural Policy Strategy for Regions, Agriculture and Trade* (QLTR-200-00394). CIEAHEM report, 13p.

EEC. 1991. *Implementation of Nitrate Directive. European Environment Commission report*, Annex 2: Code(s) of good agricultural practices. 1991.

Hazell, P.B.R. and Norton, R.D. 1986. *Mathematical Programming for Economic Analysis in Agriculture*. Macmillan Publishing Co, New York, 400p.

Heckelei, T., and Wolff, H. 2003. Estimation of Constrained Optimisation Models for Agricultural Supply Analysis Based on Generalised Maximum Entropy. *European Review of Agricultural Economics*, 30(1), 27-50.

Howitt, R.E. 1995. Positive Mathematical Programming. *American Journal of Agricultural Economics,* 77, 329-342.

Ifen. 2002. Les dépenses des Départements en matière d'environnement. *Les données de l'environnement*, n°79, 4 p.

Janssen, S., Louhichi, K., Kanellopoulos, A., Zander, P., Flichman, G., Hengsdijk, H., Meuter E., Anderson, E., Belhouchette, H., Blanco, M., Borkowski, N., Heckelei, T., Hecker, M., Li, Lansink, H., Stokastad, A., O., Thorne, G., P., van Keulen, H. and van Ittersum, M. K. 2009. Economic farm model for environmental and economic assessment agricultural systems. *Accepted for Environmental Management.*

Kanellopoulos, A., Berentsen, P., Heckelei, T., van Ittersum, M. and Oude Lansink, A. 2010. Assessing the Forecasting Performance of a Generic Bio-Economic Farm Model Calibrated with Two Different PMP Variants. *Journal of Agricultural Economics*, 61(2), 274-294

Louhichi K., Janssen, S., Kanellopoulos, A., Li, H., Borkowski, N., Flichman, G., Hengsdijk, H., Zander, P., Blanco, M., Stokstad, G., Athanasiadis, I., Rizzoli, A.E., Huber, D., Heckelei, T. and Van Ittersum, M. 2010. A generic Farming System Simulator, in *Environmental and agricultural modelling: integrated approaches for policy impact assessment* edited by F. Brouwer and M.K. van Ittersum. Springer Academic Publishing,

OECD. 2004. *Analyse of the 2003 CAP reform*. Paris.
Van Ittersum, M.K., Ewert, F., Heckelei, T., Wery, J., Alkan Olsson, J., Andersen, E., Bezlepkina, I., Brouwer, F., Donatelli, M., Flichman, G., Olsson, L., Rizzoli, A.E., van der Wal, T., Wien, J.E. and Wolf, J. 2008. Integrated assessment of agricultural systems – A component-based framework for the European Union (SEAMLESS). *Agricultural Systems,* 96, 150-165.

Chapter 21

Connecting Agri-environmental Schemes and Cross-Compliance Designs: An Exploratory Case Study in Emilia-Romagna

Fabio Bartolini, Vittorio Gallerani, Meri Raggi and Davide Viaggi

Introduction

Agri-environmental schemes (AESs) and cross-compliance (CC) are the two components of the Common Agricultural Policy (CAP) most explicitly aimed at providing environmental benefits from agriculture. A large body of literature focuses on these topics (see for example Hodge 2000, Dobbs and Pretty 2004, Osterburg et al. 2005, Helin 2008). AESs have been studied under different perspectives in Europe since the beginning of the 1990s (see for example Feinerman and Komen 2003, Dupraz et al. 2002, Latacz-Lohmann 2004). Cross compliance is a more recent field of study connected to the implementation of this policy in Europe by way of the 2003 CAP reform (Fishler reform). The literature in this field has increased significantly in recent years (see for example Varela-Ortega and Calatrava 2004, Fraser and Fraser 2005, Davies and Hodge 2006, Bennet et al. 2006). Under Regulation 1698/2005, the agri-environmental scheme design rationale has been changed, with a more clear identification of a baseline for identifying the commitments and the compliance costs in order to justify payments. Such a baseline is generally identified as the commitments of cross-compliance represented by mandatory standards. The joint implementation of AESs and CC is now to be interpreted as a major component of policy design and analysis, as it sets the boundary between positive and negative externalities.

There are two different perspectives on these issues: on one hand, the link between cross-compliance and agri-environmental schemes can be interpreted as a problem of the decision maker who would ultimately prefer a joint AESs-CC design. The joint design problem in this case may involve the impact of the CC prescription on the design of AESs, the allocation of funds to the two measures and the connected implementation of mechanisms of control and sanctions that may be able to guarantee compliance to both agri-environmental commitments and mandatory standards defined for each measure. From the farmer's point of view, the decision-making problem concerns participation in, and compliance with, the combination of the two measures. In particular, the private costs of participation in agri-environmental schemes must be added to the cost of compliance with the mandatory standards defined for each measure if they are not already implemented. These costs may be very relevant as mandatory standards are required on the whole farm even if agri-environmental schemes are applied only to a small portion of the farm.

The Italian Ministry of Agricultural Food and Forestry recently introduced a new framework regulating the mechanisms of monitoring and sanctions for cross-compliance commitments and mandatory standards when pertinent for agri-environmental commitments. This legislation seeks to tailor more effective sanction mechanisms to ensure compliance with cross-compliance and agri-environmental commitments. In spite of the increasing connection between the two policies, to the best knowledge of the authors no study deals directly with the problems raised by their joint design. In this chapter we investigate the problem from the point of view of farmers' decision making. The objective of this chapter is to investigate farmer choice under varying levels of monitoring and sanctions, with respect to compliance with the mandatory standards introduced by cross compliance and participation in, and compliance with, agri-environmental schemes under conditions of moral hazard.

The chapter is organised into three further sections: Section 2 provides a mathematical model describing farmers' behaviour when faced with the joint decision regarding participation in, and compliance with, CC and AESs. Section 3 reports a case study, including results of simulations based on the model described in section 2 and Section 4 provides a discussion.

Model

A large body of literature treats the issue of compliance with CC or AES commitments through concepts related to either contracts that reduce moral hazard, or appropriately targeted policy (see for example Choe and Fraser 1999, Fraser 2005, Wu et al. 2001, Bartolini et al. 2005, Hart and Latacz-Lohmann 2005, Bartolini et al. 2008a, 2008b). In order to investigate the effects of the linkages between CC and AES commitments, a farmer behaviour model has been developed. This model is an extension of the agent portion of the model developed by Bartolini et al. (2008a) with the inclusion of the choice to participate in AES contracts, but without searching for an optimal contract design from the viewpoint of the public regulator. Let us assume that the farmer can produce some environmental good ε. There will be some level of environmental good that the farmer is obliged to produce through CC (let us say ε^c) without any additional payment (though compliance is a requirement in order to maintain the full single farm payment). If the farmer chooses to participate in an AES contract, he/she uptakes an higher obligation ($\varepsilon^a > \varepsilon^c$) and has the right to require an additional payment for such an obligation. We represent the cost of providing the environmental good as a function $\Psi_i(\varepsilon_i)$, with $\psi_i'(\varepsilon_i) \geq 0$, $\psi_i''(\varepsilon_i) \geq 0$, $\psi_i(0) = 0$ and $\psi_i'(0) = 0$, where i represents the farm (type).

We would then represent the costs as $\psi_c = \psi\left(\varepsilon^c\right)$ for CC and $\psi_{a+c} = \psi\left(\varepsilon^{a+c}\right)$ for the costs of both the CC and the agri-environmental commitment. Several qualifications apply to this general setting. First of all, CC commitments may be not homogeneous in terms of environmental goods or technical specifications with the attached AES commitment. In addition, they may be non-homogeneous in terms of the area of application, as CC must be applied to the whole farm, while some AES requirements apply to only a partial share of the farm. Secondly, farmers may be expected to be heterogeneous with respect to compliance costs. In addition, not all farmers are obliged to adhere to CC, as not all of them receive payments. Finally, if we assume asymmetric information, farmers may be subject to self-selection or partial compliance with respect to the theoretical constraints proposed by the public administration. This may apply independently to CC and AES, or jointly.

In this chapter we focus on homogeneous AES and CC requirements, applying to the same unit of land, but with heterogeneous groups of farmers. Our focus is on farm reactions and the consequences for policy design. We model compliance with a generic set of CC prescriptions through a continuous variable $e_i^c = [0,1]$, where e^c represents the degree of compliance for farm type i. In addition, a set of compliance with AES is modelled as a continuous variable $e_i^a = [0,1]$ where e^a represents the degree of compliance and i represents the farm type. Farmer i receives a single farm payment (SFP) P_i (average per hectare as the average of the whole farm land of each farm) determined by his historical payment entitlements. If the farmer is not compliant, a sanction is imposed. Farmers can also choose to be involved in AES. In this case they receive a fixed payment P_i^a per hectare for the provision of a given amount of public good. Such a payment is calculated as compensation for the income foregone plus the additional costs, based on average conditions in the area covered by the scheme. The farmer, receiving the SFP, will provide an environmental good ε_i^c that is equal to the degree of compliance (e_i^c) multiplied by the compulsory amount of environmental good under CC (ε_i^{c*}). The decision to provide some environmental good under the AES contract is structured in the same way. This means that farmer i receives agri-environmental payments, and in return will provide an amount of environmental good ε_i^a that is equal to the multiplication between commitments of AES (ε_i^a) and the degree of compliance (e_i^a). Under these conditions we could also re-define: $e_i^c = \frac{\varepsilon_i^c}{\varepsilon_i^{c*}}$ and note that under the assumption that CC and AES are linked and presented as a continuous functions and the compliance with both AES and CC may be written as: $e_i^{a+c} = \frac{\varepsilon_i^{a+c}}{\varepsilon_i^{a+c}}$. It should be noted also that, under this formulation, while the cost of participation in the AES is based only on the differential of costs between those imposed by the AES and those imposed by the CC, the degree of compliance with AES also implies compliance with CC (and not the other way round). Sanctions for non-compliance with CC or agri-environmental commitments are calculated in the same way, namely as a function of the payment ρP, where ρ represents the share of payment subtracted as a sanction. As the punishment for non-compliance relates to the right to receive the payments, in the model we always assume that $\rho \leq 1$. Sanctions for cheating CC standards is equal to $\rho^c P^c$ and sanctions for cheating agri-environmental commitments is equal to $\rho^a P^a$. Both ρ^c and ρ^a may be treated as policy design variables, as the regulator may have the option of changing/adapting its value to encourage compliance, including by differentiating it across farms. However, as this may create political/equity difficulties we assume that it cannot be differentiated among farms. The determination of the probability of non-compliance with both CC and AES commitments depends on a number of parameters, including some random effects (e.g. mistakes or weather conditions). We simplify the problem by calculating the probability of the non-compliance being detected based on two parameters: non-compliance (directly correlated), and monitoring (inversely correlated). We assume that the probability that non-compliance is detected, if some non-compliance exists, is equal to the degree of non-compliance, i.e. $(1-e_i^c)$ for CC and $(1-e_i^a)$ for AES. Monitoring intensity (m^a) is the expectation that a farm is monitored respectively for CC and for AES. An estimate of this may be reasonably derived by the announced percentage of farms monitored each year, which may be represented by a value of between 0 and 1. We use it as a direct representation of the probability that non-compliance is detected. In other words, in any farm monitored, if non-compliant, non-compliance is detected with probability of $(1-e_i)$ for CC and $(1-e^a)$ for AES; the total probability that non-compliance is detected is equal to $m^c(1-e_i^c)$ for CC and $m^a(1-e_i^a)$ for AES.

The farmer behaviour model can be presented as a problem in which the farmer i has the option to choose one strategy among a set of four possible strategies, based on the maximizations of the expected farm profit. Formally the farmer behaviour model may be presented as:

$$\pi_i = \max\left(\pi_i^0, \pi_i^c, \pi_i^a, \pi_i^{a+c}\right) \qquad (1)$$

With:
π_i^0 = expected farm profit when the farmer is involved in neither SFP or AES;
π_i^c = expected farm profit when the farmer receives SFP and must apply CC commitments;
π_i^a = expected farm profit when the farmer participates in AES;
π_i^{a+c} = expected farm profit when farmer receives SFP, and participates in AES.
The profit function when farmer i is not involved in either SFP or AES is identified as: $\pi_i^0 = -\psi_i\left(\varepsilon_i^0\right)$. It is intuitive that without being involved in SFP or AES the optimal amount of environmental good provided by farmer i is equal to zero. The expected profit when farmer i is involved only in SFP can be determined by the sum of the profit function in case the non-compliance is not detected, and the profit function in case the non-compliance is detected, each one multiplied by the respective probability (Bartolini et al. 2008a):

Following Bartolini et al. (2008a), the first order condition with respect to the optimal level of compliance is:

$$\pi_i^c = \left(1 - \left(m^c\left(1 - e_i^c\right)\right)\right)\left(P_i^c - \psi_i\left(e_i^c \varepsilon_i^{c*}\right)\right) + \left(m^c\left(1 - e_i^c\right)\right)\left(P_i^c - \rho^c P_i^c - \psi_i\left(e_i^c \varepsilon_i^{c*}\right)\right)$$

$$\frac{\partial \pi_i^c}{\partial e_i^c} = -\varepsilon_i^{c*}\psi_i{}'\left(e_i^c \varepsilon_i^{c*}\right) = m^c \rho^c P_i^c = 0$$

(2)

which yields:

$$\psi_i{}'\left(e_i^c \varepsilon_i^{c*}\right) = \frac{m^c \rho^c P_i^c}{\varepsilon_i^{c*}}$$

The optimal level of compliance depends on the monitoring intensity, the level of sanctions and the payments. When any of the three is zero, the marginal cost of compliance (hence compliance) will be zero.

The expected profit when farmer *i* is involved in AES, similarly to the case of SFP, may be determined by two components: the profit function in case the non-compliance with respect to agri-environmental commitment is not detected, and the profit function in case the non-compliance with respect to the agri-environmental commitment is detected, each one multiplied by the respective probability:

$$\pi_i^a = \left(1 - \left(m^a\left(1 - e_i^a\right)\right)\right)\left(P_i^a - \psi_i\left(e_i^a \varepsilon_i^{a*}\right)\right) = \left(m^a\left(1 - e_i^a\right)\right)\left(P_i^a - \rho^a P_i^a\right) - \psi_i\left(e_i^a \varepsilon_i^{a*}\right)$$

$$\pi_i^a = P_i^a - \psi_i\left(e_i^a \varepsilon_i^{a*}\right) + \left(m^a\left(1 - e_i^a\right)\right)\left(-\rho^a P_i^a\right)$$

$$\frac{\partial \pi_i^a}{\partial e_i^a} = -\varepsilon_i^{a*}\psi_i{}'\left(e_i^a \varepsilon_i^{a*}\right) + m^a \rho^a P_i^a = 0$$

$$\psi_i{}'\left(e_i^a \varepsilon_i^{a*}\right) = \frac{m^a \rho^a P_i^a}{\varepsilon_i^{a*}}$$

(3)

The profit function may be simplified to:

$$\pi_i^a = P_i^a - \psi_i\left(e_i^a \varepsilon_i^{a*}\right) + \left(m^a(1 - e_i^a)\right)\left(-\rho^a P_i^a\right)$$

Under first order conditions, the optimal level of compliance with AES is determined by:

$$\frac{\partial \pi_i^a}{\partial e_i^a} = -\varepsilon_i^{a*}\psi_i{}'\left(e_i^a \varepsilon_i^{a*}\right) + m^a \rho^a P_i^a = 0$$

which yields:

$$\psi_i{}'\left(e_i^a \varepsilon_i^{a*}\right) = \frac{m^a \rho^a P_i^a}{\varepsilon_i^{a*}}$$

The optimal amount of environmental good depends on the level of monitoring intensity of AES, the level of sanctions and AES payments. As for the SFP case, when any of the three is zero, the marginal cost of compliance with the agri-environmental commitments (hence compliance) will be zero.

The expected farm profit when farmer *i* is involved in both SFP and AES can be interpreted as the sum of equation 2 and 3, plus the increase in probability to be monitored in the CC commitments as a result of the adoption of AES. As in equation 2 and 3, the expected profit is determined by two factors: the profit function in case the non-compliance with respect to CC and agri-environmental commitment is not detected, and the profit function in case the non-compliance with respect to CC and agri-environmental commitment is detected, each one multiplied by the respective probability:

$$\pi_i^{a+c} = (1-e_i^{a+c})[(1-(m^c(1-e_i^c)+m^a(1-e_i^{a+c})))(P_i^c - \psi_i(e_i^c \varepsilon_i^{c*})) + (m^c(1-e_i^c)+m^a(1-e_i^{a+c}))(P_i^c - \rho^c P_i^c - \psi_i(e_i^c$$
$$+[(1-(m^c(1-e_i^{a+c})))(P_i^a - \psi_i(e_i^{a+c}\varepsilon_i^{a+c*})) + \psi_i(e_i^c \varepsilon_i^{c*})) + (m^a(1-e_i^{a+c}))(P_i^a - \rho^a P_i^a - \psi_i(e_i^{a+c}\varepsilon_i^{a+c*}) + \psi_i(e_i^c \varepsilon_i^{c*}))]] \quad (4)$$

This expected farm profit may be simplified as:

$$\pi_i^{a+c} = P_i^c + P_i^a - \psi_i(e_i^{a+c}\varepsilon_i^{a+c*}) +$$
$$+ (m^c(1-e_i^c)+m^a(1-e_i^{a+c}))(-\rho^c P_i^c)+m^a(1-e_i^{a+c})(-\rho^a P_i^a) \quad (5)$$

As a result of the simultaneity of adoption of CC and AES in equation 5 the compliance with respect to cross-compliance (e_i^c) can be redefined as $e_i^c = \frac{\varepsilon_i^{a+c}}{\varepsilon_i^c}$ which represents the ratio between CC compliance commitments and environmental good produced as a consequence of the fact that the farmer receives SFP. Under this formulation variable e_i^c must have an upper limit $e_i^c = 1$, which implies the complete fulfilment of CC commitments. In the cases in which $e_i^c = 1$, the expected farm profit can be further simplified as:

$$\pi_i^{a+c} = P_i^c + P_i^a - \psi_i(e_i^{a+c}\varepsilon_i^{a+c*}) + m^a(1-e_i^{a+c})(-\rho^c P_i^c)+m^a(1-e_i^{a+c})(-\rho^a P_i^a)$$

$$\frac{\partial \pi_i^{a+c}}{\partial e_i^{a+c}} = -\varepsilon_i^{a+c*}\psi_i'(e_i^{a+c}\varepsilon_i^{a+c*})+m^a(\rho^a P_i^a + \rho^c P_i^c)=0$$

$$\psi_i'(e_i^c \varepsilon_i^{c*}) = \frac{m^a \rho^a P_i^a}{\varepsilon_i^{a*}} \quad\quad\quad (6)$$

Under first order condition, the optimal level of compliance with AES is determined by:

$$\frac{\partial \pi_i^{a+c}}{\partial e_i^{a+c}} = -\varepsilon_i^{a+c*}\psi_i'(e_i^{a+c}\varepsilon_i^{a+c*})+m^a(\rho^a P_i^a + \rho^c P_i^c)=0$$

Which yields:

$$\psi_i'(e_i^{a+c}\varepsilon_i^{a+c*}) = \frac{m^a(\rho^a P_i^a + \rho^c P_i^c)}{\varepsilon_i^{a+c*}}$$

The optimal level of compliance with CC plus agri-environmental commitments (with fulfils CC commitments) is dependent on the monitoring intensity of AES, and the expected reductions of CAP and agri-environmental payments.

Otherwise, when the CC commitment is not fulfilled ($e_i^c < 1$), the compliance variable can be written as: $e_i^c = \delta e_i^{a+c}$ where $\delta = \frac{\varepsilon_i^{a+c}}{\varepsilon_i^c}$ which represents the ratio between the agri-environmental commitments and CC commitments. This implies that the expected farmer profit, written in equations 5, can be simplified as:

$$\pi_i^{a+c} = P_i^c + P_i^a - \psi_i\left(e_i^{a+c}\varepsilon_i^{a+c*}\right) +$$
$$+ \left(m^c\left(1-\partial e_i^{a+c}\right) + m^a\left(1-e_i^{a+c}\right)\right)\left(\rho^c P_i^c\right) + m^a\left(1-e_i^{a+c}\right)\left(\rho^a P_i^a\right)$$

$$\frac{\partial \pi_i^{a+c}}{\partial e_i^{a+c}} = -\varepsilon_i^{a+c*}\psi_i'\left(e_i^{a+c}\varepsilon_i^{a+c*}\right) + m^c\partial\left(\rho^c P_i^c\right) + m^a\left(\rho^a P_i^a + \rho^c P_i^c\right) = 0$$

$$\psi_i'\left(e_i^{a+c}\varepsilon_i^{a+c*}\right) = \frac{\left(m^c + m^a\right)\rho^c P_i^c + m^a\rho^a P_i^a}{\varepsilon_i^{a+c*}} \qquad (6')$$

Under first order condition, the optimal level of compliance with both CC and AES is determined by:

$$\frac{\partial \pi_i^{a+c}}{\partial e_i^{a+c}} = -\varepsilon_i^{a+c*}\psi_i'\left(e_i^{a+c}\varepsilon_i^{a+c*}\right) + m^c\delta\left(\rho^c P_i^c\right) + m^a\left(\rho^a P_i^a + \rho^c P_i^c\right) = 0$$

Which yields:

$$\psi_i'\left(e_i^{a+c}\varepsilon_i^{a+c*}\right) = \frac{\left(m^c + m^a\right)\rho^c P_i^c + m^a\rho^a P_i^a}{\varepsilon_i^{a+c*}}$$

The optimal level of compliance (sum of compliance with respect to CC and AES commitments) depends on both monitoring intensity and on both levels of sanctions. It is worth noting that the linkage between AES and CC results in an increase in the "weight" of the sanctions of the CC with respect to the sanctions of the AES.

An example

The model described in the previous paragraph has been tested in an area of the Emilia-Romagna Region (Northern Italy), with the main aim of providing a numerical exemplification of the outcome of the model. The area analysed is the municipality of Argenta, which is located, in its entirety, in a nitrate vulnerable zone. As a result of this zoning, all farmers must be compliant with the commitments of the nitrate directive (directive 676/91 CEE), which stipulates compliance with a maximum level of nitrogen use (170 kg per ha). Furthermore, with the implementation of the 2003 CAP reform, restrictions due to the nitrate directive have been included as Statutory Management Requirements (SMR) in CC commitments. This implies that those farmers that receive SFP and are located in nitrate vulnerable areas must be compliant with the restriction on nitrogen use. Furthermore, in the design of AES, in particular for the input reduction measures (indeed nitrogen reduction plays a major role in this measure) the CC prescriptions have been identified as a baseline. This means that the agri-environmental payment has been tailored as compensation for the additional costs arising from the adoption of agricultural practices that go beyond CC commitments. In fact, in the justification of AESs in the 2007-2013 program the Emilia-Romagna Region has identified the nitrogen prescription for CC as the baseline for AESs (Emilia-Romagna Region 2007) for these nitrogen reduction schemes.

An average cost function for nitrogen reduction in the municipality of Argenta has been adopted: $\overline{\psi}(a) = 0.0288a^2 + 0.0814a$
where:

a = reduction of nitrogen use (kg per ha);

$\overline{\psi}(a)$ = average compliance cost.

The cost function has been derived through linear programming modelling of farmer behaviour in the area of the municipality of Argenta, through a parameterization of the amount of nitrogen available to the farm (Bartolini et al. 2007).

For the purpose of discussing the effects of the adoption of CC and AES commitments, the diversification among farms is a critical issue. In this case, it has been assumed that cost functions of different farms may be obtained as a fixed proportion of the average cost function in the area, assuming a range of plus or minus the percentage (g). For this purpose a differentiation of plus or minus 10 per cent, 40 per cent, 70 per cent with respect to the average cost function has allowed for the identification of six different farm types with respect to the cost of nitrogen reduction.

The farms are only differentiated by way of the cost function of nitrogen use reduction; other parameters, such as SFP, AES payments and both CC and AES commitments are assumed to be the same among farms. SFP has been estimated to be EUR100/ha, whereas AES payments have been estimated to be EUR70/ha, and the amount of nitrogen use reductions is 50kg/ha for CC and 100 kg /ha for AES + CC. The values of these parameters are not different from the real applications of cross-compliance and agri-environmental schemes concerning input reduction measures in Emilia-Romagna (Emilia-Romagna Region 2007, De Roest 2008, Canatossi and Ansovini 2008).

Following the prescriptions of MIPAF (2008), the percentage of payment reductions in the case of total non-compliance with CC and AES commitments is equal to 100 per cent of the payments (which means that $\rho^c = 1$ and $\rho^a = 1$). Simulations have been undertaken using GAMS software .

Figure 21.1 presents the results of simulations for the six farm types when $m^c = 0.0$ (monitoring intensity of CC) and with a sensitivity analysis of between 0 and 1 of monitoring intensity of AES.

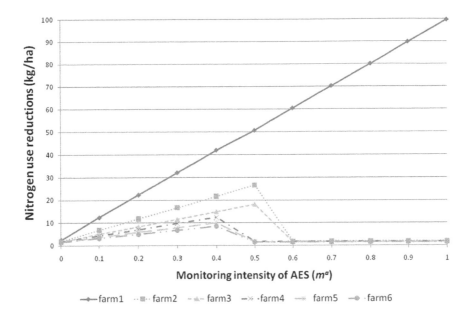

Figure 21.1 Nitrogen use reductions (SFP and AES) with $m^c = 0.0$

With a monitoring intensity level equal to 0.01 (which represents the current level of monitoring intensity),[1] the participation in AES and the compliance with CC commitments are differentiated among farm types. The participation in AES and the compliance with CC commitments changes strongly across farmers increasing the value of the monitoring intensity of AES (m^a). For all levels of m^a all farm types receive SFP, but farmers are generally not compliant with CC prescriptions, let alone AES commitments. Their compliance reflects some level of nitrogen reduction. Farmers are generally compliant with AES commitments only when there are low values of AES monitoring intensity, but they tend to cheat with both the CC and AES commitments. Only the farm type with the lowest compliance costs (farm1) is fully compliant with the CC commitments for a high value of monitoring intensity of AES ($m^a > 0.5$), but not completely with regard to AES commitments. By increasing the monitoring intensity of AES values, the nitrogen use reduction for all other farm types increases until farmers decide to remain involved in AES. For further increases of m^a (higher than 0.5) the quantity of nitrogen reduction for farmer types 2-6 are the same as in the case of not monitoring agri-environmental schemes ($m^a = 0$), as they prefer to avoid up-taking the AES.

---

    1   Value of monitoring intensity in Italy is around 1 per cent of the farm that benefited of the SFP (Catanossi and Ansovini 2008).

Table 21.1 presents the results of simulations (nitrogen reduction) in the six farm types when $m^c = 0.1$ and when $m^c = 0.5$, compared to the case of $m^c = 0.0$.

Table 21.1 Changes in nitrogen use reduction increasing monitoring intensity ($m^c = 0.1$ and $m^c = 0.5$)

m^a	farm1		farm2		farm3		farm4		farm5		farm6	
	$m^c=0.1$	$m^c=0.5$	$m^c=0.1$	$m^c=0.5$	$m^c=0.1$	$m^c=0.5$	$m^c=0.1$	$m^c=0.5$	$m^c=0.1$	$m^c=0.5$	$m^c=0.1$	$m^c=0.5$
	(kg/ha)	(kg/ha)	(kg/ha)	(kg/ha)	(kg/ha)	(kg/ha)	(kg/ha)	(kg/ha)	(kg/ha)	(kg/ha)	(kg/ha)	(kg/ha)
0	10.42	47.43	5.21	28.36	3.47	18.90	2.84	15.47	2.23	12.15	1.84	10.01
0.1	10.42	37.59	5.21	28.36	3.47	18.90	2.84	15.47	2.23	12.15	1.84	10.01
0.2	10.41	27.75	5.21	28.36	3.47	18.90	2.84	15.46	2.23	12.15	1.84	10.01
0.3	10.42	17.92	5.21	28.36	3.47	18.90	2.84	15.47	2.23	12.15	1.83	10.00
0.4	8.08	8.08	5.21	28.33	3.47	18.90	2.84	15.47	2.24	12.16	1.84	10.01
0.5	-	-	5.21	23.41	3.47	18.90	16.26	28.88	2.23	22.69	1.84	18.69
0.6	-	-	34.72	48.01	3.47	38.58	2.84	31.56	2.23	12.15	1.84	10.01
0.7	-	-	5.21	48.01	3.47	41.86	2.84	15.47	2.23	12.15	1.84	10.01
0.8	-	-	5.21	48.01	3.47	18.90	2.84	15.47	2.23	12.15	1.84	10.01
0.9	-	-	5.21	28.36	3.47	18.90	2.84	15.47	2.23	12.15	1.84	10.01
1	-	-	5.21	28.36	3.47	18.90	2.84	15.47	2.23	12.15	1.84	10.01

The effect of increasing the monitoring intensity of cross-compliance can induce farmers to have higher reductions in nitrogen use than with $m^c = 0.0$. In fact with $m^c = 0.1$ the minimum level of nitrogen reductions is double than with lower monitoring intensity, while with $m^c = 0.5$ the minimum level of nitrogen reduction is ten times higher than the level of nitrogen use reductions with lower monitoring intensity. Table 21.1 shows that the compliance under different levels of monitoring intensity (increasing ten and fifty times) has the same tendency, which is also the same as in Figure 21.1. As Figure 21.1 shows, the increasing increments of AES monitoring intensity induce farmers to increase nitrogen reduction, until they decide to remain involved in AESs. However, compared to Figure 21.1, with a CC monitoring intensity level of 0.5, the farms with low compliance costs (farmers 1 and 2) are generally compliant with CC commitments. The highest value of nitrogen reduction is particularly visible for farm 1, which with all AES monitoring intensity values, is fully compliant with CC commitments. This means that by increasing the monitoring intensity of CC from $m^c = 0.0$ to $m^c = 0.1$ the detection of CC commitment non-compliance is quite low.

Discussion

This chapter provides a theoretical analysis of farmer behaviour when cross-compliance and agri-environmental measures are linked and jointly designed, with an application to a case study in Emilia-Romagna (Italy). Through the modelling of the choices related to nitrogen use reduction, it was possible to investigate the effects of the different monitoring process designs on farmers' compliance when cross-compliance and agri-environmental schemes are jointly implemented. The results show that all farm typologies are interested in receiving SFP, but only a small portion is compliant with cross-compliance commitments. Furthermore, all farmer typologies are interested in participating in agri-environmental schemes, yet they are even more non-compliant with agri-environmental scheme commitments than with cross-compliance commitments.

The results confirm that compliance can be increased with an intensification of CC commitment monitoring. On the contrary, it is not possible to derive the same direct proportionality between AES monitoring intensification and nitrogen reductions, as a result of the additional commitments involved and the voluntary structure of agri-environmental policy.

In the monitoring and sanction conditions assumed in the chapter, the results show that farmers use agri-environmental payments to pay the cost of CC commitments. This consideration can be derived by observing that given the fixed value of the monitoring intensity of cross-compliance commitments and the increased monitoring intensity of AES, farmers tend to reduce their use of nitrogen.

Monitoring intensity in agri-environmental schemes and cross-compliance should be more accurately designed and, in order to increase effectiveness, the monitoring and environmental prescriptions themselves should be considered more explicitly as variables to be adapted to incentive compatibility criteria.

This chapter was an attempt to investigate the effects of simultaneous decisions at farm level regarding participation in cross-compliance and agri-environmental schemes. No asymmetric information between the public administrations and the farmers has been taken into account in the model, but in order to provide an optimal contract design the structure of the principal-agent model as well the inclusion of asymmetric information could provide useful improvements to the model. While the main compliance related issues are caught by the model, other elements, which reflect the simplified assumptions made with respect to the complexity of policy design, cross-compliance and agri-environmental commitments, cost structure, asymmetric information, considerations of the whole farm surface and other prescriptions, could be investigated with further research.

References

Bartolini, F., Gallerani, V., Raggi, M. and Viaggi, D. 2007. Implementing the Water Framework Directive: Contract Design and the Cost of Measures to Reduce Nitrogen Pollution From Agriculture. *Environmental Management* 40(4), 567-577.

Bartolini, F., Gallerani, V., Raggi, M. and Viaggi, D. 2005. *Contact design and targeting for the production of public goods in agriculture: the impact of the 2003 CAP reform.* Proceedings of 11th EAAE congress: *The future of Rural Europe in the global agri-food system.* Copenhagen, Denmark 24-27 August 2005, available at http://ageconsearch.umn.edu/bitstream/24559/1/cp05be04.pdf [accessed: 01 June 2010].

Bartolini, F., Gallerani, V., Raggi, M. and Viaggi, D. 2008a. Modelling the effectiveness of cross-compliance under asymmetric information: Paper tat the *Proceeding of 107th EAAE Seminar.* Luxemburg: Office for official publications of the European Communities, 291-302.

Bartolini, F., Gallerani, V., Raggi, M. and Viaggi, D. 2008b. *Effectiveness of cross-compliance under asymmetric information and differentiated compliance constraints.* Proceeding of 12th EAAE Congress: *People, Food and Environments: Global trends and European Strategies.* Ghent, Belgium 25-29 August 2008, available at http://ageconsearch.umn.edu/bitstream/44151/2/210.pdf [accessed: 01 June 2010].

Bennett, H., Osterburg, B., Nitsch, H., Kristensen, L., Primdahl, J. and Verschuur, G. 2006. Strengths and Weaknesses of Cross-compliance in the CAP, *EuroChoices*, 5(2), 50-57.

Canatossi, C. and Ansovini, G. 2008. Scattano i controlli in azienda sul rispetto della condizionalità, *L'informatore Agrario*, 5, 24-27.

Choe, C. and Fraser, I. 1999. Compliance monitoring and agri-environmental policy, *Journal of Agricultural Economics*, 50(3), 468–487.

Davies, B.B. and Hodge, I.D. 2006. Farmers' Preferences for New Environmental Policy Instruments: Determining the Acceptability of Cross Compliance for Biodiversity Benefits, *Journal of Agricultural Economics*, 57(3), 393-414.

De Roest, K. 2008. La condizionalità non è un grande costo, *Agricoltura*, Aprile, 51-53.

Dobbs, T.L., and Pretty, J.N. 2004. Agri-Environmental Stewardship Schemes and "Multifunctionality", *Review of Agricultural Economics*, 26(2), 220-237.

Dupraz P. Vanslembruck I., Bonnieux F. and Van Huylenbroeck G. (2002). *Farmers' participant in European agri-environmental policies.* Proceeding of 10th EAAE Congress. Saragozza, Spain, 28-31 August 2002, available at http://ageconsearch.umn.edu/bitstream/24799/1/cp02du57.pdf [accessed: 01 June 2010].

DG Agricultural ERR, 2007. *Metodologia di Calcolo dei sostegni delle misure dell'asse 2* [online Directorate General Agricultural Emilia-Romagna Region]. available at http://www.ermesagricoltura.it/Piano-Regionale-Sviluppo-Rurale/ Programma-di-Sviluppo-rurale-2007-2013 [accessed: 01 June 2010].

Feinerman, E. and Komen, M.H.C. 2003. Agri-environmental instruments for an integrated rural policy: An economic analysis, *Journal of Agricultural Economics*, 54 (1), 1-20.

Fraser, R. 2005. Moral hazard and risk management in agri-environmental policy, *Journal of Agricultural Economic*, 53(3), 475-487.

Fraser, I. and Fraser, R. 2005. Targeting Monitoring Resources to Enhance the Effectiveness of the CAP, *EuroChoices*, 4(3), 22-27.

Hart, R. and Latacz-Lohmann, U. 2005. Combating moral hazard in agri-environmental schemes: a multiple-agent approach, *European Review of Agricultural Economics*, 32(1):75-91.

Helin, J. 2008. *Environmental protection of agriculture -clash of policies?* Proceedings of 107th EAAE Seminar: Modelling Agricultural and Rural Development Policies, Seville. 29 January-1 February 2008, available at: http://ageconsearch.umn.edu/bitstream/6468/2/cp08he17.pdf [accessed: 01 June 2010].

Hodge, I. 2000. Agri-environmental Relationships and the Choice of Policy Mechanism, *The World Economy* 23(2), 257-273.

Latacz-Lohmann, U. 2004. Dealing with limited information in design and evaluating agri-environmental policy. Part 1, in *Multifunctional agriculture, policies and markets: understanding the critical linkages*, edited by AA VV, Rennes, France: Métropole, 33-50.

MIPAF. 2008. *Disposizioni in materia di violazioni riscontrate nell'ambito del Regolamento (CE) n. 1782/03 del Consiglio del 29 settembre 2003 sulla PAC e del Regolamento (CE) 1698/2005 del Consiglio, del 20 settembre 2005, sul sostegno allo sviluppo rurale da parte del Fondo europeo agricolo per lo sviluppo rurale (FEASR),* [online: Ministero delle Politiche Agricole Alimentari e Forestali]. Available at: http://www.reterurale.it/flex/cm/pages/ ServeBLOB.php/L/IT/IDPagina/280 [accessed: 01 June 2010].

Osterburg, B., Nitsch, H. and Kristensen, L. 2005. *Environmental Standards and their linkage to support instruments of the Common Agricultural Policy.* Paper tat the *Proceedings of the 11ᵗʰ EAAE congress: The future of Rural Europe in the global agri-food system.* Copenhagen, Denmark 24-27 August 2005. Available at: http://ageconsearch.umn.edu/bitstream/24521/1/pp01os01.pdf [accessed: 01 June 2010].

Varela-Ortega, C. and Calatrava, J. 2004. *Evaluation of Cross-Compliance: perspective and implementation.* Seminar on EU concerted Action: *Developing cross-compliance background, lesson and opportunities.* Granada, Spain 19-20 April. Available at: http://www.ieep.eu/publications/pdfs/crosscompliance/seminar4report.pdf [accessed: 01 June 2010].

Wu, J. Zilberman, D. and Babcock, B.A. 2001. Environmental and Distributional Impacts of Conservation Targeting Strategies, *Journal of Environmental Economics and Management*, 41(3), 333-350.

PART 7
Rural Development Issues

Chapter 22

Reforming Pillar II –Towards Significant and Sustainable Rural Development?

Holger Bergmann, Thomas Dax, Vida Hocevar, Gerhard Hovorka, Luka Juvančič, Melanie Kröger and Kenneth J. Thomson

Introduction

In 1997, the EU Agricultural Council adopted a set of conclusions in which it developed the basics of the concept of the European Model of Agriculture. As part of the European Strategy for Sustainable Development based on the decisions of the European Council in Göteborg (June 2001), environmental dimensions were added to the social and economic ones. In the same year, the Agricultural Council integrated environmental and sustainable development as political terms and targets into the Common Agricultural Policy (CAP), and it adopted the European Model of Agriculture and particularly the concept of the multifunctionality of agriculture as a core basis of European farming policy.

As the statement of the Finnish presidency (2006, 6) shows:

> Multi-functionality is at the heart of the European Model of Agriculture. This means that together with competitive food, fibre and energy production farming also delivers other services for society as a whole. These services, which are closely linked to food and fibre production, include safeguarding viable rural societies and infrastructures, balanced regional development and rural employment, maintenance of traditional rural landscapes, bio-diversity, protection of the environment, and high standards of animal welfare and food safety. These services reflect the concerns of consumers and taxpayers. As European farmers provide these multifunctional services for the benefit of society as a whole, which often incur additional costs without a compensating market return, it is necessary and justified to reward them through public funds. In most European countries family farms are the key element in fulfilling the objectives".[1]

---

1 For a discussion of the functions of agriculture and policies that influence the provision of goods and bads as well as environmental and social services that agriculture is likely to provide, see Bergmann and Thomson (2007).

While in most developed countries the farming sector is in decline (OECD 2006, 8), it remains vital for many remote and peripheral areas. Indeed, in such areas it is often one of the most important economic sectors, and provides incomes, employment and quality of life for both farm households and the broader public. For urban and peri-urban areas, the most important functions of agriculture are the provision of eco-system services and recreational areas, generally in the form of public goods (Weber et al. 2008).

When the European Model of Agriculture was developed in the late 1990s, there was a widely shared understanding that agricultural policy should be modified in order to support functions or roles of agriculture that go beyond the production of food and contribute to the sustainable development of rural areas. Besides the primary targets of farming within the economic development process (provision of food in the first place, and also income and employment opportunities), such roles of agriculture as the provision of eco-system services, landscapes, renewable energies and the social viability of rural communities have become more and more important (cf. Van Huylenbroeck et al. (2007, 7f).

The TOP-MARD Project

The main target of the EU FP6 research project TOP-MARD was the development of the concept of multifunctionality as helping to analyse rural development policy with a focus on the economic, social, cultural and environmental context on a territorial scale. Its approach explicitly analyses:

* regions rather than nations or individual farms
* the links between rural development and agricultural policies

public goods and services using a systems dynamics model, POMMARD. The project thus filled a gap alongside other approaches, e.g. the Roles of Agriculture (FAO 2002) and Multifunctionality within the New Rural Paradigm (OECD 2006).

The POMMARD Model

Structure and Development of POMMARD

The POMMARD model was built with the Stella© software (ISEE, 2007), and represents stocks and flows using user-defined variables, parameters, equations and time periods. It can be used to simulate the behaviour of a rural region as a whole (i.e. not individual farms or other businesses) in terms of its demography, economy, environment and Quality of Life (QoL) over a number of years (at least 15, in the case of TOP-MARD). It contains 11 modules: Land Use (see below), Agriculture, Non-Commodity Outputs or NCOs (environmental), Economy,

Investment, Human Resources (demography), Quality of Life, and Tourism, together with Initial Conditions, Scenario Controls and Indicators (i.e. major model results). Figure 22.1 depicts the model structure.

The modelling approach behind POMMARD is based on Johnson (1986) and Leontief (1953), in which dynamic regional shifts are included in a localised input-output (IO) table. The initial productionist IO approach was developed to include regional specific Social Accounting Matrices (SAM), different capitals (e.g. institutional), and Quality of Life (QoL) indicators (Bryden et al. 2008).

The primary engines of the POMMARD model are final demands by economic sector (23 in the core model) and land use by up to eight agricultural (and other, e.g. forestry) production systems. Such use, specified by shares of total regional area, determines the amounts of labour employed in these systems, and the output of farm commodities and environmental non-commodities. The regional economy is modelled via an IO table to which a "households" row and column are added, while the Investment module modifies the capacity of each sector. However, unlike many models of economic relationships, the model is partially supply-oriented, insofar as agricultural activity supplements other demand drivers.

The regional human population is modelled in some detail, e.g. four age groups and six educational levels: in and after primary (age 14), secondary (age 19), and tertiary education, respectively (age 22). These age-education cohorts are represented in the employment and migration vectors.

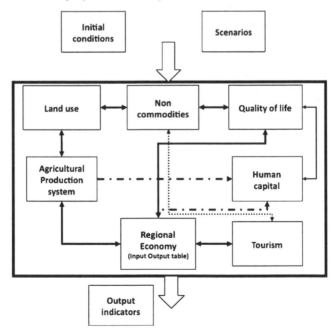

Figure 22.1   The Structure of the POMMARD Model

Source: Bergmann and Thomson (2008, 4)

The core version of POMMARD was under development throughout 2006 and 2007, Calibration in POMMARD is basically carried out by changing the most important demographic coefficients so that the whole model results in a "better" projection (see Bergmann and Thomson 2008). Such calibration was done by comparing official data between 2001 and 2007 with the results that POMMARD delivered for that period. In most cases (Germany, Scotland, Sweden), the calibration needed adaptation in the labour force participation rates, while in other cases the differences between the actual and estimated values were sufficiently small.

Output indicators employed in POMMARD

In general. the assessment of policies related to the multifunctionality of agriculture and rural development can be done with large numbers of indicators. For example, in the FP5 funded project DORA, more than 57 indicators were used, as previously done by Bryden (2002, 14f.) and Bryden et al. (2004). The EEA (European Environmental Agency) used 25 indicators to focus on the implications of policies for the status of biodiversity across Europe, and Eurofund (2008) employs more than 150 indicators to focus just on the assessment of Quality of Life.

In order to facilitate the interpretation of its results, the TOP-MARD project selected only 24 indicators, and Bergmann et al. (2007) argued that eight core indicators might be appropriate. In this chapter, the following core five categories and indicators will be used:

* Demographics – population size
* Farming – farm employment
* Economics – regional per capita income
* Population change – annual regional net migration balance
* Environmental quality – annual change in Biodiversity indicators

The case study areas

The case study areas (CSAs) that have been chosen for this comparison are:

* Pinzgau-Pongau (P-P; a tourism-dominated alpine area in North-Western Austria near Salzburg and the German border, NUTS3 Code: AT322),
* Wetterau (WE; an urbanised industrial area in the middle of Germany in the Bundesland Hessia near Frankfurt/Main, NUTS3 code: DE71E),
* Caithness-Sutherland (C&S, a remote rural peripheral area in the Far North of Scotland and a part of the Highlands and Islands, NUTS3 Code: Part of UKM61, LAU1)[2] and

---

2 Local Administrative Unit = formerly NUTS4

- Gorenjska (GK, a tourism and manufacturing dominated alpine area in the North of Slovenia near the Austrian border part of a new accession EU Member state, NUTS3 Code: SI022).

Table 22.1   Key Data for Case Study Areas, 2001

	Unit	Austria (NUTS3) Pinzgau-Pongau	Germany (NUTS3) Wetterau	Scotland (LAU1) Caithness and Sutherland	Slovenia (NUTS3) Gorenjska
Agriculture					
Number of farms	number	4,370	660	3,321	4,680*
Net farm income	EUR1,000	8.48	33.17	7.89	10.91*
Average ESU per farm	ESU	7.15	26.81	6.68	5.01*
Labour demand	head	4,510	1,408	2,325	5,420*
Farmed and Forested land	ha	176,410	36431	281,197	32,460
Demographics					
Population size	head	161,996	296,153	38,972	195,885
Under 20	head	42,361	63,847	9,177	45,457
Over 65	head	20,939	48,463	7,213	27,938
Net-migration annual flows	head	400	6,027	-100	0
population density in km²	km²	37.20	269.06	5.41	92.22*
Economics					
GVA per capita	EUR/head	22.2	33.4	10.0	9.9
GVA land use	EUR1,000	105,107	46,699	18,350	42,337
Regional employment	head	73,484	75,954	15,367	92,458

	Unit	Austria (NUTS3)	Germany (NUTS3)	Scotland (LAU1)	Slovenia (NUTS3)
		Pinzgau-Pongau	Wetterau	Caithness and Sutherland	Gorenjska
Environment					
Biodiversity indicators	none	373,757	66,359	281,193	45,252
Natural capital change	none	0	0	0	0
surface	ha	435,500	110,070	720,000	212,400*

*data for 2003

Source: Eurostat

The comparison of social, economic and ecological indicators between these four TOP-MARD CSAs reveals vast differences that are place-dependent (peri-urban, remote rural or peripheral; see Table 22.1). All areas have a lower population density than the relevant national average, are more or less rural insofar that agriculture has a large proportion of regional GVA, and are mountainous regions except for WE (Germany). The main functions of agriculture in all CSAs are to (a) produce food and fibre, (b) protect the environment, (c) ensure the social viability of rural areas, (d) guard rural culture, and (e) provide a basis for lifestyle choices (Thomson 2005).

For GK, the dominating roles are (a) and (c), and to some extent even the role of agriculture as a basis for rural development is present. On the other hand, the dominating roles of agriculture in WE are (b) to (d). The other two CSAs (C&S and P-P) can be found in between, e.g. C&S farming is basically a lifestyle choice, while in P-P it is (b) and (c).

Scenario specification and results

Scenario specification and calculation

The CAP reform of 2003 introduced decoupled "Single Farm Payments" (SFPs) and voluntary as well as compulsory "modulation". It is likely that the modulation instrument will see more use in future in that the current compulsory rate of 5 per cent will be raised. Speeches by the European Commissioner for Agriculture and Rural Development (Fischer-Boel 2008,3) indicate that:

- the common market organisations (e.g. the milk as well as the sugar market quotas) are to be phased out,

- Single Farm Payments should be paid to farmers, defined according to common sense... and
- "progressive" modulation (i.e. limiting the amount of SFPs paid to larger farms) may be introduced".

With savings used to address new challenges (e.g. climate change, bio-energy, water scarcity, biodiversity, increase social cohesion, etc.), the rural development Pillar II of the CAP will be strengthened.

It seems certain that there will be a shift in CAP expenditures towards Pillar II in order to strengthen environmental land management, rural development (including investments into the farming sector) and social cohesion (see Thomson and McGranahan 2008). The effect of this shift can be analysed with POMMARD.

Five scenarios were specified:

a. Baseline, based on EU expenditures 2001-06, including all changes that took place in 2006/7 (most prominently the introduction of SFPs, and an annual land use change defined as a trend based on the years 1991-2001),
b. "Axis 1", in which all funds being spent in Pillar II are spent in Axis 1 to improve the competitiveness of the agricultural sector,
c. "Axis 2", in which all funds being spent in Pillar II are spent in Axis 2 to provide agri-environmental goods and services as well as to support agriculture in less favoured areas.
d. "Axis 3", in which all funds being spent from 2007 onwards are spent in Axis 3 to improve the quality of life and competitiveness of rural areas.
e. "Modulation", in which Pillar I expenditures are decreased by 50 per cent and subsequently are spent in Pillar II under Axis 3.

The IO tables used referred to the year 2001 (or later in the case of the Slovenian case study area), and included EU expenditures for the years 2001 to 2006. The effects of this spending were calculated on the basis of assumptions on:

1. the economic sectors affected by each pillar's expenditures (e.g. in all four CSAs the assumption on Axis 2 expenditure was that it increases household incomes) and,
2. the leverage effect of spending under each Axis, e.g. EUR1 spent by the EU along Axis 3 attracts an additional EUR1 from the member state and EUR2 in terms of private investment.

Modelling the changes that came into effect in the year 2007 for the period 2007 to 2013 was done in a similar way, and the results were compared for each scenario by appropriate adding and subtracting of the effects that the expenditures had during the period 2001 to 2006.

All scenarios were adapted to local conditions and public expenditure patterns, to reflect the fact that in each of the CSAs the Pillar II measures are implemented with different regional coefficients and data but under common guidelines, affecting different input variables. For example, in Scotland and Slovenia, Axis 2 expenditures are shared between agri-environmental schemes and Less Favoured Area support, while in Germany the agri-environment is the target. In Austria, both schemes are characterised by a high level of support to mountain farms, underscoring the linkage of mountain farming to tourism (Dax and Hovorka 2004). Most other variables (e.g. land use change, birth rates, labour force participation rates, quality of life indicators, etc.) were estimated using time series analysis or available data from official statistical sources.

Results

Since the scales for each CSA differ to a large extent, all results in this section are calculated as a percentage of the Baseline results for the year 2015. While in Scotland and Austria the largest differences to the Baseline are up to 10 per cent, the largest effect of a scenario in the Wetterau is below 0.5 per cent, showing that in a largely urban fringe area the impact of EU policy changes is measurable but insignificant. On the other hand, in the more rural areas of P-P and C&S, the effects of policy changes are significant.

Specific case study area results: Pinzgau-Pongau

The highest increase in population size (Table 22.2) can be expected with the Axis 2 scenario, that increases the number of tourists visiting P-P, and would therefore create additional employment. On the other hand, the population would decrease with the Axis 1 and Axis 3 scenarios as an effect of the investment into sectors that need more capital per head (education, private services, etc.) compared to the additional demand for tourism labour as a result of the Axis 2 scenario. Rather surprisingly, there are no changes to agricultural labour demand in P-P over all scenarios.

Table 22.2   Scenario results for P-P. in 2015

Austria (2015)	Baseline	Axis 1	Axis 2	Axis 3	Modulation
Total Population	100.0%	99.7%	100.1%	99.8%	100.0%
Agric. Employment	100.0%	100.0%	100.0%	100.0%	100.0%
Per Capita Income	100.0%	99.4%	100.2%	99.4%	99.9%
Total Migration	100.0%	110.8%	97.0%	109.7%	101.5%
Biodiversity	100.0%	100.0%	100.0%	100.0%	100.0%

Source: Own calculations.

However, due to the fact that almost all labour in Austrian agriculture is provided through family households, provision of labour is hardly dependent on market forces over the long term, but is more determined by life-style choices and by intergenerational decisions to keep up farming (or not).

Per capita income as a measure of economic well-being over all scenarios is changed only to small amounts. The best scenario in Austria regarding this indicator is again the Axis 2 scenario in which a better environmental quality generates additional regional incomes through increased touristic demand.

Total annual net-migration is highest in the Axis 1 scenario at 111 per cent of the baseline and lowest in the Axis 2 scenario. This indicates that people tend to out-migrate less as the Axis 2 scenario significantly increases the local quality of life as well as developing new regional jobs.

The Biodiversity indicator does not change at all in P-P, because the environmental quality is good, and any measure that does not drastically change the environment has almost no effect for the region.

Overall, comparing the results of the five sets of scenario results for P-P. the most attractive option would be the Axis 2 scenario, followed by the Baseline and the modulation scenario, as in all three the population size stays stable (or increases), and per capita income levels increase or stay almost the same.

Specific case study area results: Wetterau

The WE results generally show only very small changes (<0.1 per cent) compared to the Baseline (see Table 22.3). The Axis 1 scenario would increase the population through increased investments into labour-saving technologies in agriculture, while population would decrease as the German Axis 2 measures mostly target the extensification of production systems. The highest degree of population increase can be found by measures undertaken under Axis 3, result which supports the assumption that the current LEADER measures are able to support rural viability to a small extent in the WE .

As in P-P, there are no changes to agricultural labour demand in WE over all scenarios. The same result can be found regarding per capita income. However, there is a decrease as a result of the Modulation scenario, as farm households lose a significant share of their household income, and this is only partly substituted by higher incomes of employees in non-agricultural sectors.

Table 22.3   Scenario results for WE in 2015

Germany (2015)	Baseline	Axis 1	Axis 2	Axis 3	Modulation
Total Population	100.00%	100.09%	99.99%	100.18%	100.05%
Ag Employment	100.00%	100.00%	100.00%	100.00%	100.00%
Per Capita Income	100.00%	100.00%	100.00%	100.00%	99.80%
Total Migration	100.00%	100.14%	99.86%	100.14%	100.41%
Biodiversity	100.00%	100.00%	100.00%	100.00%	100.00%

Source: Own calculations.

Total migration is negatively affected by the Axis 2 scenario, as decreased spending on economic investments in Axis 1 and Axis 2 in WE leads to a lower regional labour demand. The biodiversity indicator again shows no changes as German landscapes are highly regulated and therefore changes between the different land use categories (e.g. arable land, grassland, woodlands, etc.) are unlikely.

Overall, comparing the results of the five scenario runs for WE. the most attractive option would be the Axis 3 scenario, followed by the Axis 1 and Axis 3 scenarios as in all three the population sizes stays stable (or increases), and the per capita income increases or stays almost the same. The worst scenario seems to be the Modulation scenario in which per capita income drops while population increases somewhat.

Specific case study area results: Caithness and Sutherland   The C&S results (see Table 22.4) show the largest changes under all scenarios. Population would be significantly increased through increased investments into education and manufacturing by Axis 3 scenario, followed by a large increase effected by Axis 1 investments into machinery and other technology useful in the farming sector. Ss in the other CSAs, there are no changes to agricultural labour demand in C&S over all scenarios. Per capita income is decreased by the Axis 1 scenario by nearly 2 per cent as well as in the Modulation scenario, while it would be increase by 4 per cent in the Axis 2 scenario and by 1 per cent in the Axis 3 scenario

Table 22.4   Scenario results for C&S in 2015

Scotland (2015)	Baseline	Axis 1	Axis 2	Axis 3	Modulation
Total Population	100.0%	104.0%	100.4%	109.0%	102.9%
Ag Employment	100.0%	100.0%	100.0%	100.0%	98.3%
Per Capita Income	100.0%	98.2%	104.1%	101.0%	98.9%
Total Migration	100.0%	88.9%	83.0%	97.0%	107.2%
Biodiversity	100.0%	100.1%	100.1%	100.1%	100.3%

Source: own calculations.

Total migration is negatively and strongly affected by the Axes 1 and 2 scenarios, as decreased spending on economic investments leads to lower regional labour demand. However, the Modulation scenario would positively influence total migration by 2015, as it would be 7 per cent higher than the Baseline scenario. The biodiversity indicator sees its highest change with the Modulation scenario, probably indicating that a more diversified development approach in C&S would not only profit rural viability but also the environment.

Overall, comparing the results of the five scenario runs for C&S, the most attractive option would be the Axis 2 scenario, followed by the Axis 3 and Modulation scenario as in all three the population size increases, the per capita income increases, and the marginal change of the biodiversity indicator is significantly positive. The worst scenario under those presented would be the Axis 1 scenario, since, although it increases population size and the biodiversity indicators, it decreases the per capita income, making the regional population worse off than in the Baseline.

Gorenjska

The GK results are surprisingly similar to the results of the WE (see Table 22.5). This similarity is based on the scenario specification, as we assume that only CAP expenses are altered which represent under 10 per cent of all EU expenditures in rural areas compared to 90 per cent donated by the structural funds in Slovenia.

The Axis 2 scenario is likely to increase population size, indicating that preservation of farming and the environment in this marginal area preserves the settlement pattern. The Modulation scenario is likely to decrease population, caused by a significant number of farms being shut down. As in the other CSAs, there are no significant large-scale changes to agricultural labour demand in GK over all scenarios. However, the Modulation scenario again decreases labour demand, while the Axis 2 would increase it. Per capita income decreases by nearly 0.2 per cent in all Axis scenarios apart from the Axis 2 scenario in which increased population counteracts with the per capita income increase that is provoked by higher wages in the tourism sector than in the delivering farming sector.

Total migration is significantly negatively affected by the Modulation scenario, while all other scenarios reveal that annual net migration is higher, showing that the area becomes more attractive for potential in-migrants within each of the Axis 1 to Axis 3 scenarios. The biodiversity indicator sees its highest change with the Axis 2 scenario, suggesting that this might be a result of higher public support for environmental and spatial public goods.

Table 22.5   Scenario results for Gorenjska in 2015

Slovenia (2015)	Baseline	Axis 1	Axis 2	Axis 3	Modulation
Total Population	100.0%	100.0%	101.2%	100.0%	99.1%
Ag Employment	100.0%	100.0%	100.3%	100.0%	99.8%
Per Capita Income	100.0%	99.8%	100.4%	99.8%	99.5%
Total Migration	100.0%	101.0%	116.7%	100.5%	92.5%
Biodiversity	100.0%	100.0%	100.3%	100.0%	99.8%

Source: own calculation

Overall, comparing the results of the five scenario runs for GK the most attractive option would be the Axis 2 scenario, followed by the Axis 3 and Axis 1 scenario as in all three the population size increases, the per capita income increases and the marginal change of the biodiversity indicator is significantly positive. Probably as a sign of the not yet reached saturated development status in the other, richer, CSAs, there seems to be a need under the Slovenian circumstances first to invest into agriculture (Axis 1), the environment (Axis 2) and education/new employments (Axis 3) before a more diversified approach such as that modelled in the Modulation scenario should be chosen.

Discussion and conclusion

This chapter has presented a modelling approach that uses a holistic territorial approach to regional modelling in order to overcome the limitations of approaches that prefer a purely economic focus on questions related to rural development.

The results show that when a common specification is chosen, the results vary according to the countries of the CSA, and even more importantly – as the GK example shows – according to whether the member state is an "older" or a "newer" member.

Summarizing across the EU, the area-specific results show that:

- Axis 1 expenditure increases overall local employment more than the other three scenarios and may therefore help to ensure rural viability in farming areas. However, other components of sustainability, e.g. quality of life, and environmental quality, can be affected negatively.
- Axis 2 expenditure improves the environment as well as the quality of life in all areas, and leads to increases in local employment through multiplier effects.
- Axis 3 expenditure has positive effects in near-urbanised central European regions, but in peripheral regions is unlikely to be sustainable without continued EU support since better qualification is an additional out-migration push factor and most rural remote areas depend heavily on central governmental financial transfers to cover their cost.
- In Western European CSAs (part of the EU15), the Modulation scenario has positive effects on the local economy as well as not changing the economic position of agriculture, since with higher commodity prices farmers (even if factor prices increase as well) are likely to be compensated for loses of the SFP (a classical example that in the long term profit-seeking can have better effects than rent-seeking). The Modulation scenario in Slovenia shows that before a holistic approach to rural development can be chosen, regional pre-conditions such as those in the EU15 have to be reached.

The model results suggest that the local/regional effects of wider societal trends such as population movements, service-dominated work and commuting, and tourism diversification can be supported by EU policies but can not be reversed or even significantly changed in order to achieve more sustainability.

Furthermore, the results show that in highly developed rural areas such as C&S, P-P and WE, expenditures targeting Axis 3 are appropriate, while in GK the results suggest that, prior to extending Axis 3, steps should be undertaken to support the agri-environment through Axis 2.

References

Bergmann, H. and Thomson, K. 2008. Modelling Policies for Multifunctional Agriculture in a Remote EU Region (Caithness & Sutherland, Scotland UK), 107th EAAE Seminar in Seville (Spain), http://ageconsearch.umn.edu/handle/6596, accessed 17.02.08.

Bergmann, H., Dax, T., Hovorka, G. and Thomson, K. 2007. *Pluriactivity and Multifunctionality in Europe – a comparison between Scotland and Austria. Paper presented at the conference of the Austrian Agri-Economic Society*, Vienna, proceedings.

Bryden, J., Johnson, T., Refsgaard, K., Dax, T. and Arandia, A. 2008. Scientific Approach, Chapter 3 of the final TOP-MARD report to the EC, unpublished.

Bryden, J.M. 2002. *Rural Development Indicators and Diversity in the European Union. Conference proceeding: "Measuring Rural Diversity"*, November 21-22, 2002 Economic Research Service, Washington, DC, srdc.msstate.edu/measuring/bryden.pdf, accessed 17.02.08.

Bryden, J.M. and Hart, J.K. 2004. *A New Approach to Rural Development in Europe. Germany, Greece, Scotland and Sweden.* The Edwin Mellen Press. New York.

Dax, T. and Hovorka, G. 2004. I*ntegrated rural development in mountain areas,* in Sustaining Agriculture and the Rural Environment – Governance, Policy and Multifunctionality edited by F. Brouwer. Cheltenham, UK, p. 124-143.

Eurofund. 2008. EurLIFE interactive database on quality of life in Europe, http://www.eurofound.europa.eu/areas/qualityoflife/eurlife/index.php, accessed 17.08.08.

FAO (Food and Agriculture Organisation of the UN). 2002. *Analytical Framework – Roles of Agriculture – Socio-economic Analysis and Policy Implications of the Roles of Agriculture in Developing Countries.* Available at http://www.fao.org/es/esa/Roa/empirical_design_en.asp, accessed 18.08.08.

Finnish Presidency of the EU. 2006. *Background Paper for the Meeting of Ministers of Agriculture,* 25 July 2006. Available at http://www.eu2006.fi/eu_and_policy_areas/policy_areas/en_GB/agriculture_and_fisheries.

Fischer Boel, M. 2008. The CAP Health Check: straight ahead for responsive and sustainable farming, Speech/08/255, accessed 17.08.08.

ISEE Systems. 2007. Stella Version 9, Lebanon, NH 03766 USA. www.iseesystems.com, accessed 17.08.08.

Johnson, T.G. 1986. A Dynamic Input-Output Model for Small Regions. *Review of Regional Studies* 16 (1), 14-23.

Leontief, W.W. 1953. *Studies in the structure of the American economy.* Oxford University Press, 1953.

Organisation for Economic Cooperation and Development (OECD). 2006. *OECD Rural Policy Review* – The New Rural Paradigm Policies and Governance, Paris.

Thomson, K.J. 2005. *Decoupling and Cross-Compliance as Concepts and Instruments in Agricultural Multifunctionality.* Presented at the 93th EAAE seminar in Prague. Available at http://www.abdn.ac.uk/~pec208/index_files/Page442.htm, accessed 9th May 2010

Thomson, K.J. and McGranahan, A.D. 2008. Environment, Land Use and Amenities the New Dimension of Rural Development. *EuroChoices* 7(1), 30-37.

Van Huylenbroeck, G., Vandermeulen, V., Mettepenningen, E. and Verspecht, A. 2007. Multifunctionality of Agriculture: A Review of Definitions, Evidence and Instruments, Living Rev. *Landscape Res.* (on line) 1, 3. Available at http://www.livingreviews.org/lrlr-2007-3, accessed 18.08.08.

Weber, D., Bergmann, H. and Thomson, K.J. 2008. *Multifunctionality of Agriculture – Some remarks about the importance of different functions. UNESCO – International conference 2008 Sustainable Land Use and Water management* (Beijing, China) 8-10 October.

Chapter 23

Are Rural Development Strategies Coherent At Regional Level? Empirical Evidence From Italy

Teresa Del Giudice, Teresa Panico and Stefano Pascucci

Introduction

The 'second pillar' of the Common Agricultural Policy (CAP), formally introduced with EC Reg. No 1257/99, consists of a package of measures for rural development policy, which aims to facilitate the adaptation of agriculture to new realities and further changes in terms of market evolution, market policy and trade rules, consumer demand and preferences, and Community enlargement (Lowe et al. 2002, Dwyer et al. 2007, Thomson et al. 2010). Whereas all these changes affect not only agricultural markets but also local economies in rural areas, rural development policy aims at restoring and enhancing the competitiveness of rural areas and hence contributes to maintaining and creating employment in such areas, taking into account the need to support the multifunctional role of agriculture, the protection both of the environment and the natural and cultural heritage (Van der Ploeg et al. 2000, Marsden et al. 2001, Terluin 2003, Lowe et al. 2002, Van Huylenbroeck and Durand 2004, Dwyer et al. 2007). The reform of the CAP in June 2003 and April 2004 introduces major changes likely to have a significant impact on the economy across the whole rural territory of the European Union in terms of farm production patterns, land management methods, employment and the wider social and economic conditions in the various rural areas (Meester 2010). Accordingly, rural development policy has been further reformed to accompany and complement the market and income support policies and thus contribute to achieving relevant policy objectives. A new EC Regulation (no. 1698/2005) is the reference framework for the second pillar of the CAP. As specified below, it introduces many important changes for the implementation, programming, financial management and control framework for rural development programs (Dwyer et al. 2007, Thomson et al. 2010).

Thus, rural development policy has gained importance over time to enhance sustainable regional development paths, especially for European convergence regions,[1] which are characterized by a high percentage of rural areas with severe development gap (later referred as lagging regions, LRs). In this perspective we also aim to emphasize that different models of regional development could be introduced via rural development policy: (i) the agricultural development models, with a more specific sector-oriented approach, and (ii) the rural development models, with a more territorial (and systemic) approach[2] (OECD 1999; Van der Ploeg et al. 2000; European Commission 2003). Both development patterns imply complex strategies that involve coordination and complementarities among European Funds. To this end, EC Regulation no. 1698/2005 establishes that the EAFRD shall complement national, regional and local actions and that the assistance of the EAFRD shall be consistent with the objectives of economic and social cohesion policy. This means coordination with the European Regional Development Fund (ERDF), the European Social Fund (ESF), the Cohesion Fund (CF), the Community support instrument for fisheries, and the interventions of the European Investment Bank (EIB), and of other EU financial instruments. Therefore a complete and exhaustive analysis of agricultural-rural development policies should take into account these coordination and complementarities. In this article we will concentrate on the Regional Rural Development Plans (RRDPs), which is the part of agricultural-rural development policies financed by EAFRD, but taking into account the potential interactions with other regional development policies.

In this light this article explores 21 RRDPs of the Italian regions, in order to identify their development strategies and to evaluate whether or not they are coherent with their needs. To achieve this objective we make use of a theoretical framework according to which different kinds of agricultural-rural development models are possible (and socially desirable) for different types of regions. We identify four agricultural-rural development models: (i) a model enhancing agricultural competitiveness, (ii) producing environment services, (iii) stimulating diversification of farm and rural activities and (iv) integrated rural development. They correspond to the idea of a new structure of the CAP composed by three main pillars of intervention: (1) a *Food Market Policy*, mainly oriented to enhance competitiveness and safety of EU agri-business environment (model i) (2) an

---

1   A Convergence objective covers the Member States and regions whose development is lagging behind. Their per capita gross domestic product (GDP) is less than 75 per cent of the Community average. The phasing-out regions are those suffering from the statistical effect linked to the reduction in the Community average following EU enlargement. Hence they benefit from substantial transitional aid in order to complete their convergence process. A Regional competitiveness and employment objective is to cover the area of the Community outside the Convergence objective. The regions eligible are those coming under Objective 1 in the 2000 to 2006 programming period which no longer satisfy the regional eligibility criteria of the Convergence objective and which therefore benefit from transitional aid, as well as all other regions of the Community.

2   An interesting debate referred to the Italian context is presented by Marenco (2007) and Panico (2008).

Environmental Policy, dedicated to the support of green services (model ii) and (iii) a *Rural Development Policy*, dedicated to regional development, rural and farm diversification, button-up based approaches (model iii and iv) (Jambor and Harvey 2010).

For each of these models we may find a correspondence with a mix of the second pillar measures (EC Reg. No. 1698/2005). The total number of measures foreseen under EC Reg. No 1698/2005 represents the potential complete 'menu' from which a regional administration (and the stakeholders involved in the policy decision-making process) may select those measures which best suit their needs. Obviously, each region should pursue one or another model or a mixed one according to its socio-economic and territorial characteristics. To verify this hypothesis we use three analytical steps:

1. Principal component analysis and cluster analysis on socio-economic and territorial variables of the 21 Italian regions to identify different regional types beyond the fundamental difference between convergence and competitive regions.
2. Analysis of the RRDPs through the menu approach (Terluin and Venema 2004) to identify the strategy pursued by different regions.
3. Evaluation of the coherence between the strategy chosen by the region and their specific needs.

The article is structured as follows. Section 2 is dedicated to the theoretical background; the third to the methodological framework. Section 4 presents the results and in Section 5 it is followed by some concluding remarks.

The European agricultural model and rural development strategies: the conceptual background

In the last two decades the development of rural areas has increasingly been the focus of scientific and political debate all over Europe. The need for continuous adaptation of the European agricultural and rural model and the policies to enhance it has been stimulated by the immense changes in the European Union (Harvey 2004, 2006, Oskam et al. 2010). On the one hand, a number of 'internal' factors could be recognized as the engine of this 'adaptation process' such as the structural changes of the European economy (from an industrial-based to a service-based model), the increasing relevance of environmental issues (both at local and at global level), new relationships between health and nutrition, new life-styles and models of food consumption, renewed attention through food safety concerns, the preference of European citizens to enhance their quality of life, and an increasing demand for rurality (Van Huylenbroeck and Durand 2004, Oskam et al. 2010). On the other, the external political pressure deriving from international agreements of the EU within the WTO and the liberalization process of the global markets has necessitated a change in the political support given to agriculture and rural development strategies (Jambor and Harvey 2010).

As a consequence, the 'new' European agricultural and rural development strategy has been based progressively on two main concepts which could be summarized by the terms 'multifunctionality' and 'differentiation/diversification' (Van der Ploeg and Roep 2004, Van Huyelenbroeck et al. 2004, Brouwer and van der Heide 2009). These entail a more complex model of development based on the increasing centrality of rural areas which may be seen as a set of environmental, natural, cultural, historical and economic resources which have to be enhanced in the development process (Terluin 2003, Van Huylenbroeck and Durand 2004).

On the basis of such dynamics, the rural development strategies of the EU have been reformed in conceptual and political terms. Starting from the Cork Conference statements (Cork Declaration 1996) the objectives and priorities of rural development have been based on integrated and sustainable development in which the role of agriculture and the food sector is linked to the process of enhancing social, economical and environmental resources at the local level (Jambor and Harvey 2010). Thus the European agricultural and rural model and the related rural development strategies have gained the capacity to meet both the 'internal' and the 'external' needs of European society, on the one hand, and the constraints of the CAP, on the other.

EU Rural Development Policy is currently based on Council Regulation (EC) No. 1698/2005. This regulation provides a more strategic approach to rural development for the period 2007-2013. Its general aim is to ensure the sustainable development of rural areas focussing on a limited number of objectives relating to agricultural and forestry competitiveness, land management and environment, quality of life and diversification of economic activities, taking into account the diversity of situations, ranging from remote rural areas suffering from depopulation and decline to peri-urban rural areas under increasing pressure from urban centres.

The new rural development regulation puts in place a significantly simpler and more strategic (i.e. objective rather than measure-led) approach to rural development through the definition of three core objectives and a reorganisation of sub-objectives and measure objectives. The main changes are the following:

1. Simplification of policy implementation by introducing a single funding system (according to the principle: one fund, one programme) for rural development and the change in the programming, financial management and control framework for rural development programmes;
2. Definition of three core objectives for rural development measures (Article 4):
   - Improving the competitiveness of agriculture and forestry by support for restructuring, development and innovation;
   - Improving the environment and the countryside by supporting land management;

- Improving the quality of life in rural areas and the diversification of economic activity;

A thematic axis corresponds to each core objective, around which rural development programmes have to be built, whilst a fourth horizontal and methodological axis is dedicated to the mainstreaming of the LEADER approach.

3. Agreement of Strategic Guidelines for Rural Development, which identify European Priorities for Rural Development in order to:
   - contribute to a strong and dynamic European agro-food sector by focusing on the priorities of knowledge transfer, modernisation, innovation and quality in the food chain and priority sectors for investments in physical and human capital;
   - contribute to the priority areas of biodiversity, and preservation and development of high nature value farming and forestry systems and traditional agricultural landscapes, water and climate change;
   - contribute to the overarching priority of the creation of employment opportunities and conditions for growth;
   - contribute to the horizontal priority of improving governance and mobilising the endogenous development potential of rural areas.

Member States (MSs) should develop their rural development strategies in the light of these objectives and European priorities and, based on the analysis of their own situation, should choose the measures most appropriate to implement each specific strategy. Rural Development Programmes (RDPs) then translate the strategy into action through the implementation of these measures, which follow the four operational axes (Articles 20, 36, 52, and 63 of EC Reg. 1698/2005). Thus, each MS has prepared its rural development national strategy plan constituting the reference framework for the preparation of rural development programmes.

It seems clear that the key part of this strategy is played by the role of multifunctional and diversified agriculture as a promoter of local development processes (Van Huylenbroeck and Durand 2004, Brouwer and van der Heide 2009, Oskam et al. 2010). In this perspective the multifunctional diversified farm is not only the place where material value is created but also the organisation which could promote the immaterial welfare based on ecological equilibrium, environmental preservation, food quality and safety (Van Huylenbroeck and Durand 2004, INEA 2004). For these reasons it is important to recognise and distinguish the concept of multifunctionality (as the capacity to produce in conjunction with the primary activities a set of secondary services which have the feature of externalities such as landscape, environmental management, etc.) in the processes of diversification/ differentiation (such as the capacity of farms to implement new activity other than agriculture such as tourism, educational services, in-farm food processing, typical and local production, short chain development, organic products, etc.) to increase and broaden income sources and off-farm activities, as part of the progressive 'regrounding' of the farmer and his/her family activities through other economic sectors (Van der Ploeg et al. 2002, Van der Ploeg and Roep 2004).

Rural development strategies have to enhance this model, seeking to optimise the capacity of farms to 'create value' in the rural context by using local resources. Three main strategies to enhance this process are more evident: the first considers strategy in the dynamic of 'value creation', the process of 'deepening' farm activities to cover food processing, high quality and regional production, organic farming and short supply chains. This strategy entails a reorganisation of production, innovation and conventional asset substitution inside the farm (Van der Ploeg et al. 2002, Van der Ploeg and Roep 2004). The second strategy is based on the concept of 'broadening' which is the process of 'enlargement' of farm activities related, on the one hand, to the provision of public goods to society such as environmental management, landscape protection, rural heritage preservation and, on the other, to the production of marketable services such as tourism and recreational services (Van der Ploeg et al. 2002). The last strategy is represented by the 'regrounding' processes based on the increasing opportunity for a farmer and his/her family to develop off-farm activities and differentiate the source of income (Van der Ploeg et al. 2002).

According to the structure of the rural development policies presented in Council Regulation (EC) No. 1698/2005 we may identify a connection between the selected priorities (axes) and the type of agricultural and rural model to be supported and promoted for the near future: if the competitiveness measures (axis 1) seem to be more related to the 'deepening' strategies, 'improving the environment and the countryside by supporting land management' priority (axis 2) appears able to enhance 'broadening' farm strategies (green services), while 'improving the quality of life in rural areas and encouraging diversification of economic activity' priority (axis 3) and LEADER axis are much more related both to the 'marketable-side' of broadening strategies (agritourism, new on-farm activities, etc.) and 'regrounding' strategies (off-farm activities). More specifically, we can consider the importance attached to some measures on the menu (i.e. the share of total budget and the type of interventions supported) as the most significant 'proxy' to highlight regional rural development strategies: measures 124 and 125 could be considered an indicator of a 'deepening-oriented' strategy, measures 214 and 215 as an indicator of 'green service-oriented' strategy, measure 311 as an indicator of 'broadening-oriented' strategy while the other measures on axis 3 and the LEADER axis as an indicator of 'local development-oriented' strategy.

The methodological framework

Regional types and their socio-economic characteristics

The first step of the adopted methodological framework was to identify the different regional types. This was done using a set of socio-economic variables for the 21

Italian regions representing the main regional features related to the economic and social dynamism and competitiveness, the natural resource and the environmental endowment, the degree of development and the relevance of rural areas.

The different regions was identified and classified through an analytic technique already used and tested to determine homogeneous area systems at a sub-national level. This technique implies the identification of a range of socio-economic and geographic features related to regional administrative units, which represent the base variables. The database obtained in this way was then used for a Principal Component Analysis (PCA) in order to get a synthesis of the information detected at base level. This synthesis is represented by the Principal Components (the synthesis variables). On the basis of this information the synthesis variables were used to make a Group Analysis. It was thus possible to identify homogeneous groups of regions in relation to the main differentiation factors identified in the PCA. These homogeneous groups represent the different region typologies. The base variables used for the determination of territorial differentiation factors referred to the Context and Baseline indicators as presented and listed in the Handbook on Common Monitoring and Evaluation Framework and quantified by the Regional Administrations. The total number of base variable used is 14 related to 5 different typologies: economic, demographic, social, agricultural and environmental features (see Table 23.1).

Menu approach The menu approach is an instrument to analyse the strategies chosen from the regions and, at the same time, verify if the selected strategy are tailored to their specific needs and requirements. This part of the analysis therefore consists in identifying the chosen strategies whereas a strategy is defined through the selected measures from the second pillar menu (Table 23.1). As explained above, we need to refer to Regulation (EC) No 1698/2005, that is the actual reference framework for the second pillar of the Common Agricultural Policy (CAP). According to this Regulation, the regions have chosen their strategy for rural development, that is the measures they think best suit their rural development needs. EC Regulation No. 1698/2005 identifies three core objectives for rural development policy. Each of these may be pursued through a set of specific measures although there are some measures that suit more than one objective.

Table 23.1   Base variables used in the Principal Component and Cluster Analysis

Number	Base variable	Meaning
1	Relevance of rural-intermediate areas (% on Total regional area)	Socio-demographic features
2	Rural population (% on total population)	
3	Regional Employment Rate	Economic conditions
4	GVA/per capita (% of UE 25 mean-value)	
5	Less Favoured Areas (% on Total regional area)	Geo-economic features
6	Agricultural Land Use (% UUA/Total regional area)	Agricultural features
7	Employment Development of Primary Sector (hare of primary sector in total employment)	
8	Productivity in the primary sector (GVA/AWU)	
9	Economic Development of Primary Sector (% Total GVA)	
10	Farmers with Other Gainful Activity (% holders with other gainful activity)	
11	Relevance of intensive agriculture areas (% on Total regional area)	
12	Areas at Risk of Soil Erosion (JRC – Pasera model index)	Environmental conditions
13	Relevance of Nitrate Vulnerable Zone (% on Total regional area)	
14	Relevance of artificial land use (% on Total regional area)	

Source: own elaboration

The measures are grouped into three operational axes, each of which corresponds to one of the three objectives:

axis 1. Improving the Competitiveness of the agricultural and forestry sector;
axis 2. Improving the environment and the countryside
axis 3.Quality of life in rural areas and the diversification of the rural economy.

The fourth horizontal methodological axis, LEADER, is dedicated to mainstreaming the LEADER approach. The support granted under this axis is for:

a. implementing local development strategies to achieve the objectives of one or more of the three other axes;
b. implementing cooperation projects involving the objectives selected under point (a);
c. running the local action group, acquiring skills and animating the territory (article 59).

Finally, there is a last measure for technical assistance.

As reported in Table 23.2 the number of measures for each axis widely varies. The number of possible measures is 41: 14 for axis 1; 13 for axis 2; 8 for axis 3 and 5 for the LEADER axis. Moreover, there is an important innovation in EC Reg. 1698/2005: the balance between objectives (article 17) according to which the Community's financial contribution to each of the three objectives must cover, at least, 10 per cent of the EAFRD total contribution to the programme for axes 1 and 3, at least 25 per cent for axis 2 and at least 5 per cent shall be reserved for axis 4.

Table 23.2   Rural Development measures of the second pillar in 2007 -2013 (the menu)

Articles	Axis		Measure code	Measure title
Art.20-35	Axis 1 Competitiveness	Human capital	111	vocational training and information actions
			112	setting up of young farmers
			113	early retirement of farmers and farm workers
			114	use of advisory services by farmers and forest holders
			115	setting up of farm management, farm relief and farm advisory services
		Physical capital	121	modernisation of agricultural holdings
			122	improvement of the economic value of forests
			123	adding value to agricultural and forestry products
			124	cooperation for development of new products, processes and technologies
			125	infrastructure related to the development and adaptation of agriculture and forestry
			126	restoring agricultural production potential damaged by natural disasters
		Quality	131	meeting standards based on Community legislation
			132	participation of farmers in food quality schemes
			133	information and promotion activities

Articles	Axis		Measure code	Measure title
Art. 36-51	Axis 2 Improving the environment and the countryside	Sustainable agricultural use	211	natural handicap payments to farmers in mountain areas
			212	payments to farmers in areas with handicaps, other than mountain areas
			213	Natura 2000 payments and payments linked to Directive 2000/60/EC
			214	agro-environment payments
			215	animal welfare payments
			216	support for non-productive (agricultural) investments
		Sustainable use of forestry	221	first afforestation of agricultural land grant and premium scheme
			222	first establishment of agroforestry systems on agricultural land
			223	afforestation of non-agricultural land
			224	Natura 2000 payments
			225	forest-environment payments
			226	restoring forestry potential and introducing prevention actions
			227	support for non-productive investments

Articles	Axis	Measure code	Measure title
Art. 52-60	Axis 3 The quality of life in rural areas and diversification of the rural economy		
	Economic development	311	diversification into non-agricultural activities
		312	creation and development of microenterprises to promote economic development
		313	encouragement of tourism and developing the economic fabric
	Quality of life	321	basic services for rural population and economy
		322	village renewal and development
		323	conservation and upgrading of rural heritage
	Human capital	331	training and information measures for economic actors operating in the fields covered by axis 3
		341	skills acquisition, animation and implementation
Art. 61-65	Axis 4 LEADER		
	Implementation of local development strategies through the selection of Local Action Groups (LAGs)	411	implementation of local development strategies, competitiveness
		412	implementation of local development strategies, environment/land
		413	implementation of local development strategies, quality of life and diversification
		421	Inter-territorial and transnational cooperation
		431	running the local action groups, acquisition of skills and animation'

Source: EC Reg. 1698/2005

Table 23.3 **Number of measures selected from the second pillar menu by the Italian regions**

Region	Axis 1: max 14 measures	Axis 2: max: 13 measures	Axis 3: max 8 measures	Axis 4: max 5 measures	Total: max 40 measures
Valle d'Aosta	5	3	3	3	14
Pidmont	12	7	7	5	31
Lombardy	12	6	6	5	29
Bolzano	10	4	4	5	23
Trento	6	4	5	4	19
Veneto	14	9	7	5	35
Friuli Venezia Giulia	8	9	6	6	29
Liguria	13	9	6	5	33
Emilia Romagna	10	8	7	5	30
Tuscany	11	9	1	4	25
Umbria	13	11	6	3	33
Marche	11	10	4	3	28
Lazio	12	11	8	5	36
Abruzzo	10	8	6	5	29
Molise	12	8	5	4	29
Sardinia	12	8	3	4	27
Basilicata	11	7	5	3	26
Campania	14	10	7	5	36
Puglia	11	8	6	4	29
Calabria	9	8	6	5	28
Sicily	11	8	8	3	30

Source: Italian RRDPs 2007-2013

Second: the share of the FEARD budget among the four axes. The size of the budget indicates the relative importance of the axis. This indicator suggests the agricultural-rural development model of the region. There is, for example, a marked difference between development strategies with a large share of the FEARD budget on axis 1 with respect to another that assigns much of it to axis 3 plus the LEADER axis.

Third: the share of the FEARD budget assigned to specific groups of measures. As explained above, this refers to those measures that, for us, define a specific rural development model of the region. Measures 121 plus 123 for the competitiveness agricultural development model; measures 214 plus 215 for the environmental agricultural model; measure 311 for an agricultural development model based on the diversification of the farm economic activities; the other measures of axis 3 plus the measures of the LEADER axis for a rural development model. With respect to this third indicator we will consider the fixed share of the FEARD budget as indicative of one or another development model.

Clearly, all such indicators are important to identify the agricultural development model, i.e. the strategy chosen by the regions.

Analysis of the rural development measures selected from the menu

The socio-economic analysis

Data analysis has been carried out through a factorial analysis technique (Principal Component Analysis) followed by a hierarchical cluster analysis. As a first step, a set of variables, selected on the base of the theoretical indications, was used to identify and classify regional types. Starting from 14 variables, we extracted 5 principal components that explain the 83 per cent of the whole variance. In Table 23.4 there is the factor loading matrix, where correlation coefficient higher than 0.40 (in absolute value) were indicated. The matrix is the basis to interpret the meaning of each principal component representing the main regional differentiation factors.

Table 23.4   Factor loading matrix

Variables	Component					Communalities
	1 Socio-economic welfare	**2** Relevance of land management	**3** Intensity of agricultural process	**4** Rurality	**5** Diversification process	
Relevance of rural-intermediate areas (% on Total regional area)		**0,91**				0,88
Rural population (% on total population)				**0,87**		0,82
Regional Employment Rate	**0,90**					0,86
GDP/per capita (% of UE 25 mean-value)	**0,92**					0,90
Less Favoured Areas (% on Total regional area)		**-0,68**			0,39	0,79
Agricultural Land Use (% UUA/Total regional area)	-0,45	0,40		0,50		0,77
Employment Development of Primary Sector (share of primary sector on total employment)	-0,57			0,48		0,87
Productivity in the primary sector (GVA/AWU)	**0,70**		0,47			0,86

Variables	Component					Communalities
	1 Socio-economic welfare	**2** Relevance of land management	**3** Intensity of agricultural process	**4** Rurality	**5** Diversification process	
Economic Development of Primary Sector (% Total GVA)	-0,48			0,54	0,48	0,87
Farmers with Other Gainful Activity (% holders with other gainful activity)					0,77	0,73
Relevance of intensive agriculture areas (% on Total regional area)			0,89			0,84
Areas at Risk of Soil Erosion (JRC – Pasera model index)		0,78				0,69
Relevance of Nitrate Vulnerable Zone (% on Total regional area)			0,48			0,82
Relevance of artificial land use (% on Total regional area)			0,80	-0,46		0,93

KMO's test =0,603; Bartlett's Test of Sphericity = 169,9

The first component explains 33 per cent of the total variance and allows distinguishing regions according to their different economic welfare. As the first component increases, moving from negative to positive values, socio-economic welfare change from a condition of less development (low rate of GDP per capita, relevance of a traditional sector such as agriculture, etc.), to situation where the whole socio-economic welfare is considerable high. More information on the potential impact of agriculture on soil management is synthesized in the second component (19.9 per cent of explained variance). This variable is also a proxy of the relevance of the 'hilly' agriculture in the regional context, where higher is the risk of erosion and less intensive could be considered the agricultural processes.

The third component (12.5 per cent of the total variance) is positively correlated with the intensity of agricultural practices and the relevance of urbanised areas. It is clearly a proxy of how intensive is the agriculture inside the region. The fourth and the fifth factors show aspect referring to the type of rurality and the presence of diversification dynamics. Negative values of the fourth component (9.5 per cent of the total variance) identify those regions where the urban areas are more relevant; while, if the component value is positive, region is mainly rural. The degree of diversification can be read on the fifth component (8 per cent of the total variance).

The factor scores, that is the coordinates of the observations (the investigated regions) with respect to each of the five principal component axes, were used to group firms into clusters.

Based on agglomeration schedule six final groups were considered. Table 23.5 reports cluster centres that allow to draw the main features of each cluster and to better understand the relationship among the differentiation factors analysed in PCA.

Table 23.5　Cluster centres

Cluster	N.case	Regions	Component				
			1	2	3	4	5
			Economic welfare	Relevance of soil management	Intensity of agricultural process	Rurality	Diversi-fication process
1	4	Friuli Venezia Giulia, Liguria, Pidmont e Valle d'Aosta	0,40	-0,50	-0,41	-0,66	-1,14
2	4	Campania, Lazio, Lombardy, Tuscany	-0,06	0,57	0,50	-1,37	0,50
3	2	Trentino Alto Adige	1,86	-1,23	-0,43	0,45	1,48
4	2	Veneto, Emilia Romagna	0,81	0,18	1,60	1,24	-1,09
5	2	Umbria, Marche	0,26	2,17	-0,94	0,65	-0,38
6	7	Calabria, Basilicata, Abruzzo, Puglia, Sardinia, Sicily	-1,04	-0,36	-0,12	0,49	0,37

Source: own elaboration

The first group (FVG, Liguria, Pidmont and Valle d'Aosta) represent a regional model where high socio-economic welfare is linked to a high degree of urban population and a more specialised agricultural model. The second group is the model of 'complex' region, where the presence of relevant and sprawled urban centres is linked to intensive agriculture. Those are the region of major urban/ rural contrast. The third group is the homogeneous regional cluster of Trentino Alto Adige with a very high indicator of socio-economic welfare. Veneto and Emilia Romagna are the 'rich rural and agricultural' model while Umbria and March (group 5) could be considered as the 'hilly' model of agriculture with a high risk of soil erosion and natural resource depletions. At the end we have the group of *Mezzogiorno* regions[3] (Campania excluded) where the less development conditions are prevalent (the lowest indicator of socio-economic welfare). Even the cluster analysis confirmed that the dichotomy less/more developed regions

---

3　*Mezzogiorno* is the name of the Southern Italian regions (namely Campania, Apulia, Basilicata, Calabria, Sardinia and Sicily)

(convergence/competitiveness) seams to be the really key dominant factor of clustering of the 21 Italian regions.

Using results of the clusters we have the opportunity to look at the different regional strategies not only in terms of differentiation between 'Convergence/ Competitiveness' but also between specific agricultural/rural regional types.

Results from the menu approach analysis

In this section we focus on the selection of the measures from the second pillar menu. In Italy 21 RRDPs have been implemented, one for each of the 19 regions plus 2 RDPs for the autonomous provinces of Bolzano and Trento[4]. As can be seen in Table 3.1, these RRDPs are characterized by a large number of the measures. In all, 72 per cent of them are implemented with 28 or more measures[5]. Campania, Lazio and Veneto selected the highest numbers (36 and 35 for the latter); Valle d'Aosta and the autonomous province of Trento selected the lowest numbers (14 and 19 respectively).

Among the different axes, the measures of the first are more selected whilst those of the second are less. About 67 per cent of the regions selected 11 or more measures belonging to the competitiveness axis: Campania and Veneto selected the highest number (14); Valle d'Aosta and Trento the lowest numbers (5-6). From the measures of the environmental axis, regions selected 8 measures on average: Lazio, Umbria, Campania and Marche selected the highest numbers (10-11); Valle d'Aosta and the provinces of Bolzano and Trento the lowest (3-4). On average, convergence regions use a larger number of measures for all the axes. We have already underlined the risk that the menu consisting of numerous measures has for the fragmentation of resources. We now have some elements to be able to state that, in Italy, this risk is greater in the convergence regions. Nevertheless, to make a more complete appraisal we must focus on the distribution of expenditure among the various axes and various measures (Table 23.6). Then, if we consider the share of resources among the axes and between convergence and competitiveness regions, it emerges that no great differences exist between the two groups. Indeed, almost the same share of resources is allocated, on average, among the various axes: 37 per cent to axis 1; 44 per cent to axis 2; 9 per cent to axis 3 and 8 per cent to axis 4. This means that, considering measures according to the core objectives of European rural development policy, no specific development patterns appear among regions, not even between convergence and competitive regions. However, we *can* say that Italian regions, on average, allocate most of the EAFRD budget

---

4   Of the 19 regions, four (Calabria, Campania, Puglia and Sicily) are part of the convergence objective of the EU, one is a phasing-out region (Basilicata), another is a phasing-in region (Sardinia) while all the others are part of the competitiveness and employment objective.

5   On average 28 measures

to the competitiveness axis and environmental axis together, over and above the minimum limits fixed with the principle of equilibrium. Therefore, it seems there is a strong orientation towards a competitive-environmental services agricultural development model and a week orientation towards farm activity diversification and rural development models.

The situation becomes clearer if we look within the two groups of regions, considering at the same time single or specified groups of measures: these define specific development patterns. It is far from easy to choose limits of the budget share to define the different kinds of models.

Taking into account, at the same time, the financial equilibrium balance principle and the distribution of resources among the various axes it may be stated that a region has chosen:

- strong orientation towards the competitiveness agricultural model if the share of the resources for axis 1 is equal to or more than 40 per cent and the share for measures 121+123 is equal to or more than 20 per cent;
- strong orientation towards the environmental services agricultural model if the share of the resources for axis 2 is equal to or more than 43per cent and the share for measures 214+215 is equal to or more than 24 per cent;
- a strong orientation towards the farm activity diversification agricultural model if the share of the resources for measure 311 is equal to or more than 25 per cent;
- a strong orientation towards the rural development model if the share of the resources for axes 3 + 4 minus the share for measure 311 is equal to or more than 30 per cent.[6]

In the other situations the development model is a mixed one.

---

6   These shares were chosen since the mean values of the distribution of the relative shares budget for axes 1 and 2 are over and above the limits fixed by the financial equilibrium balance principle. For us this denotes a strong orientation towards the corresponding objectives. We thus deemed that the same should hold for the other two models.

Table 23.6 Distribution of FEARD budget shares among axes and measures in the 2007-13 RRDPs

Region	% on total budget	Axis 1	121+123	Axis 2	214+215	Axis 3	Meas.311	Axis 4	Axis 3+4	Axis 3+ 4 – 311
Valle d'Aosta	0.6	9.6	2.1	**69.4**	**31.8**	10.3	2.6	7.5	**17.8**	**15.2**
Pidmont	4.8	38.2	16.5	**44.5**	**32.5**	7.4	1.7	6.5	13.9	12.2
Lombardy	4.8	31.5	21.5	**51.6**	**25.3**	9.0	6.0	5.0	14.0	8.1
Bolzano	1.7	23.9	13.0	**62.0**	**41.1**	9.0	2.2	5.0	14.0	11.9
Trento	1.2	30.3	17.3	**52.9**	**24.9**	10.3	1.5	6.0	16.2	**14.8**
Veneto	4.9	**44.1**	**23.2**	36.9	20.7	5.0	2.4	11.0	16.0	13.9
Friuli Venezia Giulia	1.3	**43.0**	**30.5**	37.0	11.3	10.0	4.8	6.5	16.5	11.7
Liguria	1.3	**47.4**	**32.9**	23.2	14.2	6.3	3.3	20.0	**26.3**	**23.0**
Emilia Romagna	5.0	**41.0**	**26.2**	42.5	29.6	10.4	3.6	5.1	15.5	11.9
Tuscany	4.5	38.5	18.4	40.2	24.4	**10.5**	**10.5**	10.0	**20.5**	10.0
Umbria	4.1	35.2	18.6	**43.0**	**28.5**	9.0	3.8	5.0	14.0	10.2
Marche	2.5	42.2	27.6	38.8	19.9	9.0	6.7	6.0	15.0	8.3
Lazio	3.5	**47.0**	**22.8**	32.0	24.3	11.3	4.7	6.0	17.3	12.6
Abruzzo	2.0	43.0	20.2	37.0	15.9	11.0	3.9	5.0	16.0	12.2
Molise	1.0	**44.1**	**23.6**	33.8	12.1	14.1	5.0	5.0	19.1	14.1
Sardinia	6.7	28.0	13.6	**56.0**	**12.9**	1.4	0.8	13.6	15.0	**14.2**
Basilicata	4.5	26.5	11.9	**54.0**	19.9	10.0	5.6	6.0	16.0	10.4
Campania	13.1	40.0	18.3	36.0	32.4	15.0	1.8	5.0	**20.0**	**18.2**
Puglia	10.3	**40.4**	**27.0**	35.1	27.7	2.7	1.5	18.8	**21.5**	**20.1**
Calabria	7.6	**41.0**	**21.7**	**41.0**	**30.8**	10.0	5.5	6.0	**16.0**	10.5
Sicily	14.7	32.3	20.1	**52.7**	**32.6**	7.0	2.9	6.0	13.0	10.1

Region	% on total budget	Axis 1	121+123	Axis 2	214+215	Axis 3	Meas.311	Axis 4	Axis 3+4	Axis 3+ 4 – 311
Convergence (total)	**50.2**	36.8	19.8	43.1	28.7	8.9	2.9	8.4	17.3	**14.5**
Competitiveness (total)	**49.8**	37.3	20.5	44.0	23.1	8.1	4.0	7.9	16.0	11.4
RDP (total)	**100**	37.0	20.3	43.5	24.4	8.5	3.5	8.2	16.7	13.2

Source: Italian RRDPs 2007-2013 and our elaborations

Following these criteria, eight regions present a strong orientation towards the competitiveness model: Veneto, Friuli-Venezia-Giulia, Liguria, Emilia Romagna, Lazio and Molise among competitiveness regions; Puglia and Calabria among convergence regions. Seven regions present a strong orientation towards the environmental services agricultural model: Valle d'Aosta, Pidmont, Lombardy, Bolzano and Trento among competitiveness regions; Sicily among convergence regions. There are no regions strongly oriented towards the two other models. Nevertheless, there are regions that follow the mixed development model with a significant orientation towards rural development. Campania, for example, has implemented an RDP which is markedly competitiveness-oriented but a certain importance has also been attached to rural development. Of the remaining regions, Tuscany is significantly oriented towards a development model based both on environmental services and on farm activity diversification and on rural development measures.

Starting from these results we used the socio-economic classification to analyse the coherence of each regional RD strategy. The judgement was based on the principle that the type of regional strategy indicates the main problems/opportunities to be implemented. If the emerging strategy as a jointly result of the menu approach analysis met the regional model type and the stated priorities then the coherence of the RDP was considered high. On the other hand we considered not completely coherent or incoherent all the strategies not matching the regional context conditions and/or a substantial difference between emerging strategies and stated priorities. The obtained results are showed in Table 23.7.

Table 23.7 Analysis of the Italian RDPs coherence

Cluster	Regions	Type of agricultural/rural regional model	Emerging RD strategy	Stated priorities	Degree of coherence
1	Friuli Venezia Giulia		competitiveness model	mixed	intermediate
	Liguria	Developed with declining rural (mountain) areas	competitiveness model	enhance competitiveness	high
	Pidmont		environmental services model	mixed	intermediate
	Valle d'Aosta		environmental services model	integration and environment preservation	high
2	Campania		mixed	competitiveness, environmental improvements	high
	Lazio	Complex regions with urban/rural contrast	competitiveness model	mixed	intermediate
	Lombardy		environmental services model	mixed	intermediate
	Tuscany		mixed	mixed	high
3	Trento	Rural developed with integrated agriculture	environmental services model	integration and environment preservation	high
	Bolzano		environmental services model	integration and environment preservation	high
4	Veneto	Agribusiness developed with specialised agriculture	competitiveness model	enhance competitiveness	high
	Emilia Romagna		competitiveness model	network and chain development	high
5	Umbria	Green-service agriculture	mixed	mixed	high
	Marche		mixed	mixed	high

	Calabria		competitiveness model	mixed	low
	Basilicata		mixed	structural adjustment, competitiveness, rural depopulation contrast	high
6	Puglia	Less developed with rural marginality	competitiveness model	competitiveness	high
	Sardinia		mixed	mixed	high
	Sicily		environmental services model	enhance competitiveness	low
	Molise		competitiveness model	mixed	low
	Abruzzo		mixed	enhance competitiveness	intermediate

Source: own elaboration

Concluding remarks

In this study we underlined the increasing importance of rural development policy and the second pillar of CAP, over the time. They are becoming an integral part of the UE cohesion policy, particularly important for those EU regions in which rural areas and disadvantaged areas suffering of development problems, constitutes a great share of the territory (LRs). In these regions, more than in the others, the RDPs, should been aimed to the development of the whole territory and not of the specific sector, with the involvement of actors that operate in the handcraft, commerce, tourism and agricultural sectors and that should agree strategies, to share resources and costs to achieve common objectives.[7] Rural development policies, then, in these areas, should have a strong territorial approach and, with respect to the agricultural sector, should aim to the diversification of farm activities with a large attention towards the improvement of the quality of the goods and services. Agricultural development, alone, cannot solve the problems of growth and competitiveness of these areas where the scarcity of gainful productive activities generates depopulation whose consequences are hard for the same agricultural sector. Moreover, in these regions it doesn't seem that there has been the hoped integration and coordination among various EC funds for an integrated development of the rural areas. Instead, in all RRDPs it is clearly specified the demarcation among interventions financed by different funds.[8] The analysis of the 21 Italian RRDPs has highlighted that there aren't regions with rural development patterns strongly based on those measures of the 'menu' aimed to a properly local development (model iv). Generally there has been a tendency to implement RRDPs with a great number of measures, then few selective, and with a high budget shares dedicated to the measures of axis 1 and 2. If this may be justified for the competitiveness regions where agricultural and rural development have already reached a high level, it doesn't seem a good strategy for the convergence regions. In the North and Centre Italy to improve specific aspects of agricultural development (food quality, modernization of agricultural holdings, cooperation for developing of new products etc.) it could be an efficient choice to focus on these measures. Instead, to improve (bottom-up based) rural development is a more difficult task.

References

Brouwer, F. and van der Heide, M. 2009. *Multifunctional Rural Land Management.* Earthscan.

---

7   In the Italian debate an interesting note has been presented by Lanzalaco et al., 2008.

8   Panico (2008) and Sotte and Ripanti (2008) discussed this issue in the Italian context.

Dwyer, J., Baldock, D., Beaufoy, G., Bennett, H., Lowe, P. and Ward, N. 2002. Europe's rural future. The nature of rural development II, WWF/LUPG, Brussels.

Dwyer, J., Ward, N., Lowe, P. and Baldock, D. 2007. European Rural Development under the Common Agricultural Policy's 'Second Pillar': Institutional Conservatism and Innovation'. *Regional Studies*, 41(7), 873- 888

European Commission. 1996. *The Cork Declaration, A Living Countryside.* 7-9 November 1996, Cork.

European Commission. 2003. *Conclusion of the second conference on rural development, 'Planting seeds for rural futures' Rural policy perspectives for a wider Europe.* 12-14 November 2003, Salzburg.

Harvey, D.R. 2004. Policy Dependency and Reform: Economic Gains versus Political Pains. *Agricultural Economics,* 31, 265–275.

Harvey, D.R. 2006. The EU and the CAP: and agenda for review? *EuroChoices,* 5(1), 22–27.

INEA. 2004. Verso il riconoscimento di una agricoltura multifunzionale, in T*eorie, politiche, strumenti,* edited by Henke R., ESI, Napoli.

INEA. 2005. *La riforma dello sviluppo rurale: novità ed opportunità* edited by Monteleone A. ESI, Napoli.

Jambor, A. and Harvey, D.R. 2010. CAP Reform Options: A Challenge for Analysis & Synthesis. Paper presented to AES Annual Conference, Edinburgh, March 29 – 31.

Lanzalaco, L. and Lizzi, R. 2008. Government e governance come fattori critici delle politiche agricole e rurali. Paper to the XVL Convegno della Società Italiana di Economia Agraria, 25-27 settembre 2008, Portici, Italy.

Lowe, P., Buller, H. and Ward, N. 2002. Setting the next agenda? British and French approaches to the second pillar of the Common Agricultural Policy. *Journal of Rural Studies* 18, 1–17.

Marenco, G. 2007. Qualche nota su sviluppo agricolo e rurale, Rivista di Economia Agraria, LXII, n.3.

Marsden, T., Banks, J., Renting H. and van der Ploeg, J.D. 2001. The road towards sustainable rural development: issues of theory, policy and research practice. *Journal of Environmental Policy and Planning* 3(1), 75–84.

Meester, G. 2010. European integration and its relevance for agriculture, food and rural areas, in *EU policy for agriculture, food and rural areas* edited by Oskam, A., Meester, G., and Silvis, H. Wageningen, The Netherlands: Wageningen Academic Publishers, 29 – 40

OECD. 1999. Policymaking for predominantly rural regions: concept and issues, Working Party on Territorial policy in Rural Areas, Working Document DT/ TDPC/RUR (99) 2, Paris.

Oskam, A., Meester, G. and Silvis, H. 2010. *EU policy for agriculture, food and rural areas.* Wageningen, The Netherlands: Wageningen Academic Publishers.

Panico, T. 2008. Politiche di sviluppo rurale nelle regioni italiane obiettivo convergenza tra realtà e potenzialità, in *Conoscenza della realtà e politica agraria: questioni aperte per la ricerca* edited by Marenco G. Paper to the XVL Convegno della Società Italiana di economia Agraria, 25-27 settembre 2008, Portici, Italy.

Sotte, F. 2005. Sviluppo rurale e implicazioni di politica settoriale e territoriale. Un approccio evoluzionistico, in *Politiche, governance ed innovazione per le aree rurali* edited by Cavazzani A., G. Gaudio, Sivini S. ESI, Napoli.

Sotte, F., and Ripanti, R. 2008. I PSR 2007-20013 delle Regioni Italiane. Una lettura quali-quantitativa, Working paper n.6, Gruppo 2013, Forum Internazionale dell'Agricoltura e dell'Alimentazione.

Terluin, I.J. 2003. Differences in economic development in rural regions of advanced countries: an overview and critical analysis of theories. *Journal of Rural Studies* 19, 327–344.

Terluin, I.J. and Venema, G.S. 2004. Second pillar of the CAP: what can we learn from experiences with the menu approach?, *Tijdschrift voor Sociaalwetenschappelijk Onderzoek van de Landbouw*, 19(1), 6-21.

Thomson, K., Berkhout, P. and Constantinou, A., (2010). Balancing between structural and rural policy, in *EU policy for agriculture, food and rural areas* edited by Oskam, A., Meester, G., and Silvis, H. Wageningen, The Netherlands: Wageningen Academic Publishers, 377 – 392.

Van der Ploeg, J.D., Renting, H., Brunori, G., Knickel, K., Mannion, J., Marsden, T., de Roest, K., Sevilla Guzmán, E. and Ventura, F., 2000. Rural development: from practices and policies towards theory. *Sociologia Ruralis*, 40, 391-408.

Van der Ploeg, J.D., Long, A. and Banks, J. 2002. Rural development: the state of the art, in *Living Countryside. Rural development processes in Europe: the state of the art* edited by Van der Ploeg J.D., Long A., Banks J. Doetinchem, Elsevier, Amsterdam, 231.

Van der Ploeg, J.D. and Roep, D. 2004. Multifunctionality and rural development: the actual situation in Europe, in *Multifunctional agriculture: a New Paradigm for European Agriculture and Rural Development* edited by G. Van Huylenbroeck G. Durand (eds.). Ashgate, Aldershot.

Van Huylenbroeck, G., Verbeke, W. and Lauwers, L. 2004. *Role of Institutions in Rural Policies and Agricultural Markets*. Elsevier, 461 p.

Van Huylenbroeck, G. and Durand, G. 2004. *Multifunctional agriculture: a New Paradigm for European Agriculture and Rural Development*. Ashgate, Aldershot.

Chapter 24

The Generation Turnover in Agriculture: The Ageing Dynamics and the EU Support Policies to Young Farmers

Anna Carbone and Giovanna Subioli

Introduction

The measure in support of young farmers is important within the EU rural development policies because it concerns a critical issue in European agriculture: the high share of elder farmers and the lack of an appropriate generational turnover, with the well-known consequences in terms of low competitiveness and progressive set aside of resources.

This policy measure is also important because it absorbs a significant share of financial resources. In Italy, for example, between 2000 and 2006, out of a total financial support to rural development of slightly less than EUR14 billion, 826 million (about 6 per cent) were allocated to these interventions. Furthermore, with regard to Italy, in the same period of time the recipients of this support were 26,843, slightly less than the recipients of the payments for farm investments.

Nevertheless, the EU policy for young farmers and generational turnover has been widely criticized. Its efficacy has been objected (Cagliero and Novelli 2005, Carbone 2005, INEA-OIGA 2005, Sotte 2005). Recently, the influential voice of the European Parliament has joined these criticisms. On June 2008 the Parliament has approved a resolution (2007/2194 INI) that, while acknowledging the persistent problem within European agriculture, moves an open and specific criticism, not only to the scarce efficacy displayed by the CAP in counteracting the problem, but also pointing out the role that the CAP actively played in contributing to cause this situation.

This chapter illustrates the extent of senility within European agriculture and its dynamics (next section). Then illustrates the implementation of the 'measure for young farmers' in selected Italian regions over the 2000-2006 period, on the basis of the available Reports of Intermediate Evaluation (third section). The fourth section summarizes the innovations introduced by Reg. (CE) n. 1783/2003 for the 2007-2013 planning period, while the fifth discusses the European Parliament Resolution on the young farmers. Some concluding remarks are presented in the last section.

The Ageing of European Agriculture: some Empirical Evidences

The imbalanced distribution of farmers across age classes in the European agriculture is well known. The high share of elder farmers, the scarce presence of younger ones and the difficult access to the sector, are different aspects of the same phenomenon (Glauben et al. 2009, Corsi 2009). Hereafter we propose a short overview of the main updated figures on the topic.

A comparison between agriculture and the rest of the economy shows that the share of young workers is lower in agriculture (Figure 24.1). This is the case in the EU15 (approximately 35 per cent and 48 per cent, respectively) as well as in individual member States.

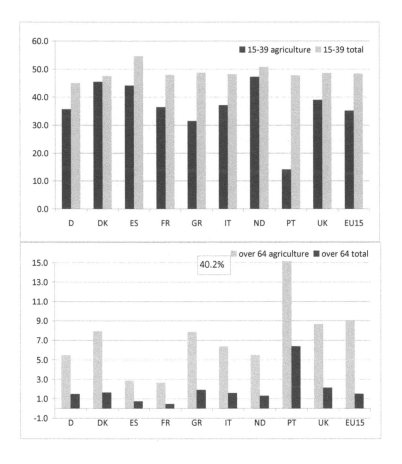

Figure 24.1   Employment by sector activity and age class, 2005-2007.

Source: Author's calculation on Eurostat Labor Force Statistics

Differences are less significant in the intermediate age class. The second graph of the figure shows that the share of the 'over 65' is, without exceptions, much higher in the primary sector (corresponding to slightly less than one tenth) than in the other sectors (less than two over one hundred). Few young farmers and many elder ones are the two faces of the same phenomenon that characterize EU agriculture.

If we move to look at the farm holders only, data show (Figure 24.2) the presence of an even wider demographic imbalance. Data from the farm structure survey are slightly different: the first age class includes persons up to 44 years old, nevertheless its share in the total is even lower, while the elder farm holders' share goes up to 30per cent in some countries. This seems to indicate the presence of significant entry and exit barriers arising from household settlements (mainly related to the use of the farmhouse and to the self-consumption of farm produce). Exit barriers, in turn, make new entries more difficult.

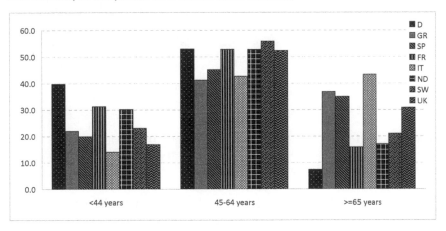

Figure 24.2   Holders by Age Class, 2005

Source Author's calculation on Eurostat data.

The data available allow for distinguishing farms by the holder's age and by economic size. So that it is possible to see that the quota of young holders in small farms (=< 1 ESU) is lower than in bigger ones, while the incidence of the elder is well higher in the first group. In particular Figure 24.3 shows that in small farms there is half the percentage of young farmers that run relatively bigger farms (>1 ESU) and this is true in every countries.[1] In the Italian literature this phenomenon is acknowledged as a virtuous circle: the larger and the more efficient are farms, the more attractive they are for young holders (Simeone 2006); in turn, the presence of young farmers makes farms more efficient and help them to increase in size

---

1   The number of Countries for which data are available here is limited to the ones shown in the figure.

over time. On the contrary, small farms are less rentable and less attractive and hence have no turnover, with old farmers that keep running few activities, with an extensive use of resources (Barberis and Siesto 1993, Mazzieri and Esposti 2005).

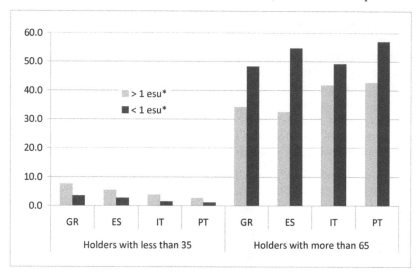

Figure 24.3　Holders by Age Class and Economic Dimension

Source: Author's calculations on Eurostat data

This demographic structure of farm holders led to a significant, ineluctable shrink of the sector (Table 24.1) (see also De Gaetano and Mazzoli 2003). Figures in Table 24.2 shows that the overall reduction is the result of two components: (i) the progressive reduction of the new entries mainly due to the loss of attractiveness of the primary sector (first row); (ii) the retirements of the elder, which is even more sizeable and determines a massive inertial effect on the demographic dynamics of the sector due to the imbalance inherited from the past periods (last row).

Furthermore, Table 24.2 also shows the presence of a relevant dynamic in the intermediate age classes. The percentage migration balances for each class show that in every country there have been significant entries in the intermediate age classes in the time span of ten-years[2] Furthermore, in some countries – notably the United Kingdom and the Mediterranean group – there have been several new entries even among the elderly.

2　For further details on the methodology see Carbone 1996.

Table 24.1 Dynamics of Farms' number and of UAA in some Eu MS countries, 1995–2005

	DK	D	GR	ES	FR	IT	ND	PT	SV	UK
Holders total	-17,090	-176,370	30,850	-213,590	-244,320	-711,100	-31,990	-128,110	-12,740	49,100
Holders %	-25.0	-31.4	3.8	-17.2	-34.0	-31.2	-29.1	-28.8	-15.3	21.8
UAA total	-18,920	0	405,580	-375,210	-385,386	-1,977,600	-40,820	-245,030	132,720	-489,660
UAA %	-0.7	0	11.3	-1.5	-1.4	-13	-2	-6.2	4.3	-3

Source: Author's calculations on Eurostat data

Table 24.2 Holders by Age Class and their demographic dynamics, 1995–2005

	DK	D	GR	ES	FR	IT	ND	PT	SV	UK
35 years*	56.1	36.1	115.8	69.9	46.0	51.3	39.7	37.0	52.3	63.2
Tm I II **	83.5	20.2	158.3	97.4	15.2	65.8	88.4	47.3	67.3	176.8
Tm II III**	1.5	-12.2	49.2	32.1	-15.4	23.5	-2.6	8.3	9.7	72.5
Tm III IV**	-22.3	-35.4	7.3	-3.2	-35.1	-14.9	-28.2	-11.4	-7.8	34.2
Tm IV V**	-38.3	-78.0	38.6	1.4	-53.3	8.5	-52.5	19.6	-13.3	46.1
% exit/entries***	425.3	118.1	437.5	694.0	259.6	1,615.0	464.5	2,288.5	462.6	639.5

*% on the same class in 1995; **Tm=(Sm/(i-1)P(i))*100, where Sm is the migratory balance of the class and P is the decade 1995-2005; ***% of over 75 (defined as exit) on under 35 (defined as entries).

Source: Authors' calculations on Eurostat data

Altogether these data indicate that the age structure of agricultural employment, and especially the ageing of farm holders, is the results of a complex set of factors, among which the most notable are: i) the presence of entry barriers, ii) the presence of exit barriers, iii) the persisting low level of factor productivity in agriculture; iv) the presence of inter-sectoral labour force movements in the intermediate age classes. This last phenomenon, in turn, is the result of: a) a change in the social consideration of agriculture, b) a migration from urban to rural areas of retired persons, in search of cheap housing and /or a more relaxed and country-like lifestyle.

The Implementation of the 'Measure for the Setting up of Young Farmers' in Italy

The discussion on the implementation of this policy measure is limited to the Italian case due to lack of information on other countries. For Italy, data are retrieved from the 'Intermediate Evaluation Reports' prepared by each sub national administration (Region), and from more detailed sample data referred to small sub-regional areas; these have been directly collected with the help of the Local Government of Lazio, Marche and Toscana Regions.[3]

Most recent available figures refer to 2000-2003.[4] Within these four years, the measure has benefited 26,843 young farmers who received, on average, EUR18,000 each, mostly as a lump sum. In order to evaluate the efficacy of this support, it is useful to compare the number of the 'new' holdings with the reduction in the total number of farms. As seen, between 1995 and 2005, Italian holdings decreased by 77,000 units per year (-31.2 per cent over the whole period). Therefore, without the 'measure for the young', at most we would have had a loss of approximately 10 per cent more every year (or 6,700 units).

In principle this figure could be significant, albeit clearly inadequate to counterbalance the exit of resources from the sector. However, we have no idea of how farmers would have behaved in the absence of the measure: would all concerned farmers have not entered the sector without the support? or would already existing farms not have had any generational turnover?

Beneficiaries of the payment – including young farmers that start a new farm and the ones that take over an existing farm – can be classified into four groups (Carbone et al. 2005),

1. Those that could not have done so without the EU payment;

---

3   The Authors acknowledge the Administrations of Lazio and Marche Region and The Grosseto Province Administrations for the provision of the data. Dott.ssa Letizia Lamoratta is also acknowledged for the data collected and the data processing.

4   The Local Governments have two or three years to complete the Final Evaluation Reports for the PORs and three years for the PSRs, so that none of them are yet available.

2. Those that could have done without the payment;
3. Those that were already informally managing their farm (in Italy the change is often formalized only when the old parent retires or dies);
4. Those that formally register as managers of the farms to have access to the payment, while management is in fact maintained by older family members.

It is evident that the measure would be truly effective only in the first case. The data do not allow each case to be detected with certainty; however they do allow interesting insights to be gained.

As shown in Table 24.3 – for those Regions that have published the Intermediate Evaluation Report – the incidence of young farmers that took over the family farm is around 70-80 per cent in most cases: this happens in Emilia Romagna, Veneto, Marche. In Tuscany this share is lower, around 40 per cent, while in other regions it can be shown that very few new farms have been started, while nothing can be said on the parental relationship between the previous and 'new young' holder. In addition, several beneficiaries of the EU payment hire land for their activities; and the share of hired land that they use is almost double the national average (more than 50 per cent vs 25.1 per cent).

Table 24.3   Turn-over in Young Farmers' Setting Up

	No. applications financed	% of turn-over	% of family turn-over
Emilia Romagna	3,696	82.1	79.2
Veneto	2,102	80	76
Piemonte	2,324	77.1	-
Calabria	1,516	76.7	-
Toscana	2,696	60	56.4
Friuli	-	-	51.5

Source: Author's calculations on Regional Evaluation Reports

Data on specific areas confirm this evidence and tell something more. In the Viterbo and Grosseto provinces (where 33 and 30 beneficiaries are located), 68 and 85 per cent respectively of the new settlements took place in family farms'. Furthermore, in about 80 per cent of the total settling (including the Marche region sample, 383 cases) the farm premium is used to hire the (family) land.

This indicates that cases 3) and 4) among those outlined above, may be the more frequent: young farmers either already settled in their farm, running it on an informal basis that simply took the occasion of the payment; or fictitious hiring contracts showing a fake transition of the management, again, just to access the payment.

Similar conclusions are presented in INEA-OIGA (2005), where 86 per cent of the sample considered is made up of young beneficiaries already settled in family holdings, or who would anyhow have settled, even without support, as they themselves stated in interviews.

Further evidence indicates that the payment has not been at the origin of significant additional settlements of young farmers. The EUR18,000 paid on average represent a very small amount if compared with the land values in our country and, hence, seems not sufficient to overcome the far most major entry barrier of the sector, which is the need to hire or purchase land. The national average value for one hectare of arable land, was of about EUR15,000 in 2002, for non-irrigated land (varying in the range 6,000-30,000), while it reached EUR25,000 for irrigated land (varying in the range 14,000-45,000).

Considering, for example, the average size of the beneficiaries' holdings in the study area, we estimated a farm value of EUR450,000 and more in the Grosseto province (equivalent to a value of EUR14,500 per hectare) and of about EUR200,000 in the Viterbo province (equivalent to a value of EUR16,200 per hectare).

The EU payment becomes even less significant when considering the need to access capital to acquire machineries, livestock, and other fixed factors of production, from buildings to computer equipment, which are necessary to start a competitive business.

The Intermediate Evaluation Reports contain further hints on the overall assessment of the appropriateness of the amount given for the settings up. The key point here is the definition of the expenses related to the settling of a new farm. If we count the administrative costs, the use of advisory services and training, the support proved by the EU can barely be sufficient in the majority of the Regions. In some cases, also the cost for improving the farmhouse, or the cost related to other minor farm improvements, were included and this turned out with a much lower percentage of the expenses covered by the payment (20-30 per cent on average).

These examples confirm that the size of the payment provided by the EU measure for young farmers offers an ineffective incentive to attract young people into the sector, and it is also insufficient to finance an increase in the competitiveness of the existing holdings through the familiar turnover within the farm.[5]

In other words, would the holdings be profitable, the turnover would happen anyway, on the contrary, non-profitable holdings are doomed to remain such: a payment of few thousand euros cannot promote a generational turnover (even if

---

5   It is worth reporting that the average farm amount allocated by the Italian law for land reorganization (441/98) was approximately of EUR280,000 (our estimate on INEA data 2005). On the other side, it should also be considered that in countries with a different land market and with a different labour cost (such as, for example, some of the new EU members), even smaller amounts, might prove of sufficient incentive efficacy.

it takes place within the family) assuring the survival of the holdings in the long period (Carbone 2005, Corsi et al. 2005).

Regulation 1783/03: the Innovations introduced for the setting up of young farmers

Given the evidence proposed, it is worth reviewing shortly the innovations introduced by Reg. (CE) 1783/03 on Rural Development for the planning period 2007-2013. The new planning period has just started; therefore it is too early to attempt an evaluation. However, it is possible to make some considerations on the novelties introduced in the measure for the settling of young farmers. Compared to the previous Regulations on Rural Development (Reg. (CE) 1257/99), the measure has been modified in three important aspects.[6]

Young People and Competitiveness

The measure is now placed under the so called First Axis, which is aimed at fostering competitiveness. Consequently, social targets such as avoiding depopulation, defending and encouraging marginal rural contexts are less emphasized; the same is true for environmental aims such as territory safeguard, landscape keeping and so on. This delimitation, if correctly understood, has important consequences in defining the areas of implementation and in the selection of beneficiaries. This means that: a) privileged territories for the payment are those where competitive agriculture is not only possible but also desirable as far as there are no economical and environmental conflicts regarding the alternative use of resources, b) the selection of applications should be more restraining than in the past.

The Business Plan

The second innovation introduced concerns the need to spell out and obtain approval of a 'business plan' as a condition to access to the funds. The importance of this aspect comes directly from binding settlement to improving the sector's competitiveness: it is clear that if new settlements should bring efficiency and competitiveness, this condition must be objectively proved. The business plan seems to be the most appropriate instrument to help reaching this goal. Furthermore it can help in a better targeting of the beneficiaries, hence reducing inefficiencies and adverse redistributive effects.

Nevertheless, it is to be said that, on the other hand, the introduction of the business plan presents some 'traps' and difficulties that it is worthwhile recalling

---

6   As a matter of fact, there seems to be a further element of novelty, that is a closer link between the measure for the young and the pre-retirement one, but the text is rather vague to this respect.

because it should be avoided. First of all there is the concrete risk that some Local Government might encounter difficulties to find the necessary competences to evaluate the business plans or that, fearing not to be able to implement the measure properly, a 'bearish' mechanism reducing the real selectivity of the criteria adopted might develop, vanishing the positive potentials of the instrument. Furthermore, the business plan might be interpreted as another bureaucratic discharge to perform, void of effective contents strictly related to what should be the entrepreneurial project of the settling young farmer and of elements proving its soundness. The real coming into effect of the above mentioned risks will obviously depend -at least partly – on the political will and on the abilities of Local Governments. Some help to this respect might come from introducing a successive check on the plan implementation as well as from a punctual definition of the contents that this plan must include, as well as of parameters to choose for the evaluation of the increased competitiveness.

The size of the payment

The third novelty concerns the increase in the size of the payment. This is settled at a maximum EUR40,000 in the form of a single premium or in the form of an interest rate subsidy (the capitalization of which should not exceed EUR40,000) but could be raised up to EUR55,000 in case of a combination of the two forms. The incentive given to young farmer that borrow money to invest in the farm project is to be judged positively also for its indirect selective effect. The payment has been considerably increased and this could be regarded, as a matter of fact, as an admission of the inadequacy of the amount previously fixed and can be considered as an improvement that may better help to overcome the entry barriers faced by newcomers, Though it is worth noting that not all the Italian Regions have chosen to fix the aid at the maximum level.

One additional novelty can be found in the implementation of the Regulation that, among others, allows the so called 'Cluster of Measures' for the setting up of young farmers. Young people who apply for the setting up aid may also apply for other measures that altogether raise the total amount of the aid and provide a wider range of sustain to the setting up. Coupled measures vary region by region but usually include: vocational training and information actions, use of advisory services, modernization of agricultural holdings, meeting standards based on community legislation and participation of farmers in food quality schemes. Many programs include preferential access and/or increased payments for these measures when included in the cluster for the setting up of young farmers.

The New Resolution of the European Parliament on Young farmers[7]

On June 5th 2008 the European Parliament has approved the resolution named 'The future of young farmers in the schedule of the present CAP reform' which comes eight years after a previous resolution on the same subject, at the initiative of Honourable Donato Tommaso Veraldi.

Starting from a description of the present status of affairs, the Resolution explicitly considers the role of the CAP in addressing the situation and starts a wide re-thinking of how the CAP should shape after 2013, having in mind possible in-progress adjustments arising from the present Health Check schedule. The outcome is an assessment of the following targets:

The generation turnover is proposed as a *condicio sine qua non* for the survival of European agriculture; therefore it asks to strengthen the instruments aimed at influencing it. According to the European Parliament, hence, one of the goals of the reformed CAP should be to reckon upon an adequate generation turnover.

In order to define suitable interventions and the related measures to be implemented, a specific analytical effort is required, both on the causes and characteristics of the phenomenon as well as on the reasons of poor efficacy that past intervention, has proved so far, taking into special consideration the different national contexts as they arise also from the enlargement process.

The document acknowledges that the topic of intergenerational turnover has been underrated and disregarded but it also assesses openly the negative effects the CAP has had on the possibility of access the sector. Therefore, specific studies are required to elaborate the nature of these effects and to quantify them exactly, in order to avoid undesirable interactive effects between measures undertaken under the first and the second pillars in the future.

Then the document describes desirable interventions both within the CAP and national policy framework. These can be distinguished in two groups: on the one hand, those aimed at strengthening existing CAP provisions; and, on the other hand, new measures proposed at community level, which are already in place in some member countries.

As for the already existing measures, the resolution outlines changes meant to strengthen their effectiveness that can be summarized as follows:

- The measure for the setting up of young farmers should be included in each Rural Development Plan, so that Member States and Local Governments should not be left free to decide the implementation.
- The measures comprised under Axes II and III, to promote infrastructures and the labour markets, should prioritize the settlement of young farmers.
- The size of the payments should be increased, especially in socially and environmentally sensitive areas (such as islands and mountains).
- More rigour should be applied in screening the requests as well as a more

---

7   This chapter is widely based on Carbone 2008.

tight definition (and enforcement) of what constitutes a 'new settlement'; this would improve the targeting of the measure and avoid inappropriate uses of the related resources.

The resolution makes one more point that, at a first sight, may seem inconsistent with what just specified, that is the necessity to extend the payment to young farmers who, though already settled in farm holdings, run a farm of suboptimal dimensions. With such payments, they could cut down the farm holding inefficiency and the uncertainty of economic results. Since estimating such condition and the related reasons can be very difficult in practice, including this new category of recipients may twart the suggested greater tightness of the criteria for eligibility.

The start-up period is estimated to be of 3 to 5 years. At the end of this period the farmer should have reached the targets reported in the 'business plan'. The length of such period stems from the production cycles, the need to integrate in the market, the variability arising from the natural and the economic environment.

The importance of the professional /vocational training is also stressed, while more professional training is suggested either through courses or other activities, study tours, or inter-regional farmers' meetings from different areas which might enable to put together/in common, improve and widespread desirable farm/ agricultural techniques and practices.

Even more interesting is to examine the innovative proposals suggested by the European Parliament in the Resolution. Hereafter, a short description and a comment on these is offered.

Very much appropriately, the Resolution recalls the general reasons of the ageing and reduction of agriculture, which stems from the comparatively low profitability of agriculture and living conditions in rural areas, which may not always be comparable to urban standards. These conditions would hold back young workers from getting into agriculture. However, this is nowadays true only in some specific contexts, and to a limited extent: while some services such as schools, transportation, health may be of lower quality, others would not; rural areas might be advantaged in terms of fewer crowds, less pollution, more space, lower cost of dwellings, a more friendly social environment. The document also recalls that being a farmer not always enjoy the social appreciation and prestige that it deserves. However this seems to be the case in countries undergoing transition where agriculture shrinks and is still associated with poverty and backwardness.

To counteract the reasons that keep the young away from agriculture, the European Parliament recalls that a wide range of interventions are necessary, that the general living conditions in rural areas improve, and farmer improves its reputation as a profession. Initiatives such as the '*European Year of Dialogue between Town and the Countryside*' could promote this change, or the institution of a European quality brand guaranteeing consumers the origin and healthiness of food as well as the existence of a European agriculture to safeguard the environment, cultural traditions and the safety of purchasing. However, these initiatives albeit commendable, could hardly make a visible difference in terms of the data mentioned above.

The document of the European Parliament also sketches out measures that can be directly implemented. For instance, the potential role of a *'bank of agricultural lands'* to be constituted on the basis of the lands released by withdrawals and early retirements which may support the settlement of young farmers. As a matter of fact, this would be an inventory of the demand and offer of agricultural land, including details of the natural, structural and infrastructural characteristics, that would facilitate trading and turnover by providing information and hence reducing transaction costs. Moreover, an institution performing this function may play an intermediary role in support of both the old farmer who wants to withdraw and the young one who is willing to take over.[8]

The startup of a new farm would proceed gradually, by acquiring the specific knowledge and participating in the network of commercial relations accumulated by the entrepreneur who is leaving. The young farmer could also rely on technicians and experts guiding her in the bureaucratic procedures as well as through the technical and economic aspects of farming.

An important role is also assigned, in the Resolution, to the interventions in the credit market, aimed at facilitating access to credit for young farmers'. The setting up of insurance patterns might – for the starting period – also protect farmers from the consequences of weather adversities, market fluctuations and other unexpected events. Finally, in order to further facilitate access to credit, Member States may adopt fiscal measures to lower the interest rate on capital.

Concluding Remarks

The data shown in Section 2 shows that the ageing of farm holders is a widespread phenomenon in the EU, with variable intensity across Member States. As a consequence, European agriculture still faces a loss of production factors among which the abandon of agricultural land and the depopulation of some rural areas causes major concerns. The structural adjustment process is distorted by the presence of many constraints and barriers that limit and prevent generational turnover. Facing these difficulties, the primary sector seems condemned to shrink progressively and to lose competitiveness.

The efforts made by the CAP to solve the problem have widely proven not to be much effective, while the general framework of the CAP – in the first phase with the market interventions and in the second with the direct payments- has had (and it is still having) a counterproductive effect. The European Parliament Resolution expressed the most updated and authoritative judgement in this direction. We argue that the measure for the setting up of young farmers has also had negative

---

8   The model could be the *Centre National pour l'Aménagement des Structures des Exploitations Agricoles (Cnasea)* which has been working in France for some years.

distributional effects, since beneficiaries have often been selected, as a matter of fact, as young person belonging to families owing a farm; whether or not intending to start running the family enterprise, whether or not already running it.

The European Parliament has highlighted the situation in the most appropriate and unequivocal way, and at the same time it has recognized the many causes and the possible solutions. It has also seen that the new Regulation on rural development introduced some innovations that seem to be going in the right direction. More could be done, however, to overcome the many barriers that are encountered in the access to the sector. Among the actions recalled in the Resolution of the European Parliament, in the Italian situation very much appropriate would be any actions to facilitate the access to the credit system and to lower the cost of loans. This would, at the same time, help to overcome one of the most stringent access barrier and help to select beneficiaries that express a real commitment to enter the sector.

It is worth remembering once more, however, that if the presence of young people could help the sector in reaching a better level of competitiveness -for as they are more dynamic, have more propensity to risk, generally represent a higher level of human capital- it is at the same time true that the general context plays a decisive role for the competitiveness of the single enterprise.

Competitiveness is less and less determined inside the holdings and is increasingly more related to the coordination of the stakeholders along the *production chain,* who interact on the territories at a local level. Consequently, in order to attract young farmers into agriculture steadily, ensuring them reasonable levels of remuneration, one cannot disregard the overall sector's competitiveness conditions. All the interventions aimed at this target, and in this respect efficacious, represent also, indirectly, an efficacious policy for the settlement of young people. On the contrary, with an agricultural and food system, on the whole non competitive, no incentive measure to entry the sector will give positive and stable results.

A very last issue it is worth pointing out relates to the most appropriate ultimate goal that should be pursued by a policy measure aimed at providing incentives to farm turnover. With the Fischler Reform the measure for the setting up of young farmers is directly linked to competitiveness and this seem to be very reasonable and appropriate in the view of a sectoral policy. At the same time the new CAP is more and more interlinked with other policies and, on the other hand, the intervention for the rural areas is increasingly conceived as only partially inscribed in the intervention for the agricultural sector. With rural policy, the EU does not only look at the sector that produces agricultural market goods, also look at a bundle of activities producing a wide set of products and services, marketable as well as not marketable. At the same time, at least in some EU countries, the rural areas have become more attractive for elder people that retire or that starts a part-time activity and wish to leave expensive and congested cities. This people can give, and often does, a contribution to the revitalization of the rural areas, but should be induced to produce those externalities and services needed. In this light, limiting the policy for the survival of the European agriculture to the measure for the setting up of young farmers is, according to us, very much reductive.

References

Agriconsulting S.p.A. 2005. Aggiornamento Rapporto di Valutazione Intermedia del Piano di Sviluppo Rurale 2000-2006 della Regione Toscana, Sito web Regione Toscana, sezione Agricoltura, tema sviluppo rurale.

Agriconsulting S.p.A 2006. Aggiornamento Rapporto di Valutazione Intermedia del Piano di Sviluppo Rurale 2000 2006 della Regione Emilia Romagna, Sito web Regione Emilia Romagna, portale Ermes Agricoltura, piano regionale di sviluppo rurale 2000-2006

Barberis, C. and Siesto, A. 1993. Agricoltura e strati sociali, Franco Angeli, Mi

Cagliero, R. and Novelli S. 2005. Insediamento giovani: un tentativo di interpretazione mediante SWOT dinamica. AgriRegioniEuropa 3, 27-31

Carbone, A. 1996. 'La presenza di giovani in agricoltura', La Questione Agraria n. 61, 143-168

Carbone, A. 2005. La misura per l'insediamento dei giovani in agricoltura: pubblici vizi e 'virtù' private. AgriRegioniEuropa, 9-10

Carbone, A., Corsi, A. and Sotte, F. 2005. La misura giovani tra nuovo regolamento sullo sviluppo rurale e prime evidenze dell'applicazione 2000-2003. AgriRegioniEuropa, 2, 9-12

Carbone, A. 2008. SOS dal Parlamento europeo: senza turnover generazionale l'agricoltura muore. AgriRegioniEuropa, 14, 41-43

Consiglio dell'Unione Europea (2003). Regolamento(CE) n. 1783/2003 del 29 settembre 2003 che modifica il regolamento (CE) n. 1257/1999 sul sostegno allo sviluppo rurale da parte del Fondo europeo agricolo di orientamento e di garanzia (FEAOG). (G.U.U.E. 21.10.2003, n. L 270)

Consiglio dell'Unione Europea. 2005. Regolamento (CE) n. 1698/2005 del 20 settembre 2005 sul sostegno allo sviluppo rurale da parte del Fondo europeo agricolo per lo sviluppo rurale (FEASR) (G.U.U.E. L 277 del 21.10.2005)

Corsi, A., Sotte, F. and Carbone, A. 2005. Quali fattori influenzano il ricambio generazionale?. AgriRegioniEuropa, 12-16

Corsi, A. 2009. Family farm succession and specific knowledge in Italy. Rivista di Economia Agraria 1, Forthcoming

De Gaetano, L. and Massoli, B. 2004. L'invecchiamento dei conduttori agricoli e le difficoltà del ricambio generazionale,in La liberalizzazione degli scambi dei prodotti agricoli tra conflitti e accordi. Il ruolo dell'Italia edited by E. Defrancesco, Franco Angeli, MI, 503-518

Glauben, T., Petrick, M., Tietje, H. and Weiss, C. 2009. Probability and Timing of Succession or Closure in Family Firms: A Switching Regression Analysis of Farm Households in Germany. Applied Economics 41(1), 45-54

INEA-OIGA. 2005. Insediamento e permanenza dei giovani in agricoltura. Rapporto 2003/2004, Roma

Mazzieri, A. and Esposti, R. 2005. Quanto sono diverse le imprese agricole 'giovani'?, AgriRegioniEuropa 2, 18-21

Ministero delle Politiche Agricole e Forestali. 2006. Piano Strategico Nazionale per lo Sviluppo Rurale, Sito web Ministero politiche agricole, sezione sviluppo rurale

Parlamento Europeo. 2008. Risoluzione 5 giugno 2008, 'Futuro dei giovani agricoltori nel quadro dell'attuale riforma della PAC'. Procedura 2007/2194 (INI), Sito web del Parlamento Europeo

Simeone, M. 2006. Le determinanti del trasferimento intergenerazionale in agricoltura: un'analisi empirica basata sulla stima di un modello probit. Rivista di Economia Agraria 4, 519-539.

Sotte, F. 2005. Affinché riprenda la riflessione strategica sul futuro della PAC. *AgriRegioniEuropa*, 1-7.

PART 8
The Future of Market Measures

Chapter 25

Phasing Out Milk Quotas: a Bio-Economic Model to Analyse the Impacts on French Dairy Farms[1]

Baptiste Lelyon, Vincent Chatellier and Karine Daniel

Introduction

For EU dairy farmers the Luxemburg agreement, decided in 2003, marked a new phase in the process of Common Agriculture Policy (CAP) reform. Direct payments were decoupled and the Single Payment Scheme (SPS) was implemented. This reform aimed to increase the competitiveness of European agriculture and to promote a market orientated agricultural sector. The new CAP reform proposed in 2008 (known as the CAP Health Check) maintains these objectives of decoupling and the removal of the milk quota system in 2015. In France, more than in some others EU member states, this decision raises questions because the government historically favoured a balanced geographical distribution of milk production through an administration of milk quotas. Moreover, for dairy farmers, these changes occurred simultaneously with an unprecedented market situation, namely high price fluctuations of agricultural raw materials.

In this context, the aim of this article is to study the implications of the abolition of milk quotas on dairy farmers' behaviour (i.e. effects on the production system, the allocation of areas to crops, and the level of intensification) with different hypothetical prices. A bio-economic model is developed and applied to four case studies: French dairy farms often have, in addition to the dairy activity, cereal or beef production. The model will allow us to highlight changes in the productive strategies of farmers in particular by studying the balance between the different productions (milk, cereals, meat) and the environmental impact at farm level (evolution of nitrogen pressure, use of purchased feed, intensification of milk production and fertilizer use).

1   This work is included in the research programmes : "Laitop" (PSDR Grand-Ouest) and "Dynamics of Dairy Territories" (coordinated by FESIA and funded by CNIEL, Credit Agricole, Groupama and Seproma)

Dairy policy setting

In France, as in all member states of the EU, milk production has been regulated since 1984 at the producer level (any excess over the authorized quantity causes a financial penalty). Milk quotas were introduced in order to control the supply of milk in a context where the storage costs of the dairy surplus became an important issue for the EU budget. Moreover, in a context marked by a modest growth in domestic consumption of dairy products and a strong competition with the countries of Oceania (Australia and New Zealand) on export markets, the authorities have been forced to progressively reduce quotas in most member states. Thus, France has lost 12 per cent of its milk production in twenty-five years and nearly half of his herd of dairy cows (due to the steady rise in the milk yield). In France, state intervention in the management of milk quotas is stronger than in most other Member States, particularly the United Kingdom, Denmark and the Netherlands where milk quotas are tradable. Indeed French authorities have adopted rules to limit the geographic concentration of milk production in regions/departments with comparative advantages: milk quotas are managed administratively in each department and they are linked to the land. A producer who wishes to increase milk production must necessarily acquire or rent hectares. The transactions of quota between producers are made by administrative decision (free attribution of volume to priority producers) and not through the market.

This regulation method for the milk supply (quota) within the EU and France is also applied in other countries such as Canada (where quotas are tradable between producers) and New Zealand (where the volumes are managed by the monopolistic cooperative enterprise, Fonterra, which provides the collection, processing and export of milk). In the United States, unlike the EU and France, milk production strongly increased over the past fifteen years (1.5 million tons of milk per year). This increase in supply, in a non limited system, allows them mainly to meet domestic demand, because exports on the world market are still relatively limited (10 per cent of the world market volume in 2008).

Twenty-five years after the implementation of milk quotas, the European Commission (2008) estimates that: "The current market outlook situation indicates that the conditions for which milk quotas were introduced in 1984 are no longer relevant." It proposes a phasing-out of milk quotas with a gradual annual increase to prepare farmers for a market without quotas post 2015. This decision also reflects some theoretical arguments against this way of regulation. Many studies (see e.g. Alvarez et al. 2006, Boots et al. 1997) show that the milk quota system is a source of inefficiencies with a non-optimal allocation of quota among producers because a high number of vulnerable and inefficient producers remain in milk production. Colman (2000) and Henessy et al. (2009) show that even if milk quotas are tradable, there are lags in adjustment and imperfections such that the theoretical optimum has not been achieved.

The removal of milk quotas by 2015 which seems to be accepted by the majority of EU Member States, raises many questions in France. These questions concern, on the one hand, the evolution in the geographical concentration of production at the national level and, on the other hand, the evolution in milk prices paid to producers. On this last point, it is clear that the milk quota system has allowed the French and European producers to benefit from stable and remunerative prices over the past two decades.

In the absence of a milk quota, the risk of a greater price volatility and lower prices exists (Bouamra-Mechemache et al. 2008), all the more so as the elasticity of demand is low in this sector. Several studies, based on partial and general equilibrium models, have already assessed the impact of the abolition of milk quotas on the price level in the EU (Kleinhanss et al. 2002, Lips and Rieder 2005, Gohin and Latruffe 2006). They showed that such a policy would lead to an increase in European milk production by 10 per cent for a diminution in prices by 26 per cent.

Methodology

In order to study the adaptation of farmers' practices in response to the abolition of the milk quota, a mathematical programming model was built. This method allows us to identify the effects of the decoupling on the production system: the allocation of land areas to different crops, the level of intensification, environmental impact.

Bio-economic model: a farm-level approach

We built a bio-economic model which takes into account the farmer's response to price variation and several technical and biological elements in order to represent as accurately as possible the functioning of a dairy farm. Mathematical Programming is a technique which enables us to represent the farm functioning in reaction to a set of constraints. It is an appropriate technique because its assumptions correspond to those of classic microeconomics: rationality and the optimising nature of the agent (Hazell and Norton 1986). This method allows us to study threshold effects and to calculate dual values of inputs (marginal yields). Farm-level modelling enables simultaneous consideration of production, price and policy information.

Any model derived from mathematical optimisation has three basic elements (Matthews et al. 2006): I) an objective function, which minimises or maximises a function of the set of activity levels; ii) a description of the activities within the system, with coefficients representing their productive responses; and iii) a set of constraints that define the operational conditions and the limits of the model and its activities. Given the objective function, the solution procedure determines the optimal solution considering all activities and restrictions simultaneously.

The model optimises the farm plan, which represents the quantities of different outputs produced and inputs used. The economic results follow from those quantities and their prices. The model is used to estimate the effects of institutional, technical and price changes on the farm plan, economic results and intensification indicators.

Many studies have demonstrated that farmers typically behave in a risk-averse way (Hardaker et al. 2004). As such, farmers often prefer farm plans that provide a satisfactory level of security even if this means sacrificing some income. For the farmer, the main issue raised by variability of price and production is how to respond tactically and dynamically to opportunities or threats in order to generate additional income or to avoid losses. Moreover, during the years 2007, 2008 and 2009, prices of agricultural commodities were subject to strong variations so that we had to take the farmer's sensitivity to price volatility. For example, the price of milk paid to the producers nearly doubled through 2007, from EUR240/t to EUR380/t before strongly decreasing to EUR220/t in April 2009. Since the beginning of 2010, milk price seems to be on an increasing trend. Prices of cereals such as wheat have followed the same fluctuations. Cereals play a special role in dairy farming because they can be both input and output.

Lambert and McCarl (1985) present a mathematical programming formulation that allows identification of the expected utility function. Their approach, which does not require an assumption of normally distributed income (unlike the E-V, MOTAD and Target MOTAD methods), can accommodate the assumption that the utility function is monotonically increasing and concave (risk-averse). Patten et al. (1988) reformulated this approach as utility efficient programming (UEP). Moreover, Zuhair et al. (1992) show that the negative exponential utility function (with constant absolute risk aversion, CARA) can better predict farmers' behaviour than cubic and quadratic functions. The CARA function is a reasonable approximation to the real but unknown utility function: the coefficient of absolute risk variation can be validly applied to consequences in terms of losses and gains for variations in annual income. The UEP method enables the model to take into account asymmetric price distribution: the skewness becomes an element of decision as well as the variation amplitude. Thus, the model maximizes the expected utility of the income as follows:

Maximize: E[U] = probability U(k, r), r varying
with: $U_k = 1 - \exp(-r_a \times Z_k)$
where Z_k is the net farm income for state k, and r is a non-negative parameter representing the coefficient of absolute risk aversion:

$r_a = (1 - \lambda)r_{min} + \lambda r_{max}$, for $0 \leq \lambda \leq 1$[a]
where λ is a parameter reflecting variation in risk preference, and r_{max} and r_{min} are

upper and lower bounds of the coefficient of absolute risk aversion (r_a). In a more detailed form, the income Z is defined by:

$$Z = \sum_a \left(T_a \times mY_a\right) \times 305 \times mP$$

$$+ \sum_a \left(aS_a \times aW_a \times aP_a\right)$$

$$+ \sum_a \left(T_a \times \left(SP_a + SPBM_a\right)\right)$$

$$- \sum_{a,p} \left(T_a \times \left(Qcf_{a,conc,p} \times cfP_{conc} \times 91.25 + I_a\right)\right)$$

$$+ \sum_c \left(X_c \times \left(Y_c \times cP_c - I_c - nQ_c \times nP + pr\right)\right) - FC$$

- The main part of the income Z is given by milk revenues: the milk quantity multiplied with Ta the total number of animal of type a (dairy cows, heifers, calves and young bulls); mYa the milk yield (litre/day) per animal by mP the milk price (EUR/litre).
- There is then the meat revenue with *aSa* the number of animals sold, *aWa* the animals' average carcass weight (kg) and *aPa* the meat price (EUR/kg). At the end of the lactation, cull cows are sold and benefit from the female slaughter premium (*SPa*) and young bulls benefit from the special premium for bovine male (*SPBMa*).
- Then we take out livestock costs as: *cfQconc,p,a* the quantity of concentrate feed ingested (kg/day/animal), *cfPconc* the concentrate feed price (EUR/kg per type of concentrate *conc*); *Ia* the specific inputs for animals (artificial insemination, medicines, herd-book and minerals).
- We add the crop revenue as: *Xc* the cultivated area (ha) for each type of crop *c* (wheat, maize (corn), rapeseed, pea, maize silage, pasture, hay and grass silage); *Yc* the crop yield (kg/ha); *cPc* the crop price (EUR/kg); *Ic* the specific crop inputs (seed, treatments and harvesting); *nQc* the nitrogen quantity (kg/ha); and nP the nitrogen price (EUR/kg).
- Finally we consider the fixed costs *FC* (electricity, water, mechanisation, buildings, rent for land, insurance, taxes and other fixed costs). These fixed costs are specific to each type of farming.

The central element in the LP model is the dairy cow. The model represents the operation of a dairy farm for a one-year period. The classical duration of lactation is 305 days, followed by 60 days of drying off. The year is divided into four seasons of 91.25 days. The fecundity rate is lower for the most productive cows, thus decreasing the number of calves per cow per year. Regarding the progeny, it is assumed that, according to the intensification level of the type of farming, 25 per cent to 35 per cent of the dairy cows are replaced per year by heifers raised on the farm (Institut de l'Elevage 2008). Concerning female calves which are not assigned to replace cows, the model can choose between: i) selling the calves at the age of 8 days; and ii) keeping the calves until 2 years old and then selling to the slaughterhouse (with the female slaughter premium).

Regarding plant production, the forage crops produced in France are mainly maize silage, grass silage, hay and pasture. All farmers aim for forage self-sufficiency; the purchase and/or sale of forage are not considered because these are activities linked to exceptional events (*e.g.*, drought or exceptional harvest) in these areas. Farmers must comply with the set-aside requirement in order to benefit from the crop premium: we use a binary variable which is 0 if the farmer does not set aside land, and 1 if he does. It is assumed that the cereals are sold at harvest time, i.e. no crop storage except for wheat used to feed the cows.

Thornton and Herrero (2001) show a wide variety of separate crop and livestock models, but the nature of crop–livestock interactions, and their importance in farming systems, makes their integration difficult. That is why, in order to precisely describe the operation of a dairy farm, this model considers four important characteristics: I) the seasonality of labour and grass production, ii) the response of crop yield to nitrogen use, iii) the non-linearity of milk yield per cow, and iv) the interaction between crop and animal production.

i) Four periods p (spring, summer, autumn and winter) are distinguished in the model. It allows for seasonal specification of grass production and grassland use (Berentsen *et al.* 2000). Seasonal variations enable us to integrate differences in the growth potential of grass during the growing season as well as the evolution of the nutrient content of grass. Moreover, we introduce seasonal labour constraints by allocating labour needs to each activity according to the work peaks (harvesting and calving). It is assumed that the farmer and his family/associates execute all the work, and thus there is no option to hire temporary labour. The model is more able to reflect temporal conditions thanks to the addition of these parameters.
For each period p:

$$\sum_a \left(\left(Wt_{a,p} \times T_a \right) + \left(Wt_{c,p} \times X_c \right) \right) + FL \leq AL_p \times AWU$$

The global working time per period (with $Wt_{a,p}$ the working time per animal; $Wt_{c,p}$ the working time per ha of crop; FL is the fixed labour) has to be lower than the labour availability per period (AL_p the available labour for each annual work unit (AWU)).

ii) Crop yield depends on the quantities of nitrogen used. Godard *et al.* (2008) formulated an exponential function, which satisfies economic requirements for attaining a mathematical optimum (the yield curve has to be concave and strictly increasing) and is consistent with its expected agronomic shape and with parameters with an agronomic interpretation.

$$Y_c \quad Ymin_c \quad Ymax_c \quad t_i \quad N_i$$

where Y is yield for each crop, and $Ymin_c$ and $Ymax_c$ are respectively the minimal and maximal yield (different according to the type of farming and its level of intensification); t_i represents the rate of increase in the yield response function to a nitrogen source i (e.g. manure, slurry, chemical nitrogen) the quantity of which is N_i. This enables us to take the increasing price of nitrogen into account and also the flow of organic nitrogen (such as manure) on the farm (Manos et al. 2007).

iii) In order to give more flexibility to the model, milk production per cow is not fixed. Farmers have the possibility to choose the milk yield per animal in a range of 1,000 litres below the dairy cow's genetic potential. It is also possible for farmers to produce beyond the genetic potential (Brun-Lafleur et al. 2009); in this case, nutritional requirements needed to produce one litre of milk are increased (from 0.44 to 1.2 energy units per litre of milk, and from 48 to 140 units of protein per litre of milk) (Faverdin et al. 2007).

iv) With these three elements, we can very accurately represent the feeding system. The quantity ingested per cow per day is determined by using i) nutritional requirements in biological unit b (energy and protein), and ii) the composition of forages and concentrate feed in equation 6 (INRA 2007). The concentrate feeds *conc* available in the model are soybean meal, rapeseed meal, wheat, production concentrate and milk powder.

For each nutrient unit b and period p:

$$\sum_a \left(T_a \left(MR_{a,b} \times 365 + mY_a \times LR_{a,b} \times 305 \right) \right) \leq$$
$$\sum_{a,c} \left(T_a \times \left(fQ_{c,p,a} \times fnc_{c,p,b} \times 91.25 \right) \right) + \sum_{a,conc} \left(T_a \times \left(CfQ_{conc,p,a} \times Cfnc_{conc,p,b} \times 91.25 \right) \right)$$

with: $MR_{a,b}$ the maintenance requirement (in energy and protein)
MYa the milk yield (in litre per animal per day)
$LR_{a,b}$ the lactation requirement (in energy and protein for one litre of milk)
$fnc_{c,p,b}$ the forage nutrient content (in energy and protein per kg of forage)
$fQ_{c,p,a}$ the forage consumption (kg) for each crop c, each period p and each type of animal a
$Cfnc_{conc,p,b}$ the concentrate feed nutrient content (in energy and protein per kg of concentrate)
$CfQcon_{c,p,a}$ the concentrate feed consumption (in kg per day per concentrate per period per animal)

The global nutritional needs for the herd must not exceed the availability in forage and concentrate feed. Moreover, the forage consumption (for each type of forage c) has to be lower than the forage production:
subject to:

$$T_a \; aS_a \; mY_a \; CfQ_{conc,p,a} \; fQ_{c,p,a} \; X_c \; nQ_c \; Y_c$$

for each type of crop c.

Consequently, in order to maximize the farm's income, the model determines the optimum for the following endogenous variables: number of each type of animal (T_a and aS_a for sale); milk yield per cow (mY_a in kg per cow per day); concentrate feed and forage consumption for each type of animal and per period ($CfQ_{conc,p,a}$ and $fQ_{c,p,a}$ in kg per animal per day per season); the crop rotation (X_c in ha); the level of nitrogen fertilisation (nQ_c for chemical nitrogen and manure, in kg); and crop yield (Y_c in kg per ha).

The model tries to offer the largest choice of technical practice for crop and animal production. That is why we choose to incorporate each "quantity variable" (as ha and kg) as endogenous variables in the model. Thus, the model has access to all possible situations, e.g.: the model can choose a full grass diet for a cow which produces 7,000 litres of milk or a full maize diet for the same cow. The model will therefore calculate the optimal quantity of input and output.

The constraints

Regarding the farm structure, the model incorporates the agricultural area, the milk quota and the available labour resources. As regards building constraints, we assume that the number of cows can increase by 10 per cent in comparison to the base year: the implementation of the programme to control pollution of agricultural origin has motivated many dairy farmers to construct new buildings with more places than required. Regarding crops, the model meets the requirements for rotation frequency and cropping pattern (Mosnier et al. 2009).

We also include three environmental measures as constraints in the model: I) the Nitrate Directive 91/676/EEC requires that farmers cannot exceed organic nitrogen application rates of 170 kg per hectare (slurry and manure); ii) farmers have to keep grasslands aged over 5 years; iii) in addition to the CAP premiums, a premium for the maintenance of extensive livestock systems or "premium for grassland" is attributed (EUR75/ha), if there is at least 75 per cent of grass in the total farm area and if the stocking rate is below 1.4 "livestock units" per hectare of grass.

Calibration: one model for four types of farming

In France, there is a high diversity of dairy farms in terms of location (mountains/ plains), intensification (intensive/extensive), feeding system (pasture, maize silage) and specialisation of production (specialized/diversified). In this context, our choice focused on the four main types in the plains regions of France: these regions are not located in the less favourable areas and do not benefit from these specific supports, we exclude the mountain areas which have a different productive system for producing milk. The data come from the annual survey of the Institut de l'Elevage (2008) with more than 600 dairy producers in the plains regions. Each type of farming is the result of the aggregation of several farms (from 20 to 45) representing similar structures and production methods.

- *The "Grass-based farm"* is a 78 ha family farm with 285,000 litres of milk quota. It produces milk with a large area of grass, which provides high fed autonomy. The milk yield per cow is low (6,000 litres per year) but the prices of milk and meat are higher thanks to a better milk composition and heavier carcasses (Normand or Montbeliarde cow). The age of first calving is 30 months and the calving period is in the spring. Cows are housed for 4 months while they consume maize. It represents eight per cent of the dairy farms in this area.
- The "Semi-intensive farm" is a 50 ha family farm with 290,000 litres of milk quota (18 per cent of the farms in the plain region). The calving period is in the autumn, which is why the use of maize is higher. The cows are more productive: Prim' Holstein with a milk yield of 8,500 litres per year.
- The "Milk + cereals farm" is a highly intensive system with 137 ha and 460,000 litres of milk quota. Each cow can produce 8,500 litres per year, and consequently the use of maize in the ration is not limited. Dairy production is the main activity on the farm, but cereal cropping is developed in parallel (wheat, rape seed, maize and pea). It represents 22 per cent of the farms in the plains regions.
- The "Milk + Young bulls farm" has 100 ha and 400,000 litres of milk quota. It is the most representative system of the area: 30 per cent of dairy farms. It has the same characteristics as the previous type, but young bull fattening activity replaces the cereal activity. The model can choose to fatten (or not) the males and buy (or not) other male calves to reach 80 young bulls. These animals are slaughtered when they are 20 months old. The young bulls benefit from the male slaughter premium (EUR80/animal) and the special premium for male bovines (EUR110/animal).

Table 25.1   Farm data for the year 2005

	Grass-based Farm	Semi-intensive Farm	Milk+cereals Farm	Milk+young bulls Farm
Share of the system in France (%)	8%	22%	30%	18%
Total area (ha)	78	50	137	100
Milk quota (litres)	285,000	290,000	460,000	400,000
Annual Work Units (no.)	1.7	1.5	2.0	2.7
Building capacity (no.)	62	37	59	122
Restocking rate (%)	0.25	0.35	0.37	0.4
Dairy genetic potential (l/year)	6,000	8,500	8,500	9,000
Max crop yield (kg/ha/year)				
Wheat	6,100	8,100	8,100	8,100
Maize	n.a.[1]	n.a.	10,000	n.a.
Rapeseed	n.a.	n.a.	3,800	n.a.
Pea	n.a.	n.a.	5,000	n.a.
Maize silage	10,200	12,200	15,200	14,200
Grass silage	8,500	8,500	8,500	8,500
Grass	8,500	7,000	6,000	6,000
Hay	8,500	7,500	7,500	7,500

	Grass-based Farm	Semi-intensive Farm	Milk+cereals Farm	Milk+young bulls Farm
Milk price (EUR/litre)	330	310	310	310
Meat price (EUR/kg)	3.0	2.6	2.6	2.6
Dairy cow carcass weight (kg)	375	325	325	325

[1]n.a.: not available

The farms of this study are located in plains areas and do not benefit from a *protected designation of origin*. Therefore, the milk processors, who collect the milk, produce cheese, yoghurt, ice cream, and liquid milk, but also butter and milk powder, which can be sold on the global market. There are no specific requirements to produce this milk in order to receive some special promotion (better milk price for giving the cows a specific food).

A calibration step is necessary: the model's results and the empirical observations have to be close. We choose the year 2005 as baseline (i.e. before the implementation of the Luxembourg agreement).

Table 25.2 gives the price level and price variation for the main inputs and outputs. With these values, we build, for each product, a random distribution of price (for 1000 states of nature k) within the range of variation and compute the model to calculate the expected utility. The use of the UEP method allows us to calculate the risk premium for each type of farming because we know the utility level.

$$E[U] = \text{probability } U(k, r) \text{ with: } U_k = 1 - \exp(-r_a \times (RP - Z_k))$$

where: U is the level of utility, r_a the coefficient of absolute risk aversion, Z the income, and RP the risk premium.

We choose an appropriate value of the coefficient of absolute risk aversion in order to calibrate the model. Bontems and Thomas (2000) show that the ratio *risk premium / income* should be around 5 per cent. Thus, the value of the coefficient of risk aversion is about 0.5 for the four types of farming. The results of the model are close to reality for the four main key criteria: income, milk yield per cow, share of cereal in total area, and share of maize silage in forage area.

Table 25.2   **Price level and price variation for inputs and outputs**

	2005 price level	Price variation (%)
Milk (EUR/litre)	0.31	10%
Meat (culled cow)	2.60	20%
Meat (young bull)	2.90	20%
Cereal crops		
Wheat (EUR/kg)	0.120	30%
Maize (EUR/kg)	0.110	30%
Rape seed (EUR/kg)	0.240	30%
Pea (EUR/kg)	0.130	30%
Concentrate feed		
Cereal (EUR/kg)	0.140	30%
Soybean meal (EUR/kg)	0.220	30%
Rapeseed meal (EUR/kg)	0.180	30%
Chemical nitrogen (EUR/kg)	0.150	30%

Results

The results of the simulations are compared to a baseline 2008 which takes into account the full implementation of the CAP Health Check measures (full decoupling of animal and crop premium and removal of set-aside). In the simulation, we assume that milk producers have the opportunity to increase their milk production up to 25 per cent compared to the baseline. This rate was fixed arbitrarily by considering that the removal of quotas would result in an increase in contracts between milk producers and processing companies. Indeed, companies or cooperatives could be encouraged to replace public regulation through certain contractual policies. The producers will be limited in their productive potential by the rules established within the framework of a contract, itself dependent on the historical milk quota. This hypothesis (+25 per cent) is retained by considering, first, that the milk quota will be increased by 5 per cent between 2009 and 2015 (following the decisions of November 2008) and, secondly, that the restructuring process will lead to a decrease in the number of dairy farms by 20 per cent over the period. In other words, the authorized volume growth is permitted with a simultaneous decrease in the number of farms.

Without the use of milk quotas, the future milk price is not predetermined. If these contractual policies permit a rigorous management of the collective supply, the producer milk price reduction could be less severe than calculated by theoretical models. Therefore the price of milk in the model (EUR280/t) is identical between the base year and the simulation, but we also test the sensitivity of dairy producers to milk and cereal price variations.

Base year results

Regarding the two farms specialized in milk production (*Grass-based* and *Semi-intensive*), they have a very similar economic dimension (EUR500 difference of income) while the structure of these two farms and their strategy are very different (Table 25.3).

The *Grass-based* farm opts for an extensive system of production: the whole area of the farm is dedicated to grassland, thus enabling it to meet the criteria of the "Grass Premium" and benefit from EUR5,900/year. With an annual milk yield of 5,250 litres per animal, the farmer chooses to remain 750 litres below the genetic potential of dairy cows. This decision has the effect of requiring a greater number of animals in order to produce the same quota. In doing so, the farmer has a greater meat revenue (the prices of milk and meat are higher thanks to a better milk composition (fat and protein) and heavier carcasses (Normand cow). This low level of production allows the farmer to apply a low cost strategy since dairy cows consume only 120 kg of purchased concentrate feed per year. Note that the variable costs for the *Grass-based* farm are lower than the *Semi-intensive* farm, however the larger size of the *Grass-based* farm (area, building) generates a higher amount of fixed cost.

The *Semi-intensive* farm applies a more productive strategy even if it has only 50 ha of total area (28 ha less than the *Grass-based* one) for an equivalent quota: it allocates nearly 25 per cent of this area for cereal production (with a EUR16,500 profit per year). Furthermore, the farmer chooses to use maize silage to feed the animals and to use a high amount of concentrate feed (890 kg/VL/year) allowing him/herr to reach a milk yield of 8,500 litres per year. It reduces the number of animals required for the production of the milk quota and thus free-up land for cereals. The dual values of land and quota are positive for both farms showing that an increase of these two inputs would result in a higher income, however the Semi-intensive farm is more highly constrained by the land factor.

The two diversified producers, *Milk+cereals* and *Milk+young bulls* farms, also have an income close to one another. Both farmers apply here a similar strategy of intensive production, the objective being to minimize the number of dairy cows in order to develop other activities. To do this, the milk yield of animals is equal to their genetic potential (8,500 and 9,000 litres of milk per cow per year). This level of production is reached through a massive use of concentrate feed. Thus, the *Milk+cereals* farm dedicates 72 per cent of land to the cereal production and the *Milk+young bulls* farm fattens 77 bulls in addition to the milk production. Lelyon et al. (2008) show that the full decoupling of the Special premium for bovine male in 2006 (EUR210/bull) encourages producers to remove the fattening activity, because the profitability of this activity is in balance with grain production. In fact, this phenomenon has seldom been observed in France because, on the one hand, producers of young bulls were often engaged in contractual relationships with slaughterhouses and, on the other hand, most farmers do not consider not using their buildings to their full capacity even if it's more advantageous from a business point of view. For these two types of farming, the dual values of land and quota are positive. The marginal yield of the land is more than twice that of the milk. Increasing the productivity of dairy cows, constrained by the genetic potential, would also allow a significant increase in income.

None of the farms studied in this base situation is constrained by the nitrate directive whose dual value is zero: nitrogen pressure (total amount of organic nitrogen produced on the farm / total area) is below the standard of 170 kg/ha.

Table 25.3 Economic and productive impact of an abolition of milk quota

	Grass-based Farm		Semi-intensive Farm		Milk+cereals Farm		Milk+Young bull Farm	
	Baseline	Quota +25 %	Baseline	Quota +25 %	Baseline	Quota +25 %	Baseline	Quota +25 %
Income (EUR)	55200	70400	54700	61300	123200	132 100	120 700	133 000
Crop area								
Cereals (ha)	0.0	0.0	11.7	8.1	99.3	97.5	20.6	21.6
Maize silage (ha)	0.0	0.0	10.4	10.9	23.5	25.3	44.0	41.3
Grassland(ha)	78.0	78.0	27.9	30.9	14.2	14.2	35.4	37.1
Premium for grassland	yes	yes	no	no	no	no	no	no
Animal activity								
Total produced milk (l)	299,200	357,800	298,700	345,500	461,400	552,800	445,970	504,400
Total sold milk (l)	285,000	356 250	290,000	344,400	460,000	551,295	400,000	500,000
Dairy cows (nb.)	57	58	35	38	54	59	50	55
Young bull (nb.)							77	60
Milk yield (l/year)	5,250	6,150	8,500	8,970	8,500	9,310	9,000	9,000
Concentrate feed (kg/cow/year)	120	420	890	1,250	1,200	2,630	1,010	1,000
Milk l/ha forage area	3,840	4,590	7,800	8,250	12,230	13,980	5,620	6,430
Organic nitrogen pressure (kg/ha)	132	135	115	126	64	70	146	142
Chemical nitrogen used (kg/year)	6,080	6,900	6,010	5,930	19,100	19,000	10,720	11,130
Working time (h/awu/year)	1,940	1,980	1,560	1,670	1,940	2,090	2,060	2,030
Economic results								
Total revenue (EUR)	142,800	164,900	132,100	145,200	315,300	340,700	301,200	313,900
Milk revenue (EUR)	85,500	106,900	81,200	96,400	128,800	154,400	112,000	140,000
Meat revenue (EUR)	33,600	34,300	16,500	18,100	23,300	25,500	102,100	85,700
Crop revenue (EUR)	0	0	12,300	8,600	106,200	103,900	21,800	22,800

	Grass-based Farm		Semi-intensive Farm		Milk+cereals Farm		Milk+Young bull Farm	
	Baseline	Quota +25 %	Baseline	Quota +25 %	Baseline	Quota +25 %	Baseline	Quota +25 %
Total subsidies (EUR)	23,800	23,800	22,100	22,100	56,900	56,900	65,300	65,300
Variable costs (EUR)	28,800	33,900	31,600	37,000	95,100	109,900	90,400	89,800
Fixed costs (EUR)	58,900	60,700	45,800	46,900	96,900	98,700	90,100	91,000
				Marginal yields				
Additional milk quota (EUR/t)	272[1]	109	187	0	217	0	193	97
Additional milk yield (EUR/l)	n.c.	2,190	300	2,180	810	4,630	320	1,925
Additional area (EUR/ha)	187	163	412	408	585	622	418	397
Additional building place (EUR/pl)	n.c.	n.c.	n.c	1,545	n.c.	2,070	n.c.	n.c.
Additional work hour (EUR/h)	n.c.	43	n.c.	n.c.	n.c.	n.c.	n.c.	n.c.

[1] n.c.: not a constraint

Abolition of milk quota: a high production potential

Regarding the economic results, a 25 per cent increase of milk production leads to an income increase (12 per cent on average, Table 25.3). However, the income increases proportionately less than the volume of milk produced due to an increase in variable costs (dairy cows, concentrate feed) and the crop-forage mix. The Grass-based farm is the one that better uses this extra volume because the substitution effects are lower than for the other farms. Thus, the quota rent of this farm is always the greatest (Table 25.4). The long run quota rents are equal to zero for the *semi-intensive* and the *milk+cereals* farms because they cannot produce all the authorised volume, while those of the *Grass-based* and *milk+young* bulls farms are positive and represent more than a third of the milk paid price. The two farms still have room of manoeuvre to increase milk production. When we aggregate the four types of farming in order to represent the whole French dairy sector, the results are consistent with the other macro level studies about the implication of the milk quota abolition (Bouamra-Mechemache et al. 2008, Cathagne et al. 2006, INRA – University of Wageningen Consortium 2002, Lips and Rieder 2005, Wieck and Heckelei 2007). The results seem to be a little high for the short and medium run quota rents compared to the Cathagne et al (2006) and Wieck and Heckelei (2007) results. At a long run scale, these results are close to those of Bouamra-Mechemache et al (2008), however their study shows that a phasing out of milk quota would lead to a very low quota rent for the producer (EUR18/t with 6 per cent increase of quota, EUR4/t with 12 per cent increase of quota and zero if quotas are removed). Lips and Ridier (2005) also show that if the milk quota disappears, French milk production could only increase by 0.8 per cent. Those results are due to the fact that the French national quota was not entirely produced for 5 years showing that the quota was not a biding limit for the dairy producer. However, many farmers deliberately choose to not produce all their quota in order to avoid paying the financial penalty and also feeding and milking the cows to finally throw the milk in the gutter.

Of course, our study is only focused on the farmer's side and we do not take into account the price and demand change in response to such an increase of volume. The main fact that our simulation demonstrates is that all the farms studied have a strong potential for milk production. The main reason for this potential stems from the fact that the growth rate of the agricultural area for dairy farms was twice that of the quota per farm over the last ten years. In France, milk quotas are linked to the land and farmers need to rent or buy additional lands to increase their quota. There was, therefore, an extensification process of milk production in France characterized by a low milk productivity per hectare of land (4,000 litres/ha against 8,800 Denmark or 11,500 in the Netherlands). In order to use these additional areas, farmers developed alternative activities such as fattening or cereals production which they can easily reduce or remove in case of an abolition of milk quotas.

However, the abolition of milk quotas creates a strong incentive for the intensification of production system: the quantity of milk produced per hectare of forage area strongly increases. This increase in level of production has a negative impact on environmental criteria. Indeed, even if the farms do not reach the maximum level of nitrogen discharge permitted by the nitrate directive, the nitrogen pressure increases. Similarly, the use of chemical nitrogen highly increases for the *Grass-based* farm and to a lesser extent in the *Milk+Young bulls* Farm: fertilization of grassland is more intensive in order to increase yield. Furthermore, the quantity of concentrate feed consumed also rise and makes farms more dependent from purchased feed and thus more vulnerable to price variations.

Table 25.4   Short run, medium run and long run quota rent estimation

	Short run quota rent	Medium run quota rent	Long run quota rent
Grass-based Farm (EUR/t)	272	215	109
(% of milk price) [1]	*91%*	*72%*	*36%*
Semi-intensive Farm (EUR/t)	187	121	0
(% of milk price)	*67%*	*43%*	*0%*
Milk+cereals Farm (EUR/t)	217	97	0
(% of milk price)	*78%*	*35%*	*0%*
Milk+young bulls Farm (EUR/t)	193	124	97
(% of milk price)	*69%*	*44%*	*35%*
France [2]			
This study (EUR/t)	209	122	34
(% of milk price)	*74%*	*43%*	*12%*
Cathagne et al. (2006)	175	104	4
	56%	*33%*	*1%*
Wieck and Heckelei (2007) [3]	178		
	62%		
Bouamra et al. (2008)		115	53
		37%	*17%*
Lips and Rieder (2005)			68.2
			22%
INRA Wageningen Consortium (2002)			108
			35%

[1] The price of milk for each type of farming is given in the table 1

[2] The average quota rent for France is calculated according to the representativeness of each type of farming in France (given in Table 1). The four types of farming considered represent 78 per cent of French dairy farms.

[3] Wieck and Heckelei give the short run quota rent for two French region (Britanny and Pays de la Loire). We give the mean of these two region in this table.

Looking at more precisely how each type of farming reacts to the removal of milk quota, we see that two farms can produce the maximum volume allowed (+25 per cent), with the same farm structure, i.e. without making any investment: the *Grass-based* and the *Milk+young bulls* farms (Table 25.3). These two farms have indeed the ability to easily increase their milk production by intensifying the production system. For the Grass-based farm, this increase in volume is mainly enabled by increasing milk production per animal (+17 per cent up to 6,150 litres/year) made possible by increasing concentrate feed. The production of the additional volume of milk is then achieved through a small increase in the number of dairy cows. To facilitate this transition, the farmer chooses to no longer feed his calves with home produced milk, but now uses milk powder (which represented an annual volume of milk of 12,600 litres or 2.1 cows). The situation of the *"Milk + Young bulls"* farm is different because this farmer can only increase a little the milk yield per animal which has already reached the limit (9,000 litres per cow per year). However, this farmer can use part of the fattening building for dairy cows. Thus, the farmer increases milk production by replacing bulls with dairy cows. Moreover, he also chooses the milk powder to feed calves and thus saves more than 40,000 litres per year.

The *"Semi-intensive"* and *"Milk+cereals"* farms do not achieve the additional 25 per cent of authorized volume. They are limited by the number of places for cows in buildings. This constraint is then lifted to enable the farmer to expand the cowshed in order to increase milk production (the cost of one place in the building is about EUR4,000 per cow: EUR330 per cow with a 12 year amortization). In this case, the *Semi-intensiv farme* needs 5 additional places to reach the threshold of +25 per cent, while leading to an increase of 6 per cent of income. The *Milk+cereals* farm needs 4 building places, thus generating 1 per cent of additional income.

To feed the additional animals, farmers change the crop rotation: the share of fodder crops increases at the expense of cereal crops (except for the Grass-based farm which had no cereals). This increase in forage area consists in equal part of an increase in the surface of grassland and maize. The full decoupling of the crops premium (which benefited maize silage) re-balances the choice between grassland and maize (Ridier and Jacquet 2002). It is also important to note that the removal of set-aside, already included in the baseline, limits the process of intensification, freeing land for fodder crops.

Among the constraints limiting the increase in milk production, the milk productivity per animal has the greatest impact on income. Indeed, when farmers manage to increase the milk yield by one litre of extra milk per cow per day, it generates an income increase from 1,900 to more than EUR4,500 depending on the farm. It is the economic gain enabled by the animal genetic level. Cows which have a higher potential can produce more milk at a lower cost: fewer animals for the same quota, thus freeing areas for other activities. Naturally, such conclusions depend on the relative prices of milk, cereals and meat.

Results are now discussed considering several hypotheses regarding price fluctuations for milk and cereals. In these simulations the price of concentrate feed is indexed to the price of cereals. It appears that maintaining milk production is always a priority for farmers, regardless of the price considered. Indeed, the costs incurred to establish a dairy activity are often too high for farmers to consider abandoning milk for cereal production. This is especially true because the agricultural area of dairy farms is often far below the threshold of profitability traditionally met in the specialized crop farms. For the *Milk+Cereal* farm, the total removal of dairy activity in favour of cereals would be, theoretically, possible in the unlikely event that the price of milk reached EUR230/t while cereals peak at EUR240/t (Figure 25.1).

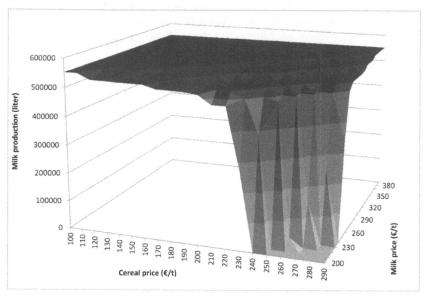

Figure 25.1   Milk production according to the milk price and cereal price (milk+cereal farm)

Nevertheless, the simulations are run with constant fixed costs. The decision could possibly be different for a farmer with no more building depreciation or who has no loans outstanding. Location in a plain region with high agronomic potential land and/or a generation change (setting up of a new farmer) could also lead to abandoning milk production since it is very time restrictive (cows have to be milked 365 days a year, twice a day).

If the farmers' strategy is to maximize milk production (regardless of the milk price), the price level of cereals has an impact on the rotation through a change in the share of cereals in the total agricultural area at the expense, or benefit, of forage areas (Figure 25.2). Regarding the feeding system, these rotation choices lead to, assuming an increase in grain prices, an increase of maize silage in the diet (and therefore to an accentuated use of concentrate feed). For the *Milk+Young bull* farm, the room to manoeuvre is more important concerning the rotation because this farmer can decide, if necessary, to reduce the number of young bulls.

When the cereal price is below the threshold of EUR130/t, cereal production is declining and even abandoned in two types of farming (*Grass-based* and

Milk+young bulls). The rise in the price of nitrogen fertilizer also causes farmers to adopt this strategy.

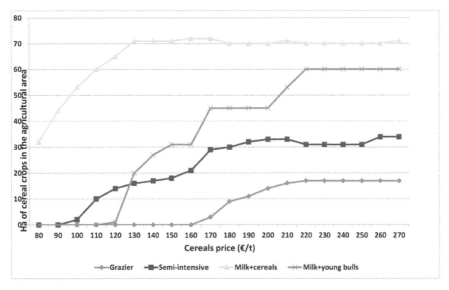

Figure 25.2   Proportion of cereals in the agricultural area according to the cereals price

Discussion and conclusion

This model, based on the mathematical programming methodology, assesses the impact of the abolition of milk quotas on the productive strategies of French dairy farms. Because we consider the interactions between types of production (both plant and animal), the main laws of biological response and the seasonality of agricultural production, this model represents, as realistically as possible, farmers' behaviour and supplies economic, technical and environmental response to the abolition of milk quotas. Based on the current construction, some improvements are possible such as to integrate other goals in the objective function (such as minimisation of labour). In a context of increased volatility in prices, the UEP method could be modified to better integrate farmers' expectations facing the direction (positive or negative) of price changes.

For dairy farmers, the abolition of milk quotas is naturally the most important issue among the various measures of the CAP "Health Check". All things being equal, and whatever the price of milk or cereals, milk producers always try to attain the maximum volume of milk production allowed. The fluctuation of cereal prices however impacts the rotation and the level of intensification. These simulations mainly emphasize that dairy farms have a high potential for increasing dairy production with constant fixed costs. However, the rise in production is mainly possible by a strong intensification of the farming system which has some negative impacts on the environment (nitrogen discharges) and the feeding self-sufficiency (more concentrate feed purchased). In the simulation, the increase in production has been limited, by hypothesis, to 25 per cent. A significant fraction of milk producers would be able to produce more if the authorization was given. This applies primarily to farmers located in areas where environmental restrictions are not too important and those for whom land is readily available. Indeed, the potential development of milk production is not homogeneous according to regions and often depends on the parallel presence of other livestock activities (pigs and poultry).

In France, more than in other Member States, the abolition of milk quotas raises significant questions. The management of the quotas allows a large state intervention in the geographical distribution of production. Moreover, milk price partly reflects the close cooperation between producers and dairy processors. In case of removal of milk quotas, companies will likely have a stronger power in the pricing of milk, the milk quality requirement, the orientation of the structure (size, intensification) and the location of the supply. Future productive strategies of French dairy farms will not only be influenced by changes in relative prices (input and output), but also by the terms of the contracts with companies.

Whatever the ways used to end the milk quota system, European and French dairy farms will continue to benefit from an interventionist agricultural policy. In New Zealand, Australia and Argentina, three major exporting countries, the dairy sector benefits from a low public support, weak border protections and a very competitive production cost per ton of milk. In the United States, a country with rapid growth of its domestic production (unlike the European Union), the dairy sector is supported by a strong market intervention. An increasing share of milk production comes from very large farms which are hardly comparable to those encountered in the European Union. The instruments used to support milk production in the European Union would therefore, despite the abolishment of milk quotas, stay quite specific in the next decade.

Acknowledgements

We would like to thank seminar participants at *"The CAP after the Fischler reform: national implementation, impact assessment and the agenda for future reform"* (109th EAAE Seminar, Viterbo, November 2008) and at the second *INRA – SFER – CIRAD* conference (Lilles, December 2008) for valuable comments on earlier versions of this chapter. This chapter also benefited from comments by J. Scott Shonkwiler visiting researchers INRA-Nantes. This research was funded by the research programmes: 'Laitop' (PSDR Grand-Ouest) and 'Dynamics of Dairy Territories in France and Europe' (funded by CNIEL, Crédit Agricole, FESIA, Groupama and SEPROMA).

Reference

Alvarez, A., Arias, C. and Orea, L. 2006. Explaining differences in milk quota values: the role of economic efficiency. *American Journal of Agricultural Economics* 88, 182-193.

Berentsen, P.B., Giesen, G. J. and Renkema, J.A. 2000. Introduction of seasonal and spatial specification to grass production and grassland use in a dairy farm model. *Grass and Forage Science* 55, 125-137.

Boots, M., Oude Lansink, A. and Peerlings, J. 1997. Efficiency loss due to distortions in Dutch milk quota trade. *European Review of Agricultural Economics* 24, 31-46.

Bouamra-Mechemache, Z., Jongeneel, R. and Requillart, V. 2008. Impact of a gradual increase in milk quotas on the EU dairy sector. *European Review of Agricultural Economics* 35, 461-491.

Cathagne, A., Guyomard, H. and Levert, F. 2006. *Milk Quotas in the European Union: Distribution of Marginal Costs and Quota Rents.* Paper to the Conference: *European Dairy Industry Model*, 1-24.

Colman, D. 2000. Inefficiencies in the UK milk quota system. *Food Policy* 25, 1-16.

Commission of the European Communities. 2008. Proposal for a Council Regulation establishing common rules for direct support schemes for farmers under the common agricultural policy and establishing certain support schemes for farmers. pp. 161. European Union.

Godard, C., Roger-Estrade, J., Jayet, P.A., Brisson, N. and Le Bas, C. 2008. Use of available information at a European level to construct crop nitrogen response curves for the regions of the EU. *Agricultural Systems* 97, 68-82.

Gohin, A. and Latruffe, L. 2006. The Luxembourg Common Agricultural Policy Reform and the European Food Industries: What's at Stake? *Canadian Journal of Agricultural Economics* 54, 175-194.

Hardaker, J.B., Huirne, R.B.M., Anderson, J.R. and Lien, G. 2004. *Coping with risk in agriculture.* Second/Ed. CAB International, Wallingford (UK).

Hazell, P.B.R. and Norton, R.D. 1986. *Mathematical Programming for Economic Analysis in Agriculture.* MacMillan, New York.

Hennessy, T., Shrestha, S., Shalloo, L., and Wallace, M. 2009. The inefficiencies of regionalised milk quota trade. *Journal of Agricultural Economics* 60, 334-347.

Howitt, R. E. 1995. Positive mathematical programming. *American Journal of Agricultural Economics* 77, 329-342.

INRA – University of Wageningen Consortium. 2002. *Study on the impact of future options for the milk quota system and the common market organisation for milk and milk products.* V. Réquillart coordinator, European Commission,.

INRA. 2007. *Alimentation des bovins, ovins et caprins : Besoins des animaux – Valeurs des aliments.* Editions Quae, Versailles.

Institut de l'Elevage 2008. *Les systèmes bovins laitiers en France: Repères techniques et économiques*. Institut de l'Elevage, Paris.

Kleinhanss, W., Bertelsmeier, M. and Offermann, F. 2002. *Phasing Out Milk Quotas: Possible Impacts on German Agriculture*. Institute of Farm Economics and Rural Studies. Braunschweig.

Lambert, D. K. and McCarl, B. A. 1985. Risk Modeling Using Direct Solution of Nonlinear Approximations of the Utility Function. *American Journal of Agricultural Economics* 67, 846-852.

Lelyon, B., Daniel, K. and Chatellier, V. 2008. Decoupling and prices: determinant of dairy farmers' choices? A model to analyse impacts of the 2003 CAP reform. Paper to the 12th Congress of the European Association of Agricultural Economists – EAAE 2008.

Lips, M. and Rieder, P. 2005. Abolition of Raw Milk Quota in the European Union: A CGE Analysis at the Member Country Level. *Journal of Agricultural Economics* 56, 1-17.

Manos, B., Begum, M. A. A., Kamruzzaman, M., Nakou, I. and Papathanasiou, J. 2007. Fertilizer price policy, the environment and farms behavior. *Journal of Policy Modeling* 29, 87-97.

Moro, D., Nardella, M. and Sckokai, P. 2005. Regional distribution of short-run, medium-run and long-run quota rents across EU-15 milk producers. Paper to the EAAE Congress. Copenhague.

Office de l'Elevage 2008. «Le marché des produits laitiers, carnés et avicoles en 2007,» Office de l'élevage, Paris.

Patten, L. H., Hardaker, J. B. and Pannell, D. J. 1988. Utility Efficient Programming for whole-farm planning. *Australian Journal of Agricultural Economics* 32, 88-97.

Ridier, A. and Jacquet, F. 2002. Decoupling direct payments and the dynamic of decisions under price risk in cattle farms. *Journal of Agricultural Economics* 53, 549-565.

Thornton, P. K. and Herrero, M. 2001. Integrated crop-livestock simulation models for scenario analysis and impact assessment. *Agricultural Systems* 70, 581-602.

Wieck, C. and Heckelei, T. 2007. Determinants, differentiation, and development of short-term marginal costs in dairy production: an empirical analysis for selected regions of the EU. *Agricultural Economics* 36, 203-220.

Zuhair, S.M.M., Taylor, D. B. and Kramer, R.A. 1992. Choice of utility function form: its effect on classification of risk preferences and the prediction of farmer decisions. *Agricultural Economics* 6, 333-344.

Chapter 26

The Role of the 'Integrated Production' Scheme in the Fruit and Vegetable CMO

Gabriele Canali

From Integrated Pest Management (IPM) to Integrated Production (IP)

Since their introduction into the market, chemical products used for pest control (the so-called pesticides or agro-chemicals and now also agro-pharmaceuticals) have played a major role in the increase of productivity of land, labour and other agricultural inputs. The importance of these inputs is especially relevant in the fruit and vegetable sector, and their use is generally more intense in this sector than in others.

Over time the increased awareness of consumers about possible health risks due to residues in these fresh products has pushed researchers, extension services as well as farmers, to try to reduce the use of agro-chemical products.

Conventional production typically relies on a number of applications of agro-chemicals based on the calendar ('calendar approach'), i.e. the applications of these inputs occur every fixed number of days in order to prevent any growth of pests. This approach does not consider the effects of fluctuations of climatic conditions on the growth of pests and therefore on the probability of great damage.

At the opposite extreme of the calendar approach, we have organic production where use of synthetic pesticides is forbidden in order to eliminate the possibility of harmful residues in the final product and to reduce or eliminate negative effects on the environment. In this case, however, crop yields are generally much lower while labour requirements are generally higher, resulting in a necessarily higher production cost and market price.

In the last few decades a different approach has been developed, Integrated Pest Management (IPM), implying a more rational and scientifically based approach to pest control and to the decisions about when, how and how much agrochemicals should be applied to crops. It is quite clear that IPM has been developed and introduced both in order to address the health (residues) and environmental concerns of consumers and citizens, and with the aim of reducing the use of these agrochemicals (and possibly the overall production cost).

IPM utilizes a number of different kinds of approach and tools for controlling pests: from the use of natural substances and application of biological enemies to control pest growth, to specific mechanical intervention as a substitute for some applications of agrochemicals, to the application of sophisticated prediction models based on biological pest data and meteorological information in order to simulate pest growth and identify more precisely when a specific chemical application is really required.

Over time IPM has evolved in different directions, and for many different reasons; in general from the 'simple' IPM approach recent attention has been paid to issues like soil management, (e.g. reduce soil erosion, maintain or increase the percentage of organic matter), water management, biodiversity protection, but especially reduction of the level of residues of agrochemicals even well under the Maximum Reside Level (MRL) defined by law.

All these different methods, which are partial modification and integration of the more traditional IPM scheme, have been generally defined as Integrated Production (IP) schemes in order to differentiate them from the IPM approach which is focused only on the management of pest control activities.

It must be noticed that these IP schemes, as well as IPM, do not rely on any public regulation at the EU level, differently from the case of organic production. The absence of this common and public standard, has allowed, if not pushed, regional/national authorities first and private certification bodies and retailers later, to introduce different definitions and certification schemes for the IP.

In particular a key role has been played, at the European level, by the possibility of supporting specific cropping techniques that are aimed at reducing the use of chemical inputs, among the agri-environmental measures introduced for the first time with EU Regulation n. 2078 on 1992. Since then, each EU member country has had the possibility, and (political) need, to define what it consider to be a 'relevant' reduction of the use of pesticides and chemical fertilizers, and therefore to define (one or more) IP schemes.

Table 26.1 shows the list of last data available at the EU level about the application of these policy tools, i.e. support of production techniques requiring less chemical inputs. From the total amount of money spent and from information about the agricultural area committed to these techniques, we have eliminated organic production in order to obtain an estimate of EU support to IP schemes.

Table 26.1 **Agri-environmental measures for crops, organic production excluded, in EU25 member countries (2005).**

	Total amount committed (000 EUR)	Number of hectares under contract	Average amount per hectare (EUR)
Austria	554.446	5.857.487	95
Belgium	32.642	264.632	123
Cyprus	76	5.623	14
Czech Republic	100.436	944.621	106
Denmark	21.107	250.830	84
Estonia	17.796	442.574	40
Finland	273.295	2.074.226	132
France	393.793	7.572.723	52
Germany	516.147	4.987.870	103
Greece	38.758	199.012	195
Hungary	166.287	N.A.	N.A.
Ireland	251.418	1.695.000	148
Italy	209.464	1.362.346	154
Latvia	2.571	19.672	131
Lithuania	0	0	-
Luxembourg	12.236	144.793	85
Malta	277	N.A.	N.A.
Netherlands	37.973	162.014	234
Poland	1	74.178	0
Portugal	93.033	630.181	148
Slovakia	20.513	274.043	75
Slovenia	25.727	194.420	132
Spain	157.318	2.712.678	58
Sweden	190.330	2.554.749	75
United Kingdom	40.149	273.533	147
EU25	3.155.793	32.640.102	97

Source: EU Directorate General for Agriculture and Rural Development, Rural Development in the European Union, Statistical and Economic Information – Report 2007.

Overall more than 32.6 million hectares have been involved by these commitments, i.e. more than 20 per cent of EU utilized agricultural land, and the average amount paid per hectare has been EUR97.

As already mentioned, the sector where IP schemes are absolutely more important is fruit and vegetables (F&V). In order to analyse the effects of the application of these different IP schemes in this sector in Italy, we summarize, first, the most important characteristics of the Common Market Organization and some key characteristics of the F&V sector; secondly, with reference to the Italian case, we identify some issues arising from the actual use of different IP schemes in order to draw some conclusions about possible useful developments.

The Common Market Organization for fruits and vegetables and IP schemes

The fruit and vegetable sector in the European Union accounts for 17 per cent of the value of agricultural output and EU27 produces about 8.3 per cent of total world production (average 2003-2005). At the EU25 level, more than 1.4 million farms produce fruit and vegetable (a total of over 9.7 million) and about 660,000 of them specialize in this. In Spain, Greece and Italy the role of this sector is particularly relevant: its share of the value of total output is respectively equal to 31 per cent, 28 per cent and 25 per cent (average 2003-2005). With respect to quantities, however, Italy is the most important producer with 16 million tons of products, on average, in the period 2003-2005 (Canali 2007).

The new Common Market Organization (CMO) for the fruit and vegetable sector was approved in June 2007 (EU Reg. n. 1182/2007, published on 26 September 2007), as a partial substitution and modification to previous regulations (especially n. 2200, 2201 and 2202/1996).

This reform has introduced a few major changes in the CMO, like the inclusion of this sector, with some specific provisions, in the Single Payment Scheme (SPS), and full decoupling for all aid previously granted to producers of a few processed F&V (e.g. tomatoes, peaches and plums, for example).

However, still consider crucial are the objectives of competitiveness of the sector and sustainability though many intervention tools were already present in previous Regulation, and especially the support of Producer Organizations (POs). Among other things, POs should help farmers to organize and to concentrate their supply of fresh products in order to better-satisfy the increasing demands by large retailers in Europe as well as in other foreign markets, and to have some possibility of bargaining with them from a less unfavourable position.

This has been also clearly stated by the EU Commissioner for Agriculture during the presentation of the reform proposal:

> "It is no secret that the retail sector, now highly concentrated, has an astonishing power to set prices. ... supermarkets seem to have the fruit and vegetable sector in a particularly strong arm-lock. It is through Producers Organisations that individual producers can stand up to the retail giant (Fisher Boel 2007)".

On the other hand, POs should also help farmers to apply the best available growing, preserving and packaging technologies, also with the aim of becoming both more competitive and increasingly sustainable from an environmental point of view.

With reference to these objectives, again, the Commissioner has been explicit: … my central aims with this reform plans are as follows:

> "- I want to help make the sector more competitive and market oriented, for the sake of sustainable production.
> - I want to help reduce the income problems caused by crisis.
> - I want to encourage people in the European Union to eat more fruit and vegetables.
> - I want to help extend the sector's efforts at caring for the environment.
> - And I want to simplify policy where possible. (Fischer Boel 2007)."

Therefore competitiveness, sustainability and care of the environment are among the most important objectives of the reform, together with the traditional one of farmers' income protection, and with the (relatively) new ones of encouraging consumption of fresh F&V and simplification of the policy.

With respect to other policy tools, it must be noticed that as a consequence of the adoption of the SPS in this sector, another provision has been introduced with the reform of the CMO, i.e. 'cross compliance', or the respect of mandatory environmental standards which becomes compulsory for farmers receiving direct payments.

Moreover, POs must also spend at least 10 per cent of their expenditure in each Operational Programme, the technical tool for obtaining public support from the CMO, for environmental measures.

Really, since 1992, i.e. since the introduction of specific support measures (Reg. 2078/1992) for low-impact agriculture, i.e. the adoption of IPM or IP schemes (as well as organic production), POs have started to promote and support with technical assistance, these new production technologies in the fruit and vegetable sector. In other words, since environmental measures have been maintained also in the new 2007-2013 rural development programme, and since POs still have an incentive to promote this adoption, there are good reasons to believe that 'some sort' of IP will remain in place at least in this sector.

The point is to demonstrate if this approach is coherent with the objectives of CMO of the F&V sector, and with the more general ones of the CAP; in particular:

1. are these *different IP schemes* the best way to help consumers to be more confident in F&V quality, and therefore to promote consumption?
2. Are these *different IP schemes* useful in order to promote competitiveness of the F&V sector?
3. And in particular, are these *different IP schemes* useful in order to shift, at least to some extent, market power from large retail chain to farmers through POs?
4. Is there any other way to apply IP technology without increasing too much production costs?

5. Are these *different IP schemes* useful in terms of simplification?
6. What are the real effects of these *different IP schemes* on sustainability?

In order to try to answer to these questions, in the following paragraph we will illustrate the case of application of these tools in the F&V sector in a specific region of Italy, Emilia-Romagna, where the production of fruits and vegetables is absolutely relevant.

The application of IP schemes in the F&V sector

The case of Emilia-Romagna

In Italy there has been a fairly widespread use of IP schemes in many regions especially with reference to the production of vegetables and fruit, both for processing and for consumption as fresh products.

In Italy, the implementation of IPM started in a few leading regions, Emilia-Romagna and Trentino-Alto Adige, under a strong push by local authorities and with a strong support from public technical assistance services (Bertazzoli et al. 2004). In Emilia-Romagna, in particular, according to recent estimates (Galassi and Mazzini) 65 per cent of the area used for cropping fruit and vegetables, is managed according to the regional IP scheme.

Moreover, Emilia-Romagna has also introduced a regional law (n. 28, 1999) with the aim of promoting the use of this regional IP standard, identifying products obtained in this way with a specific label 'Qualità Controllata' (Controlled Quality), and supporting it with public money destined for promotion activities. According to the regional IP scheme, products must be produced following detailed pest (and promotion) management rules. The objective is to promote more environmentally friendly farming systems, reduce exposure to risk due to residues of pesticides, and to promote consumption of these products with positive effects on the sector and on competitiveness.

However, after almost a decade, the results of this approach are mixed, even in this region, which plays a leading role at the national level. With reference to the fruit sector, for example, in 2006 the total production obtained respecting the regional IP production scheme was equal to 623,000 tons, i.e. about 40 per cent of total regional production of fruit. But even more interesting, and controversial, it is the fact that only 1/3 of this amount was marketed using the label *Qualità Controllata*. The same percentage for vegetables is slightly more than 14 per cent (Fanfani and Pieri 2008).

The previous data shows quite clearly that F&V growers, even with the availability of good technical assistance, and with the direct income support granted by agri-environmental measures of Rural Development policies, apply *this* regional IP scheme only on a relatively limited share of their land. Moreover, even the possibility of identifing these products with a specific label doesn't seem to work properly: the adoption percentage seems to be very low.

In order to identify possible causes we can also consider that this region had a share of 10 per cent of the Italian export of fruit in 2007 and the value of regional export in the same year was EUR486 million, while the value of the regional production, evaluated at farm prices, was EUR680 million. In other words, a very large share of this region's production of fruit is exported; in this case, a 'regional' label simply does not work. Moreover, any communication activity related to this IP scheme in these foreign markets would be too costly and ineffective. Over time, almost every Italian region has introduced its own IP certification scheme, and sometimes its own label; therefore, even at the national level, it is very difficult to realize any effective communication and promotion activity, and consumers know very little about these regional IP schemes and their (regional) labels.

One could argue that another possible positive effect of this certification could be to grant better access to large retail chains, in Italy as well as in other European countries. From interviews with buyers from retail chains and members of F&V producer cooperatives, it emerges clearly that retailers simply 'use' this certification as a prerequisite, without any possibility for farmers or POs to obtain any price premium. Moreover, they consider this certification insufficient for any specific differentiation strategy, and almost no use has been made of the *Qualità Controllata* label at the retail level for fresh products.

The interactions between different IP schemes: issues and implications

In this paragraph we compare different kinds of IP standards: regional standards defined by retail chains, national standards and a hypothetical IP standard defined at the EU level (Table 26.2).

Table 26.2   Main positive and negative effects of alternative approaches to IP standards.

Effects on:	Regional IP standards	Retailers' standards (vs. regional IP only)	National IP standard substituting regional/PL standards	EU IP standard
Farmers	(+) access to income support measure (agri-environmental measures of the Rural Development policy). (+) possible positive effect in terms of market access (-) increase in production cost because of the need to apply different IP schemes (-) limitation of pesticides available with negative effect on production costs (-) no price premium	(+) market access to large retail chain (-) further loss of market power (also because of barriers to entry in different chains) (-) increase of production costs (-) increase of production uncertainty	(+) grant market access (+) great simplification of production activity (+) decrease of production cost (w.r.t. different IP schemes for different costumers) (+) greater possibility to obtain some price premium	(+) simplification of production activity (+) decrease in production cost (only one standards) (+) fair competition among products from different country of origin (+) greater possibility to obtain some price premium
Retail chains	(+) with respect to conventional production, lower risks (in terms of MRL) (-) impossibility to use this tool for differentiation strategies between retail chains (-) impossibility to communicate to consumers	(+) increase in market power w.r.t. farmers and POs. (+) better control over farmers and reduction of risk (-) increase of costs also for control activities (-) very limited possibility to communicate to consumers the reduction of health risks (-) limited role for product differentiation based on IP	(+) reduction of transaction costs and cost for product control activities (+) increased possibility to differentiate successfully, at least at the national level, IP products from conventional and organic ones (-) (limited) loss of market powers vs. farmers (and POs)	(+) reduction of transaction costs and cost for product control activities (+) increased possibility to differentiate successfully IP products from conventional and organic ones (-) (limited) loss of market powers vs. farmers (and POs)

Consumers	(+) less risks with respect to conventional products (-) higher price	(+) less risks with respect to conventional products (-) higher price	(+) clearness and uniformity of IP contents (national products) (+) greater competition for IP products	(+)clearness and uniformity of IP contents (ALL EU products) (+) greater competition for IP products
Environment	(-) lower adoption of these IP schemes due to their inefficiency w.r.t. unique IP scheme	(+,-) uncertain effects of the restrictions (-) lower adoption of these IP schemes due to their inefficiency w.r.t. unique IP scheme	(+)positive effects if IP schemes are scientifically based (-) lower adoption of these IP schemes due to their inefficiency w.r.t. unique IP scheme	(+) increase of the probability of a scientifically based approach (+) increased diffusion of this IP scheme due to its effectiveness.

Starting from the experience of regional IP schemes, large retail chains have developed new chain-specific IP standards, based on the common 'idea' of IP but with quite different requirements. In general these private standards differ from the regional ones because of their requirements in terms of Maximum Residue Limits (MRL): they tend to require the limitation of pesticide residues to a percentage (e.g. 50 per cent) of the legal MRL. In other words, retailers seem to work simply and only on the MRL and in general they label these products with their own brand, i.e. they use them for their *private label* (PL).

In order to differentiate their products in this way, retail chains need to develop many new activities inside their structure, as, for instance, their own control systems, which can be completely internal to the firm or partly external. The reduction of health risks (i.e. food safety) through the reduction of residues and the increased intensity of control activity, is by far the most important message that they try to communicate to consumers, in order to obtain more value added and/ or to increase total sales through their PL products. However it is very difficult to transfer this message to consumers without possible (even strong) negative effects on consumption of products obtained through conventional agriculture which are sold at the same time in the same shops.

But at the farm level, all these different IP schemes generate a related increase in costs of production and commercialization, without bringing any economic benefit for farmers beside, to some extent, market access. In fact farmers need to comply, contemporaneously, with many quite different IP schemes for the same product; every IP scheme requires a different technological approach, which implies different pest management approaches, different documents, etc.

Moreover F&V growers must also satisfy, in order to enter different retail chains in different EU markets, an increasing number of other private standards like EurepGap (created in 1997), BRC (in 1998) and IFS (in 2001), to name a few examples (Duponcel 2006).

As a result, not only do production costs tend to increase quite a lot at the farm level, but also products obtained from the same farm cannot be sold to different customers simply on the basis of price and other commercial conditions, since the different IP schemes (and other private standards) represent barriers to entry into specific marketing channels without any specific and clear benefit for final consumers.

Since the application of different standards is widespread, they indirectly represent barriers to changing customers: once the farmer has chosen to fulfil one specific IP scheme, it is quite probable that if at the end of the production process the farmer decides to change customer, i.e. to sell to another retail chain, this will not be possible since he/she will not have the possibility of obtaining the required different certification(s). Therefore these conditions clearly contribute to generating and/or to reinforcing some oligopsonistic power by large retailers. This is exactly the opposite of one of the main objectives of the CAP for the F&V sector and in general.

If farmers want to maintain the possibility of selling their products to different buyers up to the end of the production process, they must apply to their entire production system the most restrictive rules of all the different IP and other private standards they may want to reach. This choice, of course, generates both an increase in production costs and a strong limitation on available technologies with the possibility of reduction in quantity and/or quality of the products obtained in the fields.

Finally, some regional and private IP schemes also introduce relevant limitation with respect to pesticides that can be used; so even if according to national and EU laws and regulation a specific pesticide can be legally used for pest control in a specific crop, it is possible that the same product cannot be used because of 'some' evaluations by, regional authorities and/or retailers. In many cases this implies that in order to control some specific pest on a specific crop, only very few, and sometimes only one pesticide(s) can be used.

Of course this is another major problem for farmers from the point of view of market power as well as technical feasibility of some production processes in economic terms: without, or with very limited, possibility to protect crops from some pests, some production activities, at least in some geographical areas, would not be possible anymore.

From the point of view of consumers, the adoption of different IP standards makes it more difficult to understand what the real contents of these different IPs are, and therefore what their value is. Therefore these different IP schemes tend to become useless if the final objective is to communicate important characteristics to consumers and to have them recognised and properly valued.

From the point of view of effectiveness in reducing the negative impact of intensive agriculture on the environment, it is very hard to say what the final output could be. These IP schemes focus especially on the level of residues at the end of the production process but do not necessarily deal with other issues which can be even more relevant from an environmental point of view in specific areas, such as, for example, soil and fertility management, water management, CO_2 emission and/or immobilization, preservation of biodiversity. Moreover, different standards deal with these issues different ways, increasing the difficulty for end consumers to understand what is at stake and what to buy.

With the recent CAP reform started in 2003, a new possibility has been granted to Single Member Countries of the EU: to define and introduce new quality systems (and labels) at the national level (the new Art. 24 of Reg. 1257/1999 as modified by Reg. n. 1783/2003), and to support farmers switching to production of products satisfying the requirements of the new quality system.

The Italian Ministry of Agriculture, together with regional authorities, has started to develop a possible new national IP scheme as a 'national quality system', which will be followed by a new specific label. It is quite clear that this could be a good move in terms of simplification if this new national IP scheme is successful in replacing other regional IP schemes, and if the real contents are clearly stated, not only with respect to their application by farmers but also with respect to communication to consumers.

Interestingly, large retail chains are starting to understand, according to statements made recently by many buyers of F&V, that their approach to PL based upon private IP schemes is not effective: it does not seem to be able to positively differentiate their product, while the cost is quite high, not only for farmers, but also for themselves. For this reason they seem to be very interested, now, in accepting the introduction of a new national IP standard as a tool for simplification and reduction of transaction costs. This new position supports the idea that retail chains may have realised that their oligopsony power with respect to F&V growers is by far less important if compared to the possibility of reducing transaction costs and/or improving their competitiveness through improved chain integration and through product differentiation strategies that in the F&V sector could be based, more successfully, on other quality characteristics.

Towards a new EU certification scheme?

The implementation of a national IP scheme will contribute to simplification, reduction of production and transaction costs for farmers, and reduction of transaction costs for retail chains, and an increase in competitiveness within the food chain.

However, it is clear that when farmers, individually or through their cooperatives and/or POs, sell their products to retailers or other economic agents in EU markets, the problem of different IP and private standards remains, even if partly simplified.

Therefore it seems clear that a new certification scheme defining a common 'integrated production' standard at the EU level, like that for organic products, would be a more reasonable solution, given the evolution of the idea of 'integrated production', and the effective use of several different certification schemes, all of them more or less closely related to it, and given all the implications of the present situation on the effectiveness of the EU agricultural policy, and especially the CMO for fruit and vegetables.

Also according to evidence on these issues (i.e. Boccaletti, forthcoming.; Govindasamy et al. 2001; Grolleau and Caswell 2006; Ventura-Lucas et al. 2002; Weaver et al. 1992), this IP quality scheme should have, at least, the following characteristics:

a. a certification scheme defined at the EU level;
b. the certification should be based upon clear principles, priorities and limitations, but it should also allow some specified flexibility in technical application to different crops in different geographical areas;
c. this quality system should imply the use of a specific label;
d. this certification should be easy to communicate to consumers, as well as retailers and other economic agents of the food chains;

e. there should be a clear distinction, both in principles and in practice, between environmental cross-compliance measures, and 'sustainable production' ones.

Among other positive implications, this new quality system would also allow the possibility for large retailers to further differentiate their PL products using other characteristics of interest to consumers.

With specific reference to the possibility and need to introduce a new 'name' for identifying and communicating the real contents of this production management system, one possibility would be to define it as 'sustainable production' or 'green production', as compared to the conventional or traditional production on one side, and 'organic production' on the other extreme.

This name would be quite easy to understand for consumers, even if the proper definition from a technical point of view would be more difficult and crucial.

Starting from the evolution from IPM to the IP, it seems fairly clear that the following issues must be addressed in order to define a 'sustainable product', even if this is far from being enough for the specific definition:

1. sustainable use of agro-chemical products for crop protection;
2. sustainable use of water;
3. soil management: crop rotation, fertilization, prevention of erosion, conservation of organic matter, etc.;
4. sustainable use of machinery and energy;
5. complete and verifiable documentation of the whole production process;
6. efficient and independent monitoring and control system;
7. clear and effective system of safety measures for workers.

The 'Green Paper on agricultural product quality: product standards, farming requirements and quality schemes' published by the Commission of the EU last October 15, 2008, in order to open the discussion on these topics with all stakeholders of the EU, could be a very good opportunity to start to elaborate and share a possible positive solution for this issue.

References

Boccaletti, S. Review of empirical studies on environmentally responsible food choice, in Household Behaviour and Environmental Policy, OECD, (forthcoming).

Bertazzoli, A., Giacomini, C. and Petriccione, G. Il sistema ortofrutticolo italiano di fronte ai nuovi scenari competitivi. INEA. Studi e Ricerche, Edizioni Scientifiche Italiane, Roma, 2003.

Canali, G. 2007. La nuova OCM ortofrutta e la sua applicazione in Italia, Working Paper, Gruppo 2013 serie, www.gruppo2013.it, n. 4, July 2007.

Commission of the European Communities, Green Paper on agricultural product quality: product standards, farming requirements and quality schemes, COM(2008) 641, 15 october.

Duponcel, M. 2006. Role and importance of producer organisations in the fruit and vegetable sector of the EU. Paper to the CAL-MED second workshop, Washington Dec. 7-8, 2006.

EU Directorate General for Agriculture and Rural Development, Rural Development in the European Union, Statistical and Economic Information – Report 2007, .

Fanfani, R. and Pieri, R. 2008. Il sistema agro-alimentare dell'Emilia-Romagna – Rapporto 2007, edited by Maggioli.

Fisher, BM. 2007. Time to shape up: a new deal for fruit and vegetables in the EU. Speech at the Meeting with Agricultural Committee at the European Parliament, Brussels, 24 January, 2007.

Galassi, T. and Mazzini, F. 2008. Prospettive per la difesa integrate nel contesto della nuova normative. Paper to the Conference: Uso sostenibile degli agrofarmaci, residui e produzione integrata (Sustainable use of agro-pharmaceuticals, residues and integrated production). 23rd October 2008. Catholic University, Piacenza (Italy).

Govindasamy, R., Italia, J. and Adelaja, A. 2001. Predicting Willingness-to-Pay a Premium for Integrated Pest Management Produce: a Logistic Approach, Agricultural and Resource. Economics Review, 30(2).

Grolleau, G. and Caswell, J.A. 2006. Interaction Between Food Attributes in the Markets: the Case of Environmental Labelling, Journal of Agricultural and Resource Economics, 31(3)

Ventura-Lucas, M.R., Ferro Godinho M.d.L. and Fragoso, R.S. 2002. The evolution of the Agri-Environmental Policies and Sustainable Agriculture. Paper to the Xth EAAE Congress: Exploring Diversity in the European Agri-Food System. Zaragoza (Spain)

Weaver, R.D., Evans, D.J. and Luloff, A.E. 1992. Pesticide Use in Tomato Production: Consumer Concerns and Willingness-to-Pay. Agribusiness An International Journal, 8(2).

Chapter 27

Red Meat Producers' Preferences for Strategies to Cope with the CAP Reform in Scotland

Cesar Revoredo-Giha and Philip Leat

"Beef marketing. It is not possible to keep a suckler cow for a year for the value of 0.9 of a suckled calf with employed labour and paying any rent. The retailers will import 3rd world beef if price goes to economic levels. Therefore the beef production business will steadily melt away to a much smaller national herd. At present time it is best to take the SFP [single farm payment], reduce production and wait and see what happens. We have to try, where practicable, to market direct to our customers and exclude the big retailers. This is difficult, but we're easing into it"

Comment of a North-East Scottish cattle producer after answering the FOODCOMM[1] questionnaire.

Introduction

The introduction of the Single Payment Scheme (SFP), agreed in 2003, represented a structural change in the way farmers are supported. Accordingly, the possible responses to the reform are expected to be wider than those predicted by any typical economic model based on historical information, as farmers have to consider a number of new or increasingly important variables in their decision making. These may include a retirement strategy, how to invest the SFP, possible succession plans, whether to cross-subsidise their production, and in what way their lifestyle might change, etc.

Furthermore, the Scottish Executive's document 'A Forward Strategy for Scottish Agriculture: Next Steps' (SEERAD 2006) recognised that farmers may go through a transition period after the Common Agricultural Policy (CAP) reform, and during this period they will evaluate how to react to the changes in support.

The purpose of this chapter is to analyse farmers' strategic preferences for coping with the reform. We concentrate the analysis on red meat producers, specifically cattle and sheep producers, due to their importance for Scotland's

---

1   FOODCOMM is the acronym of the EU-funded project 'Key factors influencing economic relationships and communication in European food chains'.

agri-food economy. Within Scotland the beef and sheep sectors are major parts of the agricultural economy, representing 27 percent and 10 percent respectively of agricultural output in 2005, with beef being the largest single part of the farming industry (Scottish Executive 2006a). In total there are approximately 13,300 holdings with beef cattle and 15,800 with sheep (Scottish Executive 2006b). Whilst production is spread across the country, there are particular concentrations of cattle in the South and South West of Scotland as well as the North East, whilst for sheep there are concentrations in the South and South West and the Highlands.

Instead of using a mathematical model to forecast farmers' actions towards the SFP, the chapter presents the results of a survey. The analysis of economic agents' (e.g., farmers) intentions as forecasts of future behaviour has long been subject to criticisms due to differences between intended and actual behaviour (e.g., Manski 1990). Nevertheless, in contexts with high uncertainty, where the actual outcomes may depend on a high number of unobservable variables such as those already mentioned, the survey of intentions is probably one of the few methodologies able to capture what future events might be. In this sense, the results of farmers' intentions surveys regarding CAP reform may inform policy development because they provide information about the options that farmers are considering during the transition period. Additionally, they provide information about how different groups of farmers may react to CAP reform (e.g., sheep or cattle producers, breeders or finishers, regional preferences etc.) and therefore they help identify whether some of the strategies are related (i.e., farmers consider them as part of a package) and what variables might explain their potential uptake.

The chapter is structured as follows. First, as a background, we briefly review the main changes introduced by the Midterm Review of the Common Agricultural Policy for the red meat sector with emphasis on the Scottish sector. Next, we proceed with the empirical section that starts describing the applied survey and construction of the statistical database. Then, we report the statistical results and also discuss the implications. Finally, we present some conclusions from the analysis.

CAP reform and red meat production in Scotland

The SFP is part of a package of measures as part of the CAP Reform ((EC) Regulation 1782/2003). It replaced direct support schemes such as the Arable Area Payments Scheme (AAPS), Beef Special Premium Scheme (BSPS), Extensification Payment Scheme (EPS), Sheep Annual Premium Scheme (SAPS), Slaughter Premium Scheme (SPS), Suckler Cow Premium Scheme (SCPS) and also associated payments like the LFA Supplement on sheep (Scottish Executive 2005).

In Scotland the SFP is calculated on the basis of a business' track record of farming activity under pre-reform subsidy schemes and the land used to support the relevant payments (i.e., historic or reference period approach). Thus, the calculation is the average of the farming activity expressed as a financial value divided by the land area to arrive at a number and rate of Payment Entitlements (Scottish Executive 2005).

In order to receive the SFP, farmers and crofters must maintain their land in good agricultural and environmental condition and respect regulations relating to public, animal and plant health, environmental protection and animal welfare. Regarding the agricultural conditions, the land must be used for arable land or permanent pasture (including common and shared grazing but excluding areas used for non-agricultural uses, e.g. buildings, permanent crops, forests, fruit, vegetables, table potatoes). Environmental conditions come as part of the cross-compliance conditions, and they comprise protecting the soil from erosion, maintaining organic matter levels in the soil and the soil structure; and ensuring a minimum level of maintenance for, and avoiding the deterioration of, habitats. The animal and plant health, environmental protection and animal welfare requirements are also part of the cross-compliance arrangements, deriving from a number of European Commission Directives and Regulations (i.e., 18 directives in total), and are known as Statutory Management Requirements. Examples of these directives are: Birds Directive, Habitats Directive, Groundwater Regulations and Identification and Registration of Livestock. Additionally, in order to receive the SFP the land must have been at the claimant's disposal for at least 10 months.

The funds allocated to the SFP are subject to the practice called 'modulation' which consists of the reduction of payments to make the funding available for a range of measures designed to assist rural development. The current modulation rate is equal to 3.5 per cent. The funds from the modulation exercise are matched by the Treasury and the total is available to be spent in Scotland.

The EC Regulation allows retention of a maximum of 10 per cent of payments under each of the relevant sectors to establish a national envelope (i.e. a ring-fence sum of money) to address the protection or enhancement of the environment or for improving the quality and marketing of agricultural products. This is applied in Scotland in the form of the national envelope for the beef sector called the Scottish Beef Calf Scheme (SBCS). The aim is to provide an incentive for the retention of cattle in more peripheral areas both for environmental and social reasons. It is important to note that this is not extra money but a redistribution of the resources allocated to a sector (e.g., beef sector).

In 2004, SEERAD published an analysis of the impact of the introduction of the SFP and the national envelope for the beef sector (Scottish Executive 2004). Unfortunately, a similar study for the sheep sector is not available. The results of the analysis show that the introduction of the SFP negatively affects breeder-producers as they are no longer able to factor future unclaimed subsidy payments into the price that they receive for their calves. Finishers, on the other hand, gain as their SFP is based on past subsidy claims irrespective of the prices they pay for store cattle. In contrast, the national envelope for the beef sector improves the position of breeder-producers at the expense of finishers. Additionally, the analysis indicates that the introduction of the SFP and the national envelope for beef have regional implications. Thus, the measures have a positive effect on the North East region and a negative one on the North West islands (Shetland, Orkney and Eileanan an Iar). The impact on the South West is mixed with negative effects in the Borders, Clyde Valley and Ayrshire and positive consequences in Dumfries and Galloway.

Due to the uneven distribution of effects of the measures associated with CAP reform (e.g., regional or breeder versus finishers), it is expected that farmers' strategies in response to the changes will also differ. In this sense, one might expect that in those areas where the CAP reform is envisaged to have more detrimental effect there might be greater inclination to adopt measures that reduce the unfavourable effects. Certainly, it is important to note, that the choice of measures is also constrained by producers' resources (e.g., human and non-human capital) and their willingness to change).

Empirical analysis

This empirical section comprises two parts. The first provides a description of the survey, the additional variables subsequently added to the database of survey responses and an overview of the sample. The second part presents and discusses the statistical results.

Survey

The data used in this study were collected through a postal survey undertaken between April and June 2006. The questionnaire comprised three sections. The first enquired about farmers' marketing problems; the second explored specific issues within the red meat supply chain with the purpose of providing a snapshot of how developed collaborative supply chains are in the sector and possible challenges ahead. The last section, which provides the core information for this chapter, dealt with possible farmers' responses for coping with the CAP reform.

The third part of the questionnaire enquired about farm characteristics and included two questions regarding farmers' intentions. The first question asked farmers about whether they expected their production to increase, decrease or remain the same in the future. The second question asked them to choose amongst a number of strategies that they would consider for coping with the effects of the CAP reform. The question presented the farmers with the following alternatives: no change in production; finishing animals at lower weight; changing production seasonality; cutting costs; producing higher quality; exiting production; reducing the scale of operation; investing to expand production; and diversifying to non-livestock enterprises. In addition, farmers could provide a different alternative. Separate answers were considered for cattle and sheep production.

The survey sample was designed to be representative of the Scottish beef and sheep producer sector (i.e., red meat producers). In order to exclude 'spare time holdings', the sample considered only farms with sizes of 1 or more Standard Labour Requirement (SLR). The SLR is a measure of farm size based upon the labour input required (1 SLR equates to 1,900 hours of labour input required per year).

According to the June 2005 Scottish Agricultural Census, the number of beef and sheep producers in Scotland with more than 1 SLR was 5,481. From this universe 1,778 producers were selected to produce a target sample that was representative by region and farm size. The sample considered 14 Scottish regions (Shetland, Orkney, Eileanan an Iar, Highland, NE Scotland, Tayside, Fife, Lothian, Scottish Borders, East Central, Argyll and Bute, Clyde Valley, Ayrshire, Dumfries and Galloway) and 4 farm size groups (farms from 1 SLR to 2 SLR, farms from more than 2 SLR to 3 SLR, farms from more than 3 SLR to 4 SLR, and farms with more than 4 SLR).

The survey questionnaire was mailed to the 1,778 producers, obtaining an overall response of 34.4 per cent after two mailing waves. From the 611 farmers of the resulting sample, 16.1 per cent were found to be cattle specialists, 27.3 per cent sheep specialist, with the remainder being producers of both cattle and sheep.

Most farmers engaged in the production of cattle were found to be exclusively breeders (53.2 per cent) or breeders and finishers (40.2 per cent), with only a small percentage being only finishers (6.6 per cent). These percentages were different in the case of sheep producers, where most were engaged in both breeding and finishing (55 per cent), followed by exclusively breeders (40.6 per cent) and only 4.3 per cent being purely finishers.

Regarding the marketing channel used, the survey showed that producers use a variety of marketing channels for their output. However, the two most common channels found for both cattle and sheep production were livestock auctions (the average percentages of the output marketed through this channel was 58.5 for cattle and 65.5 for sheep), followed by marketing directly through processors and abattoirs, where the average percentages of marketed output were 26.2 for cattle and 14.9 for sheep.

The information provided by the survey was complemented by information from the Agricultural Census describing the production of each one of the farms of the sample, and with further information from the Farm Structure Survey. Unfortunately, the information from the latter was only available for 59 per cent of the sample.

Results and discussion

The tree diagrams presented in Figure 27.1 (comprising Panels 1.a and 1.b) summarise the responses obtained from the producers regarding their strategies to cope with the CAP reform. Panel 1.a refers to their strategies with respect to their cattle production whilst Panel 1.b refers to their sheep production.

The different responses considered in the study were divided into two main groups: those that only indicated a change or not in the level of production, labelled 'status-quo' (to indicate a relatively passive response rather than no response at all) and those that indicated a sort of active response that we labelled 'strategy'. The status-quo responses were: no change in production (N1); exiting production (N2); reducing scale of operation (N3). The strategies were: finishing animals at lower weight (A); changing production seasonality (B); cutting costs (C); producing higher quality (D); investing to expand production (E); and diversifying to non-livestock enterprises (F).

Additionally, we deployed the term 'pure strategy', when a farmer chose only one strategy and 'mixed strategy' when the farmer chose more than one strategic response. It is important to note that the fact that the farmer had chosen more than one strategy does not mean necessarily that it is a strategy comprised of several parts. This is due to the fact that in the questionnaire farmers were asked to choose all the possible alternatives that they were considering to cope with the CAP reform. Therefore, from a group of chosen alternatives, farmers might end up applying only some of their indicated strategic responses.

The figures present the number of cases under each category and percentages with respect to the preceding major categories. For instance, in the case of the pure strategy categories, they present three percentages: from left to right, the first percentage is with respect to the total number of farmers in the group, the second is with respect to the total number of farmers applying at least one strategy and the third one is with respect to the number of cases with a pure strategy.

For both cattle and sheep, the analysed responses corresponded to those of producers that were specialist or mixed producers. For instance, the analysis of cattle production only considered responses of producers that were cattle specialists or were producers of both cattle and sheep, and excluded specialist sheep producers that might have some cattle. The classification of specialists and mixed producers was provided by SEERAD.

Figure 27.1 (panels 1.a and 1.b) indicates that regarding cattle production, approximately 39 per cent of the 485 farmers are not considering any strategy in particular (i.e., status-quo). In the case of sheep production that percentage is 48.6 per cent of the 313 farmers. These large percentages may indicate some degree of uncertainty about future economic conditions, which makes it difficult to envisage a more precise strategy. In addition, these high figures may also be interpreted as indicating that some farmers are in some sense willing to avoid restructuring their business by subsidising them with the proceeds from the single farm payment.

Twelve per cent of the cattle producers chose one strategy, whilst in the case of sheep this percentage was 15 per cent. Within the strategies, the most commonly chosen was 'to improve the quality of the production' for both cattle and sheep production. In the case of cattle it represented 44.8 per cent of the total number of pure strategies and 39.6 per cent in the case of sheep. This strategy is followed by two other pure strategies: reduction of costs (13.8 and 22.9 per cent of the number

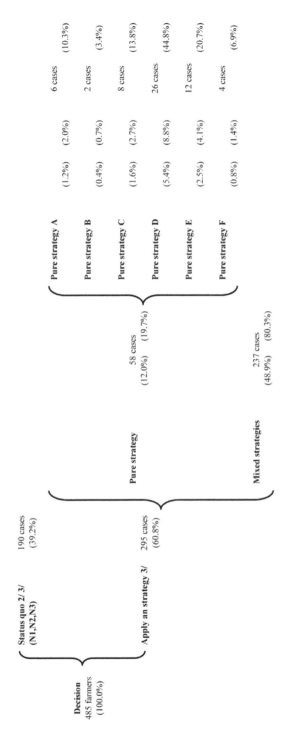

Figure 27.1a   Strategies chosen by farmers regarding their cattle production

Notes:

1/ It only considers cases of cattle specialist producers or cattle and sheep producers

2/ Farmers that chose one of the following answers: (i) will not introduce any change, (ii) reduce the scale of their operation, (iii) exiting production or (iv) did not answer.

3/ The strategies are as follows: N1=No change in production; A=Finishing animals at lower weight; B=Changing production seasonality; C=Cutting costs; D=Producing higher quality; N2=Exiting production; N3=Reducing scale of operation; E=Investing to expand production; F=Diversifying to non-livestock enterprises.

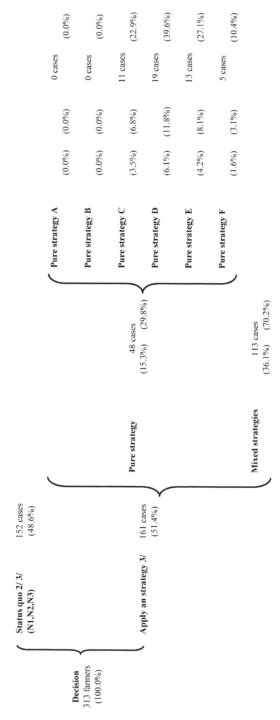

Figure 27.1b   Farmers' Preferences for Strategies to Cope with the CAP Reform

Notes:

1/ It only considers cases of sheep specialist producers or cattle and sheep producers

2/ Farmers that chose one of the following answers: (i) will not introduce any change, (ii) reduce the scale of their operation, (iii) exiting production or (iv) did not answer.

3/ The strategies are as follows: N1=No change in production; A=Finishing animals at lower weight; B=Changing production seasonality; C=Cutting costs; D=Producing higher quality; N2=Exiting production; N3=Reducing scale of operation; E=Investing to expand production; F=Diversifying to non-livestock enterprises.

of pure strategies for cattle and sheep respectively) and investment to expand production (20.7 and 27.1 per cent for cattle and sheep respectively).

The percentage of farmers indicating more than one response (i.e., mixed strategies) was also significant (48.9 per cent in the case of cattle producers and 36.1 in the case of sheep producers). It is important to note that within the number of cases considering at least one strategy, the case of mixed strategies was the most common (80.3 per cent for cattle and 70.2 per cent for sheep).

Given the importance of mixed strategies pointed out in panels 1.a. and 1.b, we proceeded to analyse the degree of relationship between pairs of strategies. Tables 8.5 and 8.6 measure the degree of association between all the categories for cattle producers and sheep producers.

The tables are comprised of two parts: the upper part of the table presents the Chi-square test of the degree of independence between the categories. The null hypothesis is that the categories are independent, therefore when rejected it indicates a degree of association between categories. The lower part of the table corresponds to the contingency coefficient that measures the degree of association of two categorical variables.

It is important to note that in contrast to the Pearson correlation coefficient, which measures the degree of association for continuous variables, the contingency coefficient does not have a maximum value of 1 (although the minimum is equal to 0). In fact, as can be seen in Tables 8.7 and 8.8, the values in the diagonal are 0.71 instead of 1. Additionally, the Chi-square tests (upper part of the tables) are used to verify whether the contingency coefficients are statistically significant different from zero (i.e., the categories are not independent). Those coefficients that are significantly different than zero appear highlighted in the table.

Table 27.1 Degree of association between red meat producers' strategies to respond to CAP reform – Cattle Production

χ^2 with 1 degree of freedom is 6.64 with $\alpha=0.01$

Strategies

Strategies	N1	A	B	C	D	N2	N3	E	F
N1	485.0	4.3	**8.9**	0.9	**8.1**	3.2	**61.5**	**22.6**	**16.8**
A		485.0	**15.7**	0.3	2.2	0.2	0.1	2.3	0.1
B			485.0	**37.2**	**12.8**	1.0	1.8	2.7	**7.9**
C				485.0	**14.0**	1.9	0.0	2.1	0.2
D					485.0	3.7	0.0	**12.7**	1.8
N2						485.0	0.1	1.4	1.7
N3							485.0	**20.5**	**20.4**
E								485.0	1.2
F									485.0

Contingency coefficient (i.e., degree of association between pairs of strategies)

Strategies

Strategies	N1	A	B	C	D	N2	N3	E	F
N1	0.71	0.09	**0.13**	0.04	**0.13**	0.08	**0.34**	**0.21**	**0.18**
A		0.71	**0.18**	0.02	0.07	0.02	0.01	0.07	0.02
B			0.71	**0.27**	**0.16**	0.04	0.06	0.07	**0.13**
C				0.71	**0.17**	0.06	0.00	0.07	0.02
D					0.71	0.09	0.00	**0.16**	0.06
N2						0.71	0.02	0.05	0.06
N3							0.71	**0.20**	**0.20**
E								0.71	
F									0.71

E   0.71   0.05

F   0.71

Notes: The strategies are as follows: N1=No change in production; A=Finishing animals at lower weight; B=Changing production seasonality; C=Cutting costs; D=Producing higher quality; N2=Exiting production; N3=Reducing scale of operation; E=Investing to expand production; F=Diversifying to non-livestock enterprises.

Cattle production shows 13 significant correlations and sheep production only seven. Of these correlations, the only one that looks puzzling is the significant degree of association between N1 and N3; this is between no change in production and decrease in the scale of production, which can be interpreted as farmers' uncertainty about the direction in the change of production scale. Another significant association was that one between no change (N1) in production and investing to expand production (E). Whilst, this answer might sound contradictory, the farmer choosing these options might have been considering N1 as a short term option and E as a long term one.

Regarding the category 'no change in production', in the case of cattle producers it is associated with change in production seasonality, the production of higher quality, investing to expand production and with diversifying to non-livestock enterprises. In the case of sheep producers it is associated with only the first three mentioned categories.

In both cattle and sheep, the strategy of increasing the quality of production is correlated with no change in production, change in seasonality, cost reduction and investment to expand production.

One of the results of the IMCAPT project (SAC 2006) was to show the dispersion in cost efficiency amongst farmers and the possibility of improving profitability by reducing the gap between them. It appears from panels 1.a and 1.b, and from the association analysis, that cost reduction is considered as a strategy by some farmers (although it is not the most common strategy) and it also appears associated with the strategy of changing the seasonality of production for both cattle and sheep.

The strategy of diversifying to non-livestock enterprises is only correlated with other categories for cattle producers. It is related to no change in production, decreasing production and with change in seasonality.

In addition, to the association analysis, a frequency distribution analysis was performed to figure out the diversity of strategies followed up by producers. The results indicated that cattle producers chose a greater range of possible strategy combinations than sheep producers. It is important to note that if we group those farmers choosing 'producing higher quality', despite whether they are planning to maintain or decrease their scale of production (i.e., cases D, N1D, N3D), then in the case of cattle they are 37 per cent of the total number of farmers that selected the option and 46 per cent in the case of sheep.

In the case of cattle producers, the two most common combinations that include 'producing higher quality' are with 'investing to expand production' (18.5 per cent of the total of cattle farmers that chose producing higher quality), 'cutting costs' (16.2 per cent) and 'diversification to non-livestock activities' (11.2 per cent). In the case of sheep producers, the combination 'producing higher quality' and 'cutting costs' is the most common (19.4 per cent of the total of sheep farmers that chose producing higher quality), followed by the combination with 'diversification to non-livestock activities' (11 per cent) and with 'investing to expand production' (6.4 per cent).

Amongst the other strategies (i.e., those not including improved production quality) the most important for cattle was cutting costs and diversify to non-livestock activities followed by expand production through investment. The same three strategies were found in the case of sheep producers, but the cutting cost strategy was the most important, closely followed by diversify to non-livestock activities and by expand production through investment.

Table 27.2   Degree of association between red meat producers' strategies to respond to CAP reform – Sheep Production

χ^2 with 1 degree of freedom is 6.64 with $\alpha=0.01$

Strategies

Strategies	N1	A	B	C	D	N2	N3	E	F
N1	313.0	0.1	2.4	0.7	**10.8**	2.2	**40.2**	6.4	3.6
A		313.0	**7.5**	3.4	2.6	0.1	1.1	0.2	1.5
B			313.0	**11.6**	**14.4**	3.0	0.8	**8.7**	4.2
C				313.0	**13.8**	0.8	0.9	0.2	0.8
D					313.0	1.7	0.7	3.5	2.0
N2						313.0	0.0	0.4	4.9
N3							313.0	6.0	6.0
E								313.0	0.2
F									313.0

1

Contingency coefficient (i.e., degree of association between pairs of strategies)

Strategies

Strategies	N1	A	B	C	D	N2	N3	E	F
N1	0.71	0.02	0.09	0.05	**0.18**	0.08	**0.34**	0.14	0.11
A		0.71	**0.15**	0.10	0.09	0.02	0.06	0.03	0.07
B			0.71	**0.19**	**0.21**	0.10	0.05	**0.16**	0.12
C				0.71	**0.21**	0.05	0.05	0.03	0.05
D					0.71	0.07	0.05	0.11	0.08
N2						0.71	0.00	0.04	0.12

1

N3	0.71		
E		0.14	0.14
F		0.71	0.03
			0.71

Notes: The strategies are as follows: N1=No change in production; A=Finishing animals at lower weight; B=Changing production seasonality; C=Cutting costs; D=Producing higher quality; N2=Exiting production; N3=Reducing scale of operation; E=Investing to expand production; F=Diversifying to non-livestock enterprises.

Table 27.3 presents a regression analysis of the two most common answers for cattle production and for sheep production, i.e., no change in production and producing higher quality. A discrete choice regression, i.e., logit model, was used to identify some variables that may be associated with the decisions behind the two mentioned categories.

The dependent variable in the logit model is a dichotomous variable (i.e., dummy variable) that takes the value of one when the category is selected (e.g., no change in production) and zero otherwise. Thus, the model measures the probability of choosing the response coded with a 1. The logit model is given by equation (1) (where the sub-index t represents observations):

$$(1) \qquad P_t = F(I_t) = F\left(\sum_{i=1}^{k} \beta_i X_t \right) = \frac{1}{1 + \exp\left(-\sum_{i=1}^{k} \beta_i X_t \right)}$$

Where P_t is the probability of choosing the category (e.g., producing high quality), I_t is an index equal to $I_t = \sum_{i=1}^{k} \beta_i X_t$, β_i are the model parameters and x_t are the explanatory variables, $F(\bullet)$ is the logistic distribution (i.e., cumulative) function and k is the number of parameters including the intercept (i.e., $x_{1t} = 1$).

Regarding the interpretation of the coefficients, a positive (negative) coefficient indicates that the variable has a positive (negative) effect on the probability of choosing the category.

Table 27.3 Logit analysis of selected farmers' responses

Dependent Variable		Intercept	Shetland	Orkney	Highland	Tayside	Lothian	Scottish Borders	Clyde Valley	Ayrshire	Dumfries and Galloway	Farm size (sh)	Squared Farm size (sh)	Breeder and finisher	Finisher	χ^2 2/	Sig. 3/
I. Cattle production																	
No change in production	Coef.	0.043						-0.755			0.413	-0.088		-0.817		28.386	***
	t 1/	0.228						-1.556			1.681	-2.104		-3.806			
Concentrate on high quality	Coef.	-0.822			-0.781	-1.103				1.017		0.204	-0.021	0.338		32.213	***
	t 1/	-2.692			-2.137	-2.162				2.897		1.541	-1.805	1.648			
II. Sheep production																	
No change in production	Coef.	-0.296	-1.244					0.582			0.721	-0.084			0.773	12.647	**
	t 1/	-1.363	-1.594					1.486			1.775	-1.780			1.252		
Concentrate on high quality	Coef.	-1.417		1.569			-1.379		1.064			0.175	-0.008			13.553	**
	t 1/	-4.755		1.786			-1.317		2.098			1.772	-1.350				

Notes: 1/ Asymptotic t statistic; 2/ Value of the log likelihood ratio test under the null hypothesis that all the coefficients except the intercept are equal to zero; 3/ * significant at 10 per cent, ** significant at 5 per cent, *** significant at 1 per cent.

The variables chosen in the analysis were dummy variables representing the different regions (shown in Table 27.3); a continuous variable, the standard labour requirement (SLR) representing the farm size (SLR was also incorporated in the regressions in a non-linear way by including a quadratic term in addition to the linear one); and dummy variables for whether the farmer was a specialist or had mixed production and whether a breeder or finisher. All the variables were tried in the four regressions and only those that were significant (or approximately) were kept in the table.

As measured by the likelihood ratio test -which indicates whether the explanatory variables provide some explanation of the variance, in addition, to that already explained by the intercept- the cattle production equations were more significant than those for sheep. In the former the likelihood ratio test rejected the null hypothesis at 1 per cent, as opposed to 5 per cent for the latter.

With respect to cattle production, the probability of choosing no change is production is affected negatively (i.e., reduces the probability) if the farm is located in the Scottish Borders and positively if it is in Dumfries and Galloway. In all the other regions, the effect was not statistically significant and different from that captured by the intercept. Farm size has a negative effect on the probability, indicating that the bigger the farm the less probable it is that changing production was selected. A similar result was found with respect to whether the farmer is a breeder and finisher.

Regarding the answer of concentrating on high quality production for cattle, location in the regions of Tayside and Lothian had a negative impact on the probability of selecting this strategy, whilst location in Ayrshire showed a positive effect. The regional effects for this regression were quite significant with asymptotic t statistics above two. Farm size entered into the regression in a non-linear way. The signs showed that the probability of selecting the quality strategy increases with farm size but decreases beyond a certain point. Whether the farmer was a breeder and finisher also had a positive impact on the probability of concentrating on higher quality.

For sheep production, the results indicated that farms from Shetland are less likely to change their production whilst those in the Scottish Borders, and Dumfries and Galloway showed the opposite effect. Similar to the response for cattle, farm size affected negatively the probability of choosing no change. It is important to note that the quadratic term was not statistically significant. If the farmer is a finisher this had a positive effect on the probability of not introducing production changes.

Finally, farms in Orkney and the Clyde Valley showed a greater tendency to concentrate on high quality production, whilst Lothian farms showed just the opposite. In terms of farm size, the results showed a non-linear effect that increases the probability until some point after which a decrease occurs.

Conclusions

The results show that the nature of adjustment is still uncertain, reflected in the high numbers of farmers that do not know what strategy to follow, or that will maintain the same production levels despite the reform (a reflection of this situation is the fact that beef cow numbers post reform, i.e., 2004 to 2006, decreased only by about 1.2 per cent). However, a sizable percentage of farmers indicated their intention to concentrate their production on high quality output. The latter response opens the possibility of performance-enhancing strategies, which not only improve production quality, but also may influence the cohesion/ relationships and communication along the red meat supply chain.

Overall, the analysis shows that a large proportion of farmers surveyed indicated no change in production or a decrease in production without choosing any specific strategy. This may be explained by the uncertain conditions surrounding agriculture after the CAP reform and still prevalent at the time of the survey. It may also indicate that some farmers are willing to subsidise their production (by using the SFP to fund their productive activities) in order to maintain their farming lifestyle.

It is important to note that the variety of strategies chosen is high, and this is particularly true in the case of cattle production. However, amongst the strategies –i.e., all the alternatives other than to remain unchanged, to decrease the scale of production or exit the business- the preferred option is that of concentrating on higher quality. This result is clear from the various analyses. Moreover, it might be regarded as positive in that it matches the broad goals of the Forward Strategy and because it introduces clear avenues of action. It also indicates a positive background attitude for the national envelop for beef to develop farmers' interest in producing higher quality.

Other strategies that show some importance and were also combined with 'concentrating on high quality production' were cutting costs, which is important given the cost efficiency dispersion observed in livestock production, and diversification to non-livestock activities.

The logit regression analysis was used to identify variables that may characterise those farms answering 'no change in production' and 'concentrating on higher quality'. It showed some differing regional effects (i.e., some regions have positive effects whilst others have negative effects on the probability of choosing the respective answers). Also, the farm size variable presented a negative effect on the probability of not changing production for both cattle and sheep and a non-linear effect in the case of concentrating on higher quality. Additionally, in some of the regressions, being a breeder and finisher -as in cattle- might have a negative effect on the probability of not changing production and positive effect on the probability of concentrating on high quality production. For sheep being a finisher, increases the probability of not changing production.

References

Manski, C.F. 1990. The Use of Intentions Data to Predict Behavior: A Best-Case Analysis *Journal of the American Statistical Association*, 85(412), 934-940.

Scottish Agricultural College (SAC). 2006. Implications of the CAP reform (IMCAPT), SAC, Edinburgh.

Scottish Executive. 2004. Beef National Envelope Analysis. Available at: http://www.scotland.gov.uk/Topics/Agriculture/Agricultural-Policy/CAPRef/ BNE/BNE intro.

Scottish Executive. 2005. CAP Reform. Frequently Asked Questions as of November 17, 2005. Available at: http://www.scotland.gov.uk/Topics/ Agriculture/Agricultural-Policy/CAPRef/CAPrefQandA/ SFPFAQ.

Scottish Executive. 2006a. Scottish Agriculture Output, Input and Income Statistics, 2006, Scottish Executive – National Statistics.

Scottish Executive. 2006b. Economic Report on Scottish Agriculture-2006 edition, Government Statistical Service.

Scottish Executive. 2006c. A Forward Strategy for Scottish Agriculture: Next Steps. Available at http://www.scotland.gov.uk/Resource/Doc/94965/0022832.pdf.

PART 9
The Future of Direct Payments

Chapter 28

The Single Payment Scheme in the Impact Assessment of the CAP 'Health Check'[1]

Beatriz Velázquez

Introduction

With the 2003 Reform, the way support is granted to farmers has changed radically. The vast majority of payments have been decoupled from the volume of production, and made conditional to the respect of environmental, food safety and animal welfare standards (cross-compliance). Some degree of coupling has been allowed to Member States (MS) in order to avoid abandonment of production, under well defined conditions and within clear limits. Other important elements of the reform include the strengthening of the rural development policy with more EU money and new measures, the introduction of a mechanism of financial discipline, and the revision of market policies in the dairy, arable crop, rice, and other small sectors.

Although first appraisals of the 2003 reform indicated that a fundamental reform of the CAP for the remaining horizon of the 2007-2013 financial perspectives was neither necessary nor desirable, experience gained so far from reform implementation had put in light adjustments needs that were not foreseen when the reform was carried out (European Commission 2007).

The 'Health Check' of the Common Agriculture Policy (CAP) had the objective of assessing the experience of the 2003 CAP reform, and of introducing necessary adjustments to the policy. The policy issues addressed in the Health Check (HC) covered the three basic regulations of the CAP: i) direct support to farmers[2] ii) the Single Common Market Organisation (CMO),[3] iii) the Rural Development (RD) Policy.[4]

This chapter is based on the comprehensive impact assessment of potential changes to the CAP that accompanied the HC legal proposals tabled by the Commission on 20 May 2008 (European Commission 2008). It focuses in

---

1 European Commission, Directorate for Agriculture and Rural Development, Brussels, Belgium. The views expressed in this chapterr does not necessary reflect the official views of the European Commission.

2 Council Regulation (EC) No 1782/2003.

3 Council Regulation (EC) No 1234/2007.

4 Council Regulation (EC) No 1698/2005.

particular on the analysis of different options for adjustment of the Single Payment Scheme (SPS).

The functioning of the different implementation models is presented in the first paragraph. The second paragraph describes the situation before the Health Check putting in evidence the link between payments distribution and structural variables. Open issues and objectives for adjusting the system are presented in the third paragraph. The economic impact of alternative options is analysed in paragraph 4. The last paragraph contains brief concluding remarks.

Basic facts of the Single Payment Scheme

The introduction of the Single Payment Scheme (SPS) aimed at allowing farmers to follow market signals applying a system as simple as possible from an administrative point of view, and compatible with WTO requirements for Green Box payments.

In implementing the SPS, Member States (MS) could opt for a historic model (payment entitlements based on individual historic reference amounts per farmer), a regional model (flat rate payment entitlements based on amounts received by farmers in a region in the reference period) or a hybrid model (mix of the two approaches, either in a static or in a dynamic fashion). The new Member States[5] (EU12) could choose to apply the Single Area Payment Scheme (SAPS), a simplified area payment system, for a transitory period until end 2010 (2011 for Bulgaria and Romania).[6]

In order to receive payments, farmers have to activate their SPS entitlements by matching them with a corresponding number of eligible hectares. In the historic model the number of payment entitlements corresponds to the number of hectares that generated support payments in the reference period. Eligible land not used to activate entitlements remains as 'naked land'. On the other hand, in the regional implementation the number of payment entitlements broadly matches the number of eligible hectares. As can be seen in Figure 28.1, the percentage of 'naked land' is higher, on average, in MS applying the historic model.

---

5   Cyprus, Poland, Slovenia, Slovakia, Check republic, Malta, Latvia, Lithuania, Estonia, Hungary, Bulgaria and Romania.

6   Denmark, Germany, Luxembourg, Finland, Sweden, UK - England and UK - Northern Ireland apply the hybrid model (some static, some dynamic). Belgium, Ireland, Greece, Spain, France, Italy, the Netherlands, Austria, UK – Scotland and UK – Wales apply historic the model. Malta and Slovenia apply the regional model. The rest of the EU12 apply the SAPS. The single area payment of the SAPS is a flat rate payment per hectare at MS level. It is calculated by dividing the annual national financial envelope of the MS by the agricultural area under SAPS in a given year.

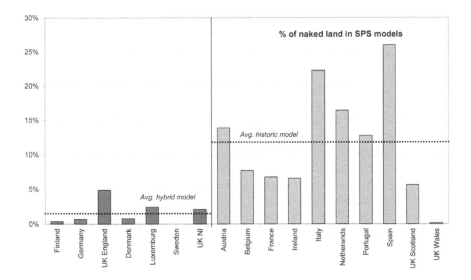

Figure 28.1 'Naked land' in historic and regional/hybrid SPS models (2006)

Source: European Commission – DG Agriculture and Rural Development

Whatever the choice of model, both historic and regional approaches have a fixed reference in payments and in the area to which these payments correspond. However, the two models, and their variants, differ significantly in terms of distribution of support, whether this is fixed in one shot (static) or gradually (dynamic).

The historic model keeps the previous level of support that farmers received, and leaves redistribution issues to be dealt with through modulation.[7] Thus this SPS model uses the *farm as the fixed reference* for the allocation of payment rights (entitlements).

The regional model, driven mainly by equity arguments since redistribution was significantly scaled down because of the limited extent of modulation, addresses issues of redistribution of support through the SPS. As a result, this SPS model uses the *area as the fixed reference* for the allocation of entitlements.

In budgetary terms, MS choices led to an almost even split between historically-based and regionally-based support. Both approaches achieved the objective of WTO compatibility, by introducing fixed references for the payments farmers receive. Although the initial implementation of the regional model proved to be more complex, once in place both models have had similar implementation rules.

---

7   Modulation is the transfer of funds from payments in Pillar 1 to Pillar 2, and involves farmers across the EU but not the smallest ones.

The situation before the Health check

The distribution of support to farmers across the EU, which resulted form the application of the 2003 reform, has been often criticised.

It has been recalled that the distribution of direct payments among farms is unequal since 80 per cent of beneficiaries receive roughly 20 per cent of payments. This was confirmed by data from the Clearance Audit Trail System (CATS) for 2006, which showed that in the EU15 80 per cent of beneficiaries received almost 20 per cent of payments, while in the EU10[8] the corresponding figure is slightly above 20 per cent (Figure 28.2). Differences in payment levels reflected the different production structures across Europe as well as the level of previous support that generated such payments.

The average payments also differed widely between MS whether the variable used as reference was the payment per area (hectare) or whether the payment was calculated per beneficiary.

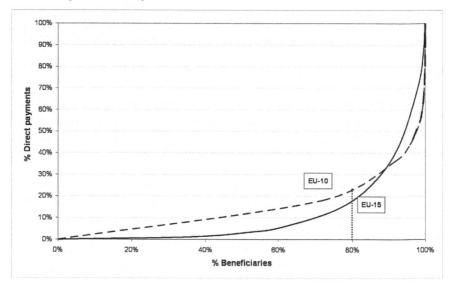

Figure 28.2 Distribution of direct payments among beneficiaries

Source: CATS data (2006 budget year)

In Figure 28.3 average farm payments was calculated in a simplified way to facilitate comparison, based on the expected amounts of payments once all MS would fully have implemented CAP reforms and new MS would have been

---

8   Cyprus, Poland, Slovenia, Slovakia, Check republic, Malta, Latvia, Lithuania, Estonia, Hungary.

wholly integrated into the CAP (2016). It is worth noting that MS with the highest per hectare payments would have been among the ones with the lowest rate of payment per beneficiary; on the other hand, the country with the highest payment per beneficiary (Greece) would be placed around the EU average if area would have been taken as the reference variable.

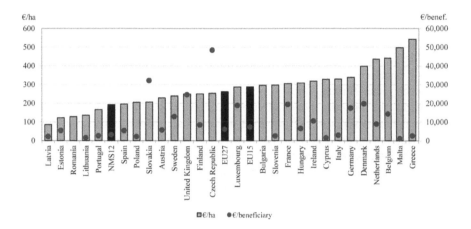

Figure 28.3 Average direct payment per MS in EUR per hectare and EUR per beneficiary

Source: DG Agriculture and Rural Development, calculations on the basis of Eurostat (UAA), CATS, and Regulation (EC) No 1782/2003 (national ceilings)

The distribution of payments per hectare based on Farm Accountancy Data Network (FADN), on structures for 2004 with payments simulated for 2009, showed that the value of payments per hectare in EU15 would have varied depending on what SPS model were implemented. In MS applying the historic model, the value of payments per hectare would have varied more than in MS implementing the regional model, both dynamic and static.[9]

What has been discussed previously reflects the issue of support equity and its relation to production structures. Being aware of the complexities of this concept, it might nevertheless be worth noting the difficulties in solving the gap between 'perceived' and actual equity issues. To take an oversimplified example, it would be possible that for some a fair distribution of payments would be a '20-20' distribution (20 per cent of farmers receive 20 per cent of payments). But the way to arrive at such a distribution might be entirely unrealistic: either by having every

---

9   See detailed analysis in Annex F 'Microeconomic (FADN) analyses' of the HC Impact Assessment in: http://ec.europa.eu/agriculture/healthcheck/ia_annex/f1_en.pdf

farmer with the same aid regardless of his farm area, or by somehow forcing every farmer to adjust their farm area in the way that they received the same payments.

Issues and objectives

Two aspects of the SPS implementation indicated that there was space, and need, for adjustment.

First, under the legislation in force before the Health Check there was no provision for MS to make adjustments to their SPS models. Even so, certain Member States could have been willing to introduce adjustments based on the experience gained through implementation.

Second, the historic model allowed farmers to be market-oriented while keeping their past support level, whereas the regional model redistributed support to farmers in a way that payments per hectare were similar within regions. As time passed, the historic reference period for these payments became more distant and individual differences in SPS values increasingly harder to justify.

One main objective pursued by the Health Check in the area of direct payments was to give Member States the possibility to modify their chosen SPS and SAPS models. Other objectives were to address concerns about the equity in the distribution of payments among farmers, to preserve transfer efficiency, market orientation and sustainability of farming, to limit administrative burdens, and to simplify the system where possible.

Options and expected impacts

In the light of the above mentioned objectives, five policy options were analysed in the Impact Assessment which accompanied the proposals tabled by the Commission in May 2008 (Table 28.1).

Table 28.1   Policy options for the SPS model

Option	SPS model	Description
0	Status quo – baseline	No review possibility for MS; no changes to historic and hybrid/regional models
1	EU-wide flat rate per eligible hectare	The same flat-rate payment entitlement per eligible hectare applies to all EU MS
2	SAPS for all MS	The Single Area Payment Scheme of new MS becomes the model for all EU MS
3	Regional flat rate per hectare	Move towards regional flat-rate entitlements applied to all eligible area
4	Regional flat rate per entitlement	Move towards regional flat-rate entitlements based on current entitlements

The starting point for understanding the impact of a move towards flatter rates of payments was to assess the impact of various SPS options on the distribution of payments. The analysis focused mainly on the economic and social sphere, although attention was also devoted to environmental and administrative implications.

Option 0 – Status quo.   The two broad SPS models, historic and regional, differ widely in terms of equity/redistribution.

The historic model has not redistributed support between farmers, and thus asset values (especially land) of the farms have been little or not at all affected. Part of direct support has been captured ('capitalised') in the value of land. Therefore, any redistribution of support would have also affected land values.[10] Capitalisation of support in land values would have been the higher, the less 'naked land' (eligible land not currently used to activate entitlements) existed. Very high transfer efficiency resulted from the fact that there has been some 'naked land', so support was better aimed at active farmers instead of landowners.

Within the regional model payments could be adjusted in certain regions according to differing natural conditions and cost structures. However, the redistribution of support among farmers in this model could have had an effect on asset prices, inducing structural responses of farms. Its implementation could have led to increased capitalisation of aid in the value of land, resulting in a somewhat lower transfer efficiency of direct support since some of it may have had benefited non-farming landowners. This would have had different impacts on Member States depending on their ownership structure. As can be seen in Figure 28.4 ownership structure varies widely across EU.

Experience with the implementation of regional model showed that in most cases the number of beneficiaries increased substantially already during the first

---

10   For a detailed analysis of the impact of the SPS payments in land markets see 'Study of the Functioning of Land Markets under the Influence of Measures Applied under the Common Agricultural Policy', European Commission DG Agri-2007-G4-14, 2008.

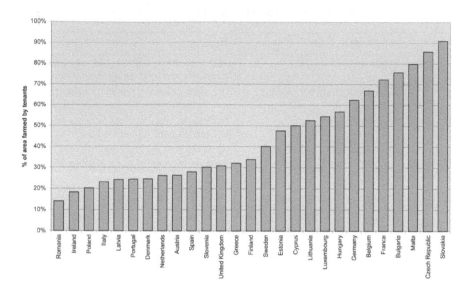

Figure 28.4 Ownership structure of EU farm land (2005)

Source: Eurostat

year of implementation. This was quite different with respect to previous claims and was due to the fact that in the regional model all beneficiaries who declared eligible hectares in the first year of implementation received entitlements. On the contrary, in the historic model only those who received support in the reference period received entitlements. As a result, the regional model led to a redistribution of payments not only between farmers but also from 'old' beneficiaries to newcomers.

Due to the higher degree of capitalisation of support in the regional model, it is to be expected that more support ends up with (possibly non-farming) landowners that under the historic model. Results of a 6th Framework Research Programme project[111] indicated that the impact of direct payments on farm income was higher in the historic model than in the regional model. This is because the stronger capitalisation in the regional model channels a larger share of payments from active farmers to landowners when land is rented.

---

11   IV Framework Project GENEDEC, see power point presentation 'Insights from GENEDEC – Brussels meeting' by Arfini, F., Kleinhanss, W., Kuepker, B. and Jayet, P.A. (2007), in http://www.grignon.inra.fr/economie-publique/genedec.

Considering that various reforms have been completed since the introduction of the SPS, and that more sectors[12] entered into the system, the experience so far had showed that certain provisions were unnecessarily rigid and complex and could be modified in a neutral way with respect to the applied model and without hampering its functioning. For example, some of the needed changes to the system were the reduction of the number of entitlements types, the elimination of transfer restrictions for entitlements received from the national reserve, and the elimination of transfer restrictions for entitlements without land.

Option 1 – EU-wide flat rate per eligible hectare. An EU-wide flat rate per eligible hectare would have meant an equal payment per hectare across EU. Although this may have been perceived as more equitable, in some of the biggest beneficiaries from such a move, for example in the new Member States, this option would have reversed the very logic of the Accession Treaty. In particular, the need to avoid huge income discrepancies within the population, for example between the agriculture sector and the rest of the economy.

The level of support in the different Member States, once the 2003 reform was fully implemented, would be the result of a delicate balance of budgetary transfers, among which fixed national SPS ceilings for payments were defined within the framework of the 2007-2013 financial perspectives. An EU-wide flat rate would have undermined such a financial framework.

The national ceilings established in Regulation (EC) No 1782/2003 were compared with the amounts they would receive if an EU-wide flat rate had been implemented. Results showed that a move towards an EU-wide flat rate per hectare would have implied a substantial redistribution impact in all Member States. Figure 28.5 indicates the percentage change in the overall amounts each Member State would have received.

Simulations made with FADN data showed that while the change in the level of payments per farm would have been considerable in some cases, a flat (equal) support rate per hectare in the EU would have implied almost no change in the unequal distribution of support among farms at EU level. This is because such a flat rate per hectare links the distribution of payments to the distribution of land between beneficiaries, and land is no more evenly distributed between farms but tends to be just as skewed as the distribution of production in the EU (which is the historic reference for payments).

---

12   Olive oil, cotton and tobacco (2004), sugar (2006), fruit and vegetables, wine (2007).

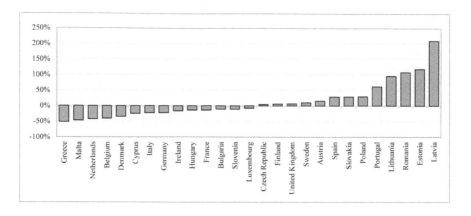

Figure 28.5 Redistribution among MS with an EU-wide flat rate

Source: European Commission – DG Agriculture and Rural Development calculations

In the case of an EU-wide flat rate payment based on all Utilised Agricultural Area (UAA), the distribution of direct payments would have corresponded to the distribution of area in the EU27 (Figure 28.6), for example, in 2005, 20 per cent of the farmers with the largest area would have counted on 8.2 hectares and 87 per cent of the EU27 Utilised Agricultural Area.

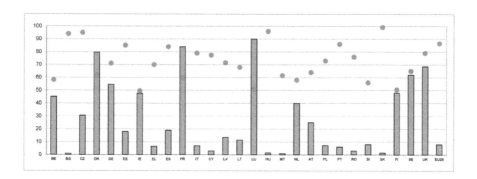

Figure 28.6 Distribution of payments for EU wide flat rate on all UAA (2005)

Source: Eurostat, Farm Structure Survey

Despite the low impact on the overall distribution, one expected effect of an EU-wide flat rate would have been the reallocation of direct payments by economic size of farms. Simulations made with FADN data showed that payments per hectare and income per Annual Worker Unit (AWU) would have decreased in the largest farms and would have increased in the lower size classes (Figure 28.7).

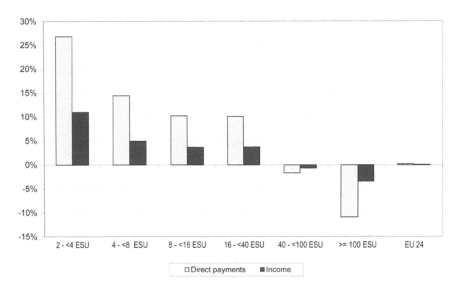

Figure 28.7   Impact on direct payments and income by class of economic size with an EU flat rate in EU24

Source: DG Agriculture and Rural Development – EU FADN simulation (size classes based on European Size Units)

Another important aspect assessed using FADN data was the extremely different impact that an EU-wide flat rate would have had on the various farm types. Both income and direct payments in milk specialists, field crops and other permanent crop specialists would have been negatively affected, whereas impacts on grazing livestock and granivore specialists, and horticultural farms[13] would have been positive. Impacts on the different types of farming differed substantially also among Member States[14] (Figure 28.8).

---

13   Note that although the impact on horticulture seems large in terms of direct payments per farm, the average area in horticulture is very small, so the impact on income is small.

14   This effect regards Member States applying the historic and those applying the hybrid model of the SPS since the bulk of the envelope of payments distributed on a historic basis in hybrid models tends to come from beef, sheep and dairy.

The impacts showed so far put in evidence that this option, which was analysed because it had some proponents in the context of the public debate, would have implied a fundamental reform of the SPS beyond the scope of the Health Check exercise. The exercise showed that an EU-wide fat rate would have brought no greater benefits in terms of the declared objectives for adjustments to the SPS.

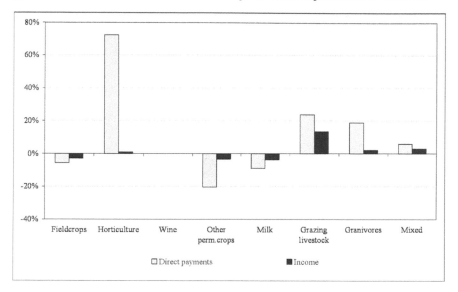

Figure 28.8 Impact of EU flat rate on different types of farming

Source: DG Agriculture and Rural Development – FADN simulation

Option 2 – SAPS for all Member States.   The Single Area Payment Scheme has been introduced in the new Member States as a preceding step to the introduction of the SPS. This was justified by the much lower levels of support to agriculture in the new Member States, the lack of previous payment and area references and the need to give new Member States time to develop the necessary control systems. As a transitional system, SAPS was designed to assist the integration of new Member States in a smooth manner, given the significant differences between general and rural economies within new Member States and with respect to EU15.

As the deadline for the expiration of SAPS got close, and new Member States considered their integration into the SPS, the possibility of extending this deadline to the end of the current financial framework, if they so wished, seemed a natural choice. Such an approach would have also allowed EU15 Member States a chance to review their SPS implementation, with the possibility to move towards a flatter rate model.

As a transitional scheme, SAPS clearly performed its intended role. Nevertheless, at times, SAPS has been mooted as a system which should be applied to all Member States, implying an inverse move from SPS to SAPS. Although such suggestions were not widespread, they contributed little to the policy debate since they failed to focus on the main difference between SPS (with the fixed entitlement reference) and SAPS (with the varying area reference).

The SAPS is a flat rate area payment per hectare at Member State level, calculated by dividing the annual national financial envelope of the Member State by the agricultural area under SAPS in a given year. This was acceptable as a transitional scheme but as a permanent scheme would have contradicted the philosophy of decoupled support because it is not based on past fixed references.

In terms of its impact on land values[15] and transfer efficiency, a SAPS for all Member States would have led to substantial impacts on land prices within Member States (due to redistribution among farmers) and higher capitalisation of support, with less transfer efficiency. Because SAPS is an area payment that did not separate payment entitlements from land, it led to a higher degree of capitalisation of support in land values. The distribution of payments among farms would have mirrored that of land and would thus have continued to be skewed.

Option 3 – Regional flat rates per eligible hectare. Regional flat rates per eligible hectare may have been perceived as more equitable than a regional model as they provide the same per-hectare support for all farmers within a region.

However, as with the EU-wide flat rate, a regional flat rate per eligible hectare would have not changed much the distribution of support between farms in the EU. Simulations made using FADN data showed that the share of direct payments received by the 20 per cent beneficiaries with the highest direct payments would have remained unchanged in the EU15 with a regional flat rate, while in the other Member States the share would have changed, although less than under option 1 (EU-wide flat rate).[16] In other words, in a number of Member States improvements towards a more equal distribution would have been expected, with a reallocation of direct support from larger to smaller economic size classes of farms.[17]

The impact of this model on different farm types was also assessed, but it seemed less pronounced than in Option 1 (Figure 28.9). Milk specialists and

---

15   The study 'Review of Transitional Restrictions Maintained by New Member States with Regard to the Acquisition of Agricultural Real Estate' provides some information on land price developments in the new Member States. See http://ec.europa.eu/internal_market/capital/docs/study_en.pdf

16   However, it has to be noted that this is also due to the fact that only the EU15 were regarded by this kind of flat rate.

17   For details see Annex F 'Microeconomic (FADN) analyses' of the HC Impact Assessment, in: http://ec.europa.eu/agriculture/healthcheck/ia_annex/f1_en.pdf

other permanent crop specialists would have been negatively affected,[18] whereas grazing livestock and granivore specialists and horticulture would have benefited. A regional flat rate per eligible hectare probably would have increased the number of beneficiaries and redistributed funds from 'old' beneficiaries to newcomers.

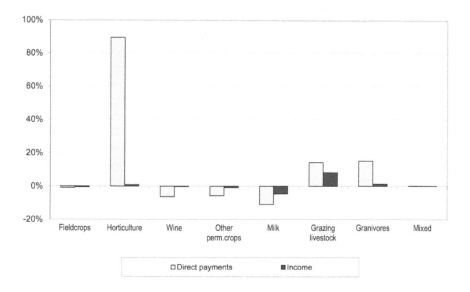

Figure 28.9 Impact of an EU regional flat rate on different types of farming

Source: DG Agriculture and Rural Development – FADN simulation

A general flat rate could have had significant redistribution impacts even within the same Member State if its farm structures differed widely. However, such impacts could have been mitigated with a targeted move towards a flat rate that takes into account such differences by harmonising payments in a regional

---

18   In the case of wine, as a reminder: whatever the model (historic, hybrid or regional), vineyards are not eligible to direct payments in the reference period used for these calculations, and had no direct payments except for the area to produce dried raisins (important production in Greece). Moreover it is not rare that olive trees are associated to wine production. Hence, farms producing exclusively wine are not affected by the negative impact on direct payment shown in the graph. Due to the reference period used, these calculations do not take into account the consequences of the Council agreement on the wine reform achieved on 19/12/2007 which will allow eligibility of all vines from 01/01/2009 and allow grape wine producers to be granted entitlements.

context.[19] This could be done by dividing the total level of reference payments of the historic model in a region by the eligible area in this region.

In a move towards a regional model, the extent to which land remains 'naked' becomes a crucial parameter. A regional flat rate per eligible hectare would have reduced the percentage of 'naked land' leading to more capitalisation of support in land values.

Option 4 – regional flat rate per entitlement. This option aimed at mitigating the undesired consequences of a redistribution of support favouring landowners, leading to new beneficiaries and still keeping the desired impact of more even payments among active farmers. The suggestion was to move towards a flat rate based on the existing entitlements, for example dividing the total value of payment entitlements within a region by the number of existing entitlements.

With respect to overall redistributive effects, the effects of a regional flat rate per entitlement would have been similar to those of a regional flat rate per eligible hectare.[20] However, there are notable differences.

A flat rate per entitlement would have not changed the degree of capitalisation of support in land values with respect to the status quo, since the amount of 'naked land' would have remained the same. The share of support of landowners and farmers would have not been affected and the transfer efficiency would be similar. However, the existing capitalisation of support would still have meant that the redistribution of support would have led to adjustments to farmers' land values under this option. Nevertheless, like in option 3, the impact on land values would have been mitigated by the possibility to design appropriate regions.

Final remarks

The historic model of the SPS and optional partially coupled support enabled a smooth transition to decoupling in Member States whose variable production structures called for progressive integration into the SPS in 2003. However, at the time of preparing the HC, it seemed appropriate to allow Member States a targeted adjustment towards a flatter rate for payments, which would have addressed concerns of unequal distribution of payments between farmers.

The legislative proposals tabled by the Commission contained four alternative ways of adjusting the SPS implementation. Member States applying the historic model may approximate the unit value of entitlements, setting a 'regional flat rate per entitlement' (option 4 in the Impact Assessment). A second alternative allowed Member States applying the historical model to change over to the regional

---

19   This seems to reflect the choice of England, where the ratio between the highest and the lowest flat rate is 7 to 1.

20   Due to technical complexities the specific analysis was not possible on the basis of FADN.

model defining a 'regional flat rate per eligible hectare' (option 3). The third alternative foresaw that Member States applying the hybrid model may review their decisions. Adjustments were optional so a last alternative was not to make any change (option 0). Option 2 of the Impact Assessment (an EU-wide flat rate) was discarded because it has been considered as going beyond the scope of the Health Check, both in distributional and budgetary terms.

The Commission proposed a transition period, with pre-established steps and an upper limit to the difference between starting and final entitlement values in order to be in line with the general principles of Community law and the objectives of the CAP allowing farmers to adapt reasonably to changes in the level of support.

The options given to Member States for adjusting their SPS models has set the pace and the dimension of adjustments, although changes in other CAP instruments had a role as well. The progressive component of modulation was considered a way of addressing the uneven distribution of payments. A stronger modulation (both basic and progressive) would probably have been more acceptable to the general society, and particularly to those in favour of a much strong rural development instruments, but it found resistance from farmers (and governments) due to leakages and implementing complexities.

Specific concerns and solutions were tackled via changes in Article 68 of Regulation (EC) No 73/2009 (ex Article 69). The new provisions allowed Member States to use part of their available SPS support to target particular sectors and regions with specific needs.

In practise, changed introduced by MS into their SPS system, the possibility and structure of Article 68, the amount of resources shifted to the second pillar through modulation and the revision of Rural Development instruments and objectives have defined the starting point for the current debate around the future of the CAP

References

European Commission. 2008. Commission Staff Working Paper, Health Check of the CAP Impact Assessment accompanying the Legislative Proposals, SEC(2008) 1885.

European Commission. 2007. *Communication from the European Commission to the European Parliament and the Council 'Preparing for the 'Health Check' of the CAP*, COM(2007) 722 final.

Chapter 29

The Fortune of Modulation in the Process of the CAP Reform

Roberto Henke and Roberta Sardone

Introduction

With the review of the Fischler reform of common agricultural policy (CAP) – the so-called Health Check – an overall picture of the adjustment process is highlighted (European Commission, 2008 and 2007). The path that has been designed shows that the Health Check represents the predictable conclusion of a process started in 2003, although the rationale and the consequences of the approach proposed are something more than a simple 'technical adjustment' of the CAP (Cooper et al. 2007).

Looking at Pillar I of the CAP, the new regulations (Reg. (EC) 73/2009) remark on the ongoing process of decoupling of direct payments, together with the increasing dismantling of the traditional market policies. In a nutshell, the Pillar I reform goes in the direction of turning the single payment, decoupled on a historical base, into a flat-rate regionalised payment, addressing the remuneration of environmental and social services through the conditions it imposes. In this way, direct payments become, at least in the intentions of the policy makers, the rewards for providing goods and services to the whole community, rather than compensation offered to farmers for the progressive dismantling of the specific supports granted to agricultural products.

As for Pillar II, the Health Check draws a picture that apparently seems to reinforce the role and objectives of the rural development policies. However, on deeper examination, one could argue that Pillar II has been considered as a sort of 'black box' where new functions and measures that are a consequence of the progressive revision of the Pillar I support can be placed: measures of land management due to the elimination of the set aside; a 'soft landing' for the milk quota regime elimination, providing specific measures for the production in mountain and marginal areas; measures aimed at the management of risk coming from further decoupling and the increasing dismantling of market policies; finally, the reinforcement of measures concerning with climatic change and the management of the natural resources. All in all, it can be maintained that it is the sector-based family of measures that tend to be enhanced, in contrast with the emphasis put on the territory-based approach of Pillar II.

Given this as a more general picture, the debate around the financial resources available for the two pillars of the CAP is mainly focused on the imbalance still existing between them, with Pillar I still representing about 75 per cent of the total expenditure for the CAP, and that in spite of the fact that the uneven distribution of resources was first highlighted at the conference of Cork in 1996 and then during the debate developed around Agenda 2000. Since then, many instruments have been discussed, all concerning the reduction of direct payments: degressivity, capping and modulation. Degressivity is simply a progressive reduction of the total amount of direct payments enjoyed by farms, never really taken into consideration so far. Capping was originally conceived as a cut to the higher brackets of direct payments that was supposed to reinforce Pillar I of the CAP, but it was then set aside. Only, it recently popped out again in the Health Check as 'upper and lower limits to support levels'. Modulation was launched with Agenda 2000 as a voluntary instrument and since then it has become one of the backbones of the CAP, being the only tool that actively shifts resources from the first to the second pillar of the CAP and that redistributes resources among Member States.

The aim of this chapter is to reconstruct the fortune of the modulation of direct payments in the CAP, tracing its evolution from its first appearance on the CAP scene with Agenda 2000 in 1999 up to the recent version of the Health Check of the Fischler Reform of June 2003. The continuous change in the rationale and implementation of the modulation underlines on one hand the 'experimental' feature of the modulation, and on the other the deep conflict between its supporters, that are in favour of the reinforcement of Pillar I at the expense of Pillar II, and those who still are in favour of a larger first pillar, although reformed and addressing the secondary functions of agriculture through conditionality.

In more detail, specific attention will be paid to the following aspects: the shift from a State-based voluntary modulation to a mandatory one (passing by a new voluntary version); the analysis of the most recent version of modulation coming with the Health Check; the theoretical and actual reinforcement of Pillar II of the CAP, given also the financial decisions adopted in 2006; finally, the redistributive effects of modulation among Member States.

Modulation from Agenda 2000 to the Health Check

Modulation was introduced in the CAP tool box with Agenda 2000, in the framework of the so- called 'horizontal regulation' (Reg. 1259/1999) and it featured for the first time a voluntary cut in the direct payments (DP) granted to farmers and conceived as a compensation for the decrease in the institutional prices. The horizontal regulation supplied with the voluntary modulation a legislative framework for the reduction of DP on the basis of parameters connected with farm employment, total farm income and total amount of DP received by a single farmer, but in any case not exceeding 20 per cent of the total amount: the Regulation fixed these basic criteria, but each Member State was allowed to choose whether to apply modulations and how to apply it, (Dwyer and Bennet 2001, Lowe et al. 2002). The same regulation also established that revenues obtained through the application of modulations had to be channelled towards the implementation of additional measures within the former 'accompanying measures' (Regs. 2078/92, 2079/92 and 2080/92), and allowances for the disadvantaged areas, all included in Regulation 1257/99 on rural development with Agenda 2000. It is, in fact, with Agenda 2000 that Pillar II of the CAP took its current shape, and modulation became the tool with which a shift of resources from the first to the second pillar was assured (Dwyer and Bennet 2001).

At the time of Agenda 2000, the debate that had arisen around modulation became quite intense, casting light on the pros and cons of the instrument: on one hand, it was considered the first clear signal of the need to balance financial resources between the two pillars of the CAP and to limit the expenditure for direct payments. On the other hand, most experts were dubious about its effectiveness being based on a voluntary approach and having effects only on direct payments, in a stage of the CAP life when other forms of support were quite relevant and common market organisations (CMOs) were rather heterogeneous in terms of the tools implemented (INEA 2000). It is interesting to stress how such debate was somehow based on the underlying idea that modulation was a temporary measure aimed at launching a signal in favour of the rebalancing of the CAP expenditure, from a sector-based dominant criterion to a territory-based one.

Not surprisingly, after almost ten years and two in-depth CAP reforms, modulation is not only still present in the tool box, but it is still the only active instrument that shifts resources from the first to the second pillar and it is actually considered as a milestone of the new CAP.

Coming to the mid-term review of Agenda 2000 (MTR), modulation became mandatory and was originally conceived as drastically drawing on the direct payments (20 per cent cut) together with the proposal of capping (Henke and Sardone 2004). However, the resistance of Pillar I support ended up in a sensible reduction of the cut, reduced to a percentage that went from 3 per cent in 2005 up to 5 per cent in 2012, whereas the capping was abandoned. It is worth noting that the actual rate of modulation would have been lower than the theoretical 5 per cent (which, in turn, was considered much lower than the proposed 20 per cent), given the franchise established at EUR 5,000, according to which the amount of direct payments under that threshold would have not been effected by the cut of modulation. Moreover, the franchise, as will be shown in more detail later, also makes the actual cut different among Member States, and especially between Mediterranean partners and North European ones, given the different farm structure and payments' distribution (Henke and Sardone 2004, Osterburg 2006).

During 2007, another proposal of a voluntary modulation to put beside the mandatory one was discussed, but it was strongly opposed by the European Parliament, that considered it an inappropriate instrument to counterbalance the cut that resources for rural development policy had borne during the financial decisions in 2006 (Osterburg 2006). Eventually, voluntary modulation was approved, but it has been implemented only by two Member States: the UK and Portugal.[1]

Moving to the latest reform, the Health Check stresses the importance of modulation as a financial instrument for the valorisation of Pillar II, also with specific reference to the new functions generally ascribed to rural development policies in the discussion paper released in November 2007.[2]

With the 2005 financial decision, resources for Pillar II for the planning period 2007-2013 were cut down from the expected EUR88.7 million to EUR69.7 million, which was to be topped up by about EUR8 million coming from modulation set up at five per cent, according to the decisions of the Fischler Reform, so that the resources available amounted to about EUR77.7 million (Mantino 2006). This is to say that, in some way, modulation has always been counted on in the balance between the first and second pillar, so it becomes relevant the mechanism of modulation itself to know the amount of resources shifted from one pillar to the other.

In the first proposal of the new modulation (May 2008), the 'old' ideas of capping and modulation were combined together in what is called the 'progressive modulation'.[3] However, in the final version of the overall agreement on the CAP review, such proposal was further simplified, and the idea itself of capping has been clearly diminished. For the EU-15, two different rates are applied: the first one affects only payments over EUR5,000, that is composed of the current rate (five per cent) plus an additional rate that grows from two per cent in 2009 up to five per cent in 2012; the second one hits payments over EUR300,000, from 2009 on, with a rate equal to four per cent. In synthesis, the modulation rate, at its full implementation, may vary from ten per cent for the share of payments over EUR5,000, up to 14 per cent for payments over EUR300,000.

---

1   According to the new voluntary modulation, Member States have been allowed to raise the rate of modulation up to 20 per cent, as originally proposed by the MTR. The UK set its modulation at different rates according to Regions (from nine to fourteen per cent), while Portugal is at 10 per cent.

2   This document, released on November 20, 2007 to open the discussion about the Health Check proposals, explicitly quotes: 'With the CAP budget now fixed until 2013, strengthening rural development funds can only be achieved through increased co-financed compulsory modulation' (Commission of the EC 2007, p. 9). In the first proposal launched by the Commission, modulation was proposed together with a rather heavy capping of the DP, but that specific proposal was then deleted, while the progressive component of modulation was proposed.

3   See the proposal for a council regulation COM(2008) 306/4 'establishing common rules for direct support schemes for farmers under the common agricultural policy and establishing certain support schemes for farmers', chapter 2, Art. 7. Such proposal was based on a growth of the mandatory modulation rate from five per cent up to 13 per cent, gradually reached through a annual increase of two per cent. Moreover, a progressive rate of modulation needed to be added to the 'basic' one, according to different levels of direct payments, implemented as a progressive taxation system.

The implementation of the new modulation

The new modulation is rather complex not only in terms of the mechanism of the cuts implemented, but also in terms of redistribution criteria among countries as well as the use of the resources saved by modulation. Starting from the criteria of resource distribution among Member States, the new regulation distinguishes the overall resources coming from modulation into two different components: one originating from the application of the 'base modulation', that is the current five per cent rate of modulation, and one that is originated by the application of the 'supplementary modulation' rate, that is the additional modulation rate (2-5 per cent) plus the progressive component (four per cent over EUR300,000). Such distinction is important considering the utilisation of the resources. Those coming from the base modulation have been already included in the budget for Pillar II planning period (2007-2013), and so they will keep following the 'traditional' rules for the Member States distribution, that is 80 per cent is redistributed to Member States by the EU on the base of the so-called 'objective criteria': agricultural area, agricultural work, GDP per-capita, and 20 per cent stays within the Member State where the cut has been originated. On the contrary, the new components of the modulation are joined together to create a sort of 'national envelope' that will stay in the Member State and will be used for the implementation of the 'new challenges' starting in 2010. The new challenges represent the main policy tasks to be addressed in the future by the CAP, for which the resources available in the rural development plans (RDPs) are considered not enough: climate change, renewable energies, water management, biodiversity; to these the technological innovations and the soft landing for the milk sector have been added following the dismantling of the quota regime. The national development plans need to be reviewed accordingly.[4]

In the next sections a simulation of the effects of modulation as designed by the new regulation is provided. In order to calculate the cuts on direct payments, the 2006 data base organised by the DG-Agri has been used[5]. The original data set has been re-organised according to the specific topic of modulation, that is calculating all the direct payments below the threshold of EUR5,000 (the franchise proposed by the Commission) that are not hit by the modulation, and considering those above the threshold of EUR300,000.

On average, in the EU-15, 75 per cent of the farms enjoy payments below the franchise set up at EUR5,000, while only 13.4 per cent of the payments is below the same threshold. Such shares hide a high variability among the Fifteen: the Northern partners show a much lower rate of both farms and payments under

---

4   In Regulation (EC) 74/2009 a list (Annex II) provides all the possible existing measures for each Member State that can be allowed as addressing the new challenges. New measures are also possible.

5   The data set is available on the following web page: http://ec.europa.eu/agriculture/fin/directaid.

the EUR5,000 franchise, (France 34 per cent of farms and only 3.2 per cent of payments; Germany 51 per cent and 6.5 per cent respectively; The United Kingdom 49.4 per cent and 3.4 per cent) while the Mediterranean countries show a reversed situation (Greece 90.6 per cent and 51 per cent; Portugal 92.5 per cent and 28.7 per cent; Italy 91.3 per cent and 31.6 per cent) and Spain in an intermediate position.

In Table 29.1, data are organised according to the mechanism of the modulation, so that the first EUR5,000 for each farm are reported in class 0, all payments between EUR5,000 and EUR300,000 are in the second class and the share of payments over EUR300,000 are reported in the third class (II). All in all, 33.8 per cent of the total payments are into the franchise, while only two per cent of the total payments are over the highest threshold and so generate resources for the progressive modulation, with Germany being the Member State with the highest share of payments within that bracket (7.6 per cent).

Table 29.1   Direct payments submitted to modulation cut by payment brackets, 2006

	(million euro)				%		
	0	I	II	Total	0	I	II
Belgium	156.5	310.1	0.3	466.9	33.5	66.4	0.1
Denmark	222.0	698.9	3.2	924.1	24.0	75.6	0.3
Germany	1,255.1	3,411.9	383.3	5,050.3	24.9	67.6	7.6
Greece	1,237.2	379.1	0.2	1,616.5	76.5	23.5	0.0
Spain	1,841.2	2,559.0	62.7	4,462.9	41.3	57.3	1.4
France	1,650.5	5,922.3	42.9	7,615.7	21.7	77.8	0.6
Ireland	484.1	718.3	0.3	1,202.7	40.3	59.7	0.0
Italy	1,781.9	1,609.3	64.8	3,456.0	51.6	46.6	1.9
Luxemburg	8.4	23.9	0.0	32.3	26.0	74.0	0.0
Netherlands	240.0	406.9	1.8	648.7	37.0	62.7	0.3
Austria	386.5	276.5	1.6	664.6	58.2	41.6	0.2
Portugal	244.9	292.3	1.2	538.4	45.5	54.3	0.2
Finland	243.8	256.6	0.2	500.6	48.7	51.3	0.0

Sweden	237.2	429.6	2.2	669.0	35.5	64.2	0.3
United Kingdom	612.1	2,852.0	59.4	3,523.5	17.4	80.9	1.7
EU-15	10,601.4	20,146.8	624.2	31,372.4	33.8	64.2	2.0

Legend: 0 = up to EUR5,000; 1 = from 5,000 a 300,000; 2 = over 300,000

Source: elaborations on EU data 2008.

Given the figures shown in Table 29.1, the results of modulation as proposed in Regulation n. 73/2009 are presented in Table 29.2.[6] In the EU-15, modulation cuts resources from Pillar I for about EUR1,480 million in 2009, a sum that increases up to slightly more than EUR2,000 million in 2012. The only 'supplementary' modulation (additional + progressive), that is the component that finances the national envelope, has a weight on the total amount modulated that goes from about 30 per cent in 2009 up to slightly more than 50 per cent in 2012, when modulation reaches its stable state.

Table 29.2   The components of the new modulation (million euro)

	Base modulation	'Additional+Progressive' Modulation			
	2009-2012	**2009**	**2010**	**2011**	**2012**
Belgium	15.5	6.2	9.3	12.4	15.5
Denmark	35.1	14.2	21.2	28.2	35.2
Germany	189.8	91.2	129.2	167.1	205.1
Greece	19.0	7.6	11.4	15.2	19.0
Spain	131.1	54.9	81.2	107.4	133.6
France	298.3	121.0	180.7	240.3	300.0
Ireland	35.9	14.4	21.6	28.8	35.9
Italy	83.7	36.1	52.8	69.6	86.3

---

6   According to the elaborations of the EU Commission, the modulation is applied to the ceilings of the direct payments. As a consequence, the amount of the cut in each country is slightly higher than the one here calculated. Moreover, the data set is not complete since more direct payments are to be included in the overall amount after 2006. As a consequence, data utilised can be slightly underestimated.

Luxemburg	1.2	0.5	0.7	1.0	1.2
Netherlands	20.4	8.2	12.3	16.4	20.5
Austria	13.9	5.6	8.4	11.2	14.0
Portugal	14.7	5.9	8.9	11.8	14.7
Finland	12.8	5.1	7.7	10.3	12.8
Sweden	21.6	8.7	13.0	17.4	21.7
United Kingdom	145.6	60.6	89.7	118.8	148.0
EU-15	**1.038,5**	**440.4**	**648.1**	**855.8**	**1,063.5**

Source: elaborations on EU data (2008)

If one considers the share of the actual resources modulated on the total amount of direct payments, in the EU-15 it goes from 4.7 per cent in 2009 to 6.7 per cent in 2012. This can be considered a sort of 'actual rate of modulation', depending on the amount of payments within the franchise and on the distribution of payments in the payment brackets. It is just the case to highlight how in all the Mediterranean countries such actual rate is particularly lower than the theoretical one, due to the size of farms that feature a high concentration of direct payments in the lower brackets.[7]

It is worth remembering how the additional and progressive modulation components, together making the national envelope, represent the real new element of the modulation, that is the actual additional resources coming from the reform of modulation in the Health Check. These additional resources remain in the same country of origin and are constrained in the use to the address of the 'new challenges'.

The progressive component of the modulation (deriving from a cut of four per cent of the direct payments over EUR300,000) has been heavily re-tailored during the debate on the Health Check (Henke and Sardone 2008), reducing and simplifying both the thresholds (from three to one) and the rate of modulation (only one, at four per cent). In practice, this component has been reduced to a sort of symbolic measure, since it acts, at the European level, only on 0.05 per cent of farms and on 4.5 per cent of direct payments. Altogether, it generates around EUR25 million per year, adding up to EUR125 million between 2009 and 2013 (about 37.7 per cent of resources devoted to the national envelopes). Among Member States, the progressive component of new modulation hits particularly Germany (17 per cent of payments over EUR300,000) and, at a significant distance, Italy (4 per cent) and Spain (3.2 per cent). This last evidence is quite interesting,

---

7   For example, in 2010 the actual modulation rate goes from 2.3 per cent in Greece to 8.3 per cent in the United Kingdom.

given that Mediterranean countries are also those less hit by modulation thanks to the franchise: it can be interpreted as a proof of the high polarisation of direct payments in these countries.

The reinforcement of Pillar II of the CAP

The new modulation proposed with the Health Check keeps addressing the two main objectives of this instrument that is the financial reinforcement of Pillar II through a shift of resources originally placed in Pillar I. The other objective introduced in the earlier proposals that is the introduction of a progressive element of capping to direct payments enjoyed by large farms has been drastically reduced. The reinforcement of Pillar II is an 'old' task that deepens its roots in the Conference of Cork and the debate about the livelihood of the rural areas in the EU. The latter has more to do with distributive issues within the EU and the growing need to intervene to limit the concentration of direct payments in the largest farms.

The enhancement of Pillar II brings about a number of questions that are not fully addressed by the specific norms of the CAP. Firstly, the amount of resources shifted from the first to the second pillar of the CAP does not entirely follow the actual distribution of the financial support per country but rather tends to alter it. This, of course, can be a positive result in itself, and it is worth underlining here that modulation does affect the resource balance among Member States and among farms. Secondly, the shift of resources from Pillar I to Pillar II can be opposed not only by the supporters of Pillar I – those that prefer to receive money for their status of farmers and according to the historical distribution of direct payments – but also by the national governments that should add resources on their own because of the co-financing rules of Pillar II.[8]

Finally, another relevant issue is the fact that modulation progressively adds resources to the rural development programmes (RDPs).[9] Such a shift opens relevant issues which need further discussion. One has to do with the concentration of resources on specific objectives, the other with the timing of expenditure. The regulation proposal on this matter mentions as objectives specific schemes addressing the 'new challenges' facing the European rural environment: climate change, bio-energies, biodiversity, water management, 'soft landing' for the milk

---

8   On this matter, the final decision about co-financing reduces significantly the contribution of Member States compared to the EU financing, so to lighten the financial burden of Member States; however, the principle if co-financing is safeguarded.

9   It has been noted that the link between Pillar I and Pillar II is not founded on solid ground: the regional allocation of modulation depends ultimately on the historical distribution of Pillar I payments and on farm structures; moreover, resources coming from modulation often address 'accompanying measure- like' interventions, for which expenditure is simpler and whose main objectives are internal to the primary sector (Osterburg 2006).

sector. This requires that the national plans need quick review in order to embody these measures in them. Plus, a specific monitoring activity for these measures is expected, thus making more complex the whole process of monitoring and control.

Another issue has to do with the fact that, with modulation, the pick of resources shifting from Pillar I to Pillar II would be reached in 2012, when the planning period is practically over. It is then not very clear when and how those resources would be available for the programmes, especially in a totally uncertain future of the CAP and the rural development policy after 2013 (Mantino 2007).[10] Finally, considering also the national co-financing of resources coming from modulation into the RDPs, there may even be a risk of over-financing of the plans, with the institutions managing the plans (Regions, States) unable to spend in a effective way all the resources in the programmed time, even considering an extended period for them (n+x>2).

Having said that, in the next pages the focus will be on the process of reinforcement of financial resources for Pillar II coming from modulation. According to what the new regulation says, the amount of resources raised with modulation will follow two main streams: the five per cent cut (that is, the 'old' mandatory modulation, net of the franchise) would return to the EU which will redistribute the resources among Member States according to a key that combines three elements: the share of labour in agriculture, the share of UAA and the GDP per capita. All the rest of the modulation (the additional and the progressive modulation) will be used to create a 'national envelope' that will transfer resources from Pillar I to Pillar II within Member States. According to this complex mechanism, the redistributive effect of the new modulation is quite reduced, being limited only to the original five per cent cut of the direct payments and to the franchise.

The contribution of modulation to Pillar II and its effects are shown in the next tables. Table 29.3 reports the resources cut to Pillar I, that amounts, for the whole EU-15, to more than EUR11 billion.

Given the process of redistribution activated by modulation, some countries receive more resources than the actual cut (Table 29.4). This is the case for Spain, Italy, Austria, Portugal and Finland. On the other hand, Member States like France, Germany and United Kingdom are the net contributors of modulation.

Altogether, modulation brings about 20.5 per cent of additional resources to the RDPs budget, however with very variable rates among Member States: it ranges from 8 per cent in the case of Austria to more than 50 per cent in the case of the United Kingdom. The increase of the new modulation, compared to the current one, at the European Union level equals to about six per cent points (from 14.2 per cent to 20.5 per cent).

---

10   Regarding these specific issues, it is possible that the Commission takes into consideration a different schedule of fund expenditure for the future, not strictly following the n+2 module.

Table 29.3 **Modulation cuts in the planning period 2007-2013 (million euro).**

	2007	2008	2009	2010	2011	2012	2013	2007-2013
Belgium	15.5	15.5	21.7	24.8	27.9	31.0	31.0	167.4
Denmark	35.1	35.1	49.3	56.3	63.3	70.3	70.3	379.7
Germany	189.8	189.8	281.0	318.9	356.9	394.8	394.8	2,125.9
Greece	19.0	19.0	26.6	30.4	34.1	37.9	37.9	204.8
Spain	131.1	131.1	186.0	212.2	238.5	264.7	264.7	1,428.3
France	298.3	298.3	419.3	478.9	538.6	598.2	598.2	3,229.7
Ireland	35.9	35.9	50.3	57.5	64.7	71.9	71.9	388.2
Italy	83.7	83.7	119.8	136.5	153.3	170.0	170.0	917.0
Luxemburg	1.2	1.2	1.7	1.9	2.2	2.4	2.4	13.0
Netherlands	20.4	20.4	28.7	32.8	36.9	41.0	41.0	221.3
Austria	13.9	13.9	19.5	22.3	25.1	27.9	27.9	150.5
Portugal	14.7	14.7	20.6	23.5	26.5	29.4	29.4	158.7
Finland	12.8	12.8	18.0	20.6	23.1	25.7	25.7	138.8
Sweden	21.6	21.6	30.3	34.6	39.0	43.3	43.3	233.7
United Kingdom	145.6	145.6	206.2	235.3	264.4	293.5	293.5	1,584.0
EU-15	**1,038.5**	**1,038.5**	**1,478.9**	**1,686,6**	**1,894.4**	**2,102.1**	**2,102.1**	**11,341.2**

Source: elaborations on EU data 2008.

Table 29.4 Impact of additional resources coming from modulation on RDPs, 2007-2013.

	Cut	Total redistributed	Difference	RDP with base mod.	Base mod./RDP	RDP with tot. mod.	Tot. mod./RDP	Envelope/RDP
	(meuro)	(meuro)	(meuro)	(meuro)	%	(meuro)	%	%
Belgium	167.4	138.9	-28.5	418.6	19.1	477.6	29.1	12.4
Denmark	379.7	282.1	-97.6	444.7	33.3	578.7	48.7	23.2
Germany	2,125.9	1,813.6	-312.3	8,112.5	12.5	8,910.2	20.4	9.0
Greece	204.8	511.6	306.8	3,707.3	11.9	3,779.4	13.5	1.9
Spain	1,428.3	1,770.1	341.8	7,213.9	17.5	7,724.6	22.9	6.6
France	3,229.7	2,711.1	-518.6	6,442.0	24.4	7,583.9	35.7	15.1
Ireland	388.2	343.9	-44.3	2,339.9	8.9	2,476.5	13.9	5.5
Italy	917.0	1,198.5	281.5	8,292.0	10.5	8,623.1	13.9	3.8
Luxemburg	13.0	12.5	-0.5	90.0	8.3	94.6	12.7	4.8
Netherlands	221.3	234.6	13.3	486.5	32.2	564.6	41.6	13.8
Austria	150.5	316.9	166.4	3,911.5	6.7	3,964.6	8.0	1.3
Portugal	158.7	361.5	202.8	3,929.3	7.8	3,985.3	9.1	1.4
Finland	138.8	171.5	32.7	2,079.9	5.9	2,128.8	8.1	2.3
Sweden	233.7	229.0	-4.7	1,825.6	8.0	1,908.1	12.0	4.3
United Kingdom	1,584.0	1,338.8	-245.2	1,909.6	40.5	2,474.6	54.1	22.8
EU-15	11,341.2	11,341.2	0.0	51,203.4	14.2	55,274.7	20.5	7.4

Source: elaborations on EU data 2008.

Looking at the only envelope, it represents a share of 7.4 per cent on the total RDP budget, again with a wide variation among countries: from 1.3 per cent for Austria up to 23 per cent for United Kingdom. Since this is the share of the new RDO budget specifically devoted to the new challenges, it is quite clear from the last column in Table 29.4 that the new challenges will be financed with very different budgets among Member States. From this point of view, the lack of some redistribution mechanism for this component of the modulation is quite a weak point in the whole modulation process. Moreover, this component has been drastically reduced in the negotiation process, while more 'challenges' have been added, so that the actual amount of resources devoted to each of them has been clearly cut.

Conclusions

The last fifteen years of CAP reforms have seen a progressive shift of the political focus from market tools to rural non-sector tools, even though the financial resources available for the two pillars are still quite unevenly distributed. Presently, it is very hard to acknowledge Pillar I as a market oriented set of policies and Pillar II as a 'pure' territorial one, for three main reasons. In no particular order, the first is the actual dismantling of market policies (dramatically reduced after the Health Check decisions). The second has to do with the mix of policies under Pillar II, addressing different issues according to the logic of the Axes (sector-based, environmental, territorial). Finally, the third is the increasing level of integrations between the two pillars, so that it is more and more difficult to include one intervention into one or the other pillar. Moreover, in spite of the emphasis registered in all the official papers of the EU about the need to enhance Pillar II of the CAP, the results are quite disappointing, especially if one considers the budget cuts borne in 2006.

Starting from Agenda 2000 and still in the most recent CAP reform proposal in 2007, modulation is by far the main instrument that shifts resources from the first to the second pillar of the CAP, being the other policies within some CMOs (tobacco and wine). Is such a shift a good thing? The supporters of the traditional market policies would say it is not, mainly because it is a way to cut 'certain' resources to farmers that return in a much more complicated way to the rural areas but not necessarily to the farmers and certainly not to those ones that bear the cut. On the other hand, the complex network created around the rural areas, made of local institutions, non-government organisations, research institutions, groups of citizens and of farmers, environmental organisations and so on, are very much in favour of a progressive reinforcement of Pillar II that overlaps also with a more decentralised management of the CAP expenditure and with a financial involvement of the national governments (co-financing).

From a more theoretical point of view, the whole process of reinforcement of Pillar II should better be placed in a wider rethinking of the future of the CAP and the EU budget (Esposti 2007, IEEP 2007). The whole process of CAP reform seems to design two pillars whose boundaries are less and less clear and also increasingly difficult to justify. As underlined before, the objectives and the instruments of both pillars tend to become similar, while the functioning and the financing rules are still quite different.

Coming to the specific tool of modulation, it still is the only active instrument of the reinforcement of Pillar II, according to a logic that seemed to respond more to a temporary than to a stable financial mechanism. Looking at the latest version of modulation as it has been designed by the *Health Check*, it is evident how modulation is moving from a 'symbolic approach', with a marginal cut of direct payments, to a more relevant one, that ensures, altogether, quite a relevant amount of resources to Pillar II, often increasing the total resources available for the rural development policies by a considerable amount.

With regards to modulation and the new Member States, it has to be underlined that the specific features of the pillars in these countries, where the actual distribution of financial resources is rather different form the Fifteen and relatively in favour of the rural development policies. Given this picture, does modulation have a role in the new Member States? Given the redistribution effect of modulation, this could in the future become another cause of conflict among old and new Member States in terms of competition for resources (Henke and Storti 2005).

In conclusion, looking at the more general issue of the distribution of resources among pillars, two aspects need to be underlined: one has to do with the instrument of modulation in itself; the other, more general, with the increase of resources for the policies within Pillar II of the CAP. Modulation activates a flow of financial resources that basically depends on the distribution of the funds allocated for Pillar I: if the logic behind this is, on one hand, understandable and to be supported to some extent, on the other, it is hard to accept that the shape and the strength of one pillar is to depend on the other. As for the second aspect, as said earlier, the increasing overlapping of objectives, rules and mechanisms of the two pillars at this stage of the reform process would require a deeper debate about their future and, in a more general way, about the future and the scope of the financial resources of the CAP within the EU budget.

References

Dwyer, J. and Bennet, H. 2001. *Using Modulation to Support Rural Development (Background Paper)*, The Countryside Agency – Institute for European Environmental Policy (IEEP), Brussels.

Esposti, R. 2007. 'La PAC dopo il 2013', *Agriregionieuropa*, 11. Available at: http://www.agriregionieuropa.univpm.it/dettart.php?id_articolo=285.

European Commission. 2008. *Proposal for a council regulation establishing common rules for direct support schemes and establishing certain support schemes for farmers*, COM (2008) 306/4, Brussels.

European Commission. 2007. *Preparing for the 'Health Check' of the CAP reform*, Com (2007) 722, Brussels, 20 November.

Henke, R. and Sardone, R. 2004. *The reorientation process of the CAP support: Modulation of direct payments*, in G. Van Huylenbroeck, W. Verbeke, L. Lauwers (eds.), *Role of Institutions in Rural Policies and Agricultural Markets*, Elsevier, Amsterdam and London.

Henke, R. and Sardone, R. 2008. Effetti nazionali e regionali della nuova modulazione degli aiuti diretti, *Economia e Diritto Agroalimentare*, 3.

Henke, R. and Storti, D. 2005. *CAP reform and EU enlargement: effects on the second pillar endowments*, in Ortner K.M. (ed.), *Assessing Rural Development Policies of the Common Agricultural Policy*, Wissenschaftsverlag Vauk Kiel KG, Kiel.

Cooper, T., Baldock, D. and Farmer, M. 2007. *Towards the CAP Health Check and the European Budget Review*, The German Marshall Fund of the United States, GMF Paper Series, Washington D.C.

Istituto Nazionale di Economia Agraria (INEA). 2000. *La modulazione degli aiuti diretti della PAC in Italia. Prime valutazioni*, Osservatorio sulle politiche agricole dell'UE, Roma.

Lowe, P., Buller, H. and Ward, N. 2002. Setting the next agenda? British and French approaches to the second pillar of the Common Agricultural Policy. *Journal of Rural Studies*, 18(1), 1-17.

Mantino, F. 2007. Dove sta andando la politica di sviluppo rurale comunitaria? Una analisi dei possibili scenari, *Agriregionieuropa*, 11.

Mantino, F. 2006. Quali risorse per la programmazione regionale 2007-2013 dello sviluppo rurale?, *Agriregionieuropa*, 2.

Mantino, F. 2005. Financial Perspectives over 2007-2013 period and main implications for the Rural Development Policies in Europe, *Politica agricola internazionale*, 1, 2-3.

Osterburg, B. 2006. Implementing modulation in Europe's Member States: an update on the new system, which Member States will make use of voluntary modulation? Agra Europe 4th Annual Conference on Rural Development 'Putting the Rural Development Regulation into practice', London, 14 November.

Rural Europe. 2007. CAP 'Health Check' floats 13% modulation idea, n. 57.

Chapter 30

Analysis of the Effects in Italy of Alternative Hypotheses of Regionalization of the Single Payment Scheme

Maria Rosaria Pupo D'Andrea

Introduction

The Fischler reform approved in 2003 represented a turning point in the concept of the 'first pillar' of the CAP. The Single Payment Scheme (SPS), in fact, deepened the process of decoupling agricultural support that began with the MacSharry reform, removing the link between support received by producers and what they produce and linking support instead to the possession of land and to exercising an agricultural activity in the respect of cross-compliance.

The SPS can be applied in two ways. The first, *historic,* entails that each farm receives a payment equal to the average support received by the same farm in the historic reference period. In the countries that opted to apply this model there was, consequently, a 'freezing' of the distribution of support among farms. The second model entails a criterion of *regionalized* distribution, on the basis of which *all* farmers receive a flat rate payment per hectare of equal value in each of the 'regions' identified, irrespective of whether or not they had, in the past, enjoyed direct CAP payments and, if so, of their amounts.[1]

Recent reform following the Health Check of the CAP, contained in EC Regulations no. 72/2009 and no. 73/2009 (regarding the first pillar of the CAP), weaken the link between support received at the present and past levels of production (or types of production) in order to move toward a flatter rate payment.

As regards the application of the Fischler reform, Italy adopted the historic model of redistribution so as to preserve the historic beneficiaries of the first pillar of the CAP. If this model has enabled, in the short term, the principle of total decoupling to be accepted, it is not sustainable in the longer term, especially in light of the completion of the Fischler reform that entailed the almost complete decoupling of the support from *what* is produced and *how*, for all CMOs (Community Market Organizations). In fact, it would become increasingly

---

1 For a description of the SPS model applied in France, Germany, Ireland, Italy and Great Britain, the reasons on which the choices are made and the expected redistribution effects see Swinbank et al. 2004.

difficult to justify the fact that farms with the same production profile, the same internal organization and the same production techniques receive different levels of support merely on the basis of what they formerly received, or did not receive. Although it is not our intention here to discuss the justification of the support allocated through the SPS,[2] it has become increasingly clear today, and it will be even more so in the future, that there will be the need to re-examine the system of allocation of support between areas and farms in order to find a more equitable redistributive model. With reference to this, the regionalized model appears far more equitable than the system based on historic farm reference. In the context of the 'region' of reference, this model will bring about a reduction in the differences in support received by farms and consequently (depending of the percentage of regionalization) a more or less skewed distribution of support between farmers. Regionalization, nevertheless, does not resolve the problem of unfair distribution of support between 'regions' and Member States determined, once again, by the support each received in the historic reference period (Anania 2008).

The aim of the present work is to quantify the redistribution effects, both at territorial and farm level, of alternative hypotheses of regionalization based on the recent Health Check reform, in order to help the current debate on this issue in Italy. Given the prospect of a possible move to a flat rate payment from 2014 forward, the results obtained could provide a useful basis to help decision makers and stakeholders work out how to apply 'regionalization' in our country. The following section describes the decision of the Health Check regarding regionalization. Section 3 examines the working hypotheses and the methodology employed. The results of the elaborations are contained in Section 4 and the final section draws conclusions from the study undertaken.

Health Check decisions on regionalization

EC Regulation no. 73/2009, which contains the Health Check decision on direct payments, replaced the EC Regulation no. 1782/2003 of the Fischler reform. The new Regulation foresees two mechanisms:

1. regionalization;
2. approximation.

Regionalization allows Member States that adopted the historic model of the SPS (on the basis of EC Regulation no. 1782/2003), if they so wish, to move to the regionalized model from 2010. The new Regulation allows for a partial regionalization (no more than 50 per cent of the regional ceiling). Consequently, it will become possible to divide up to 50 per cent of the regional ceiling between all

---

2   For a useful discussion on the economic justification of the SPS at the time of the introduction of the Fischler reform see Sotte 2005.

the farmers whose holdings are located in the 'region' concerned, including those that on the currently applied historic model do not hold payment entitlements (because they were not beneficiaries of direct payments in the reference period), on the basis of the hectares declared by the farmer at a certain time. The remaining part (at least 50 per cent of the regional ceiling) will be divided to historic beneficiaries only (i.e. those currently-held entitlements) – in addition to what they receive on the basis of regionalized distribution – in proportion to the value of their payment entitlements matured over the historic period.

Of particular interest is the question of 'special entitlements'. These are the ones held by livestock farmers who, prior to the Fischler reform, received headage payments (for example, slaughtering premiums), to obtain which it was not necessary to declare or possess any land. Not being linked to the ownership of land, the beneficiaries of this kind of support did not necessarily have a reference area to attach to their entitlement. Consequently, the farmers that hold special entitlements 'without land' would be severely damaged by the flat rate redistribution based on the number of hectares declared at a given date.[3]

Approximation allows, on the other hand, reducing the differences in the value of current entitlements in the 'region' of reference. This mechanism operates, therefore, only for currently-held entitlements. As the implications of approximation fall outside the scope of this work, we shall concentrate here solely on the mechanism of regionalization.

Hypotheses adopted and methodology used

On the basis of indications contained in EC Regulation no. 73/2009, preliminary hypotheses on how to divide Italy into 'regions' have been formulated. Three alternative hypotheses of 'regions' have been considered:

- 20 administrative Regions,[4]
- 4 territorial 'macro-regions' (Northern Italy, Central Italy, Southern Italy, the Islands);
- Italy as a single 'region'.

The analysis was carried out on the assumption that regionalization would take place in 2006, leaving everything else unchanged. Consequently, the reform of the CMOs for wine and fruit and vegetables, which stipulated that relative support would be included in SPS from 2008, was not considered. As a result, as regards the effects of regionalization, the positive redistributive effects in favour of historic

---

3   For an estimate of the effects of the regionalization on special entitlement in Germany and England see Swinbank et al. 2004, DEFRA 2005, DEFRA 2007a, 2007b.

4   The 20 administrative Regions into which Italy is politically divided. Each Region, in turn, is divided in Provinces.

wine and fruit and vegetables producers (and areas) and the negative effects on the other producers (and areas) are overestimated. In the same manner, the analysis does not consider the rise in the value of entitlements for beet producers as the relative reform of the sugar CMO comes in force. This leads to an overestimation of the possible gains for beet producers (and areas) and an underestimation of the possible losses for the same producers (and areas). The opposite is true for other producers (and areas). Similarly, no account is taken of the transfer to rural development of 50 per cent resources of tobacco CMO to take place in 2010. Here too, this leads to an overestimation both of the negative effects for tobacco producers (and areas) who will suffer less than expected and the positive effects for other producers (and areas), who will gain less than expected. Finally, in this work no account is taken of the other decision of the Health Check reform with reference to minimum thresholds, modulation and inclusion of some direct payments in the SPS.

Once divided the national ceiling between the 'regions', the next step was the calculation of the flat rate payment in each of the alternative hypotheses of 'region', according to three different assumed percentages of support subject to regionalization: 10 per cent, 50 per cent and 100 per cent. The last threshold is put forward as a reference point for an extreme regionalization scenario, one which has not been considered in the Regulation but could, nevertheless, become relevant when decisions are taken further along the road. The calculation of flat rate payment per hectare has been carried out by dividing the part of the regional ceiling subject to regionalization by the UAA (Utilized Agricultural Area) of the 'region' considered.[5]

Finally, the calculation of flat rate payments has been carried out in two ways, first taking all entitlements into account including the special entitlements that in some way are included in the regionalization, and secondly excluding these from the ceiling subject to regionalization.

The analysis at the farm level was conducted considering the 14,100 farms which were part of the FADN-ITALIA sample in 2006.

The amount of support that each farm of the FADN sample would receive on the basis of the UAA and the flat rate payment in each of the considered hypotheses was calculated.

As regards this, it needs to be said that in the farm analysis there are only nine scenarios evaluating the effects of regionalization, because of the impossibility of obtaining information on which farms hold special entitlements from the FADN sample.

Moreover, it was considered useful to work with the values of the single payment that had not already been affected by the cut of modulation. This allowed us to take account of the fact that regionalization involves a change in the support received by each farm that could influence the amount of resources drained off by

---

5   The Regulation speaks of eligible hectares. Based on the definition contained in Article 34 (2) this corresponds to the UAA.

the modulation (by changing the distribution of farms that fall below or above the franchise of EUR5,000).

Results of the analysis

Expected effects of regionalization at territorial level

The redistributive effects of regionalization are directly linked to the crop land uses on the basis of which historic support was calculated. The larger the region and, therefore, presumably wider the crop land use in the historic reference period (and, thus, the support per hectare received by farms in the reference period on the basis of which the single farm payment was calculated), the greater will be the effects of redistribution (Anania 2008). In the same way, the greater the percentage of regionalization, the more substantial will be the transfer of resources, because the larger will be the share of support to be redistributed on the basis of the *overall* area of *all* farms, regardless of what each farmer receives today.[6]

The total amount of the transfer of resources at territorial level, therefore, will be linked to the 'distance' between per hectare payment received in the past. In general terms, the effects of regionalization will tend to privilege crop land uses which had little or no support in the past (fruit and vegetables except those for processing, vineyards, large-scale livestock rearing) and penalize crop land uses that in the historic reference period were favoured through higher support (milk, olive oil, tobacco, rice but also tomatoes for processing). Consequently, the loss or gain in each administrative Region (and, in this context, in each Province) will depend on the crop land uses and per hectare related support in the reference period used for the calculation of the single payment compared to the average for the 'region'. The objective of the analysis carried out in the following pages is to quantify these effects.

In the territorial analysis we shall reflect on the hypothesis of 50 per cent regionalization, which is the maximum envisaged in the Health Check reform.

For each Province it was possible to calculate the difference ('losses' or 'gains') in absolute and percentage terms of the new amounts of resources deriving from the hypothesis of regionalization considered with respect to the status quo defined by the value of resources attributed in 2006. Moreover, in each administrative Region it was possible to estimate the amount of resources redistributed internally (the amount of support transferred from those Provinces who lose out to those who gain) and the amount of resources that the administrative Region loses or gains

---

6   The studies on the impact of the proposal of regionalization contained in the Health Check are limited to Member States that in 2004 chose to use the historic model to apply the SPS. For more information see Anania 2008, Anania and Tenuta 2008, Frascarelli 2008, Chatellier 2007, 2008, DEFRA 2005, Welsh Assembly Government 2004.

with the increase in size of the 'region' considered, moving from administrative Region, to territorial macro-region to the third hypothesis, Italy as single 'region'.

The first scenario considered is one in which the 'region' is defined as administrative Region. In this first simulation, the net balance for the administrative Region is naturally zero, insofar as the redistribution can only take place within the Region itself and not between one Region and another. One can witness, however, a significant redistribution between different areas (Provinces) inside the Region.

In the scenario of 50 per cent regionalization with special entitlements included, the Regions within which the greater transfer of resources is recorded (i.e. resources transferred from certain Provinces to other Provinces of the same administrative Region) are Apulia and Lombardy followed by Calabria and Veneto. In relative terms, that is in relation to the historic support for each Region, the highest percentage of support redistribution is recorded in Abruzzo, where 50 per cent regionalization would imply a move from one Province to another of 13 per cent of historic regional resources overall. The Regions with the least redistribution between Provinces in this case are Basilicata (0.4 per cent of historic resources) and Trentino Alto Adige (1.13 per cent).

Yet, when special entitlements are excluded from regionalization, it appears that the redistribution of resources is higher in some Provinces, on account of the greater heterogeneity of the crop land uses, that gave birth to historic support without special entitlements. A case in point is Trentino Alto Adige.

The second scenario is that of 'region' defined as territorial macro-regions. In this case, there is redistribution both between administrative Regions and between Provinces within each macro-region. Within the macro-region Northern Italy witnesses a redistribution of resources from Lombardy and Veneto towards other Regions (Table 10.1). As regards Central Italy as macro-region, the Regions penalized by regionalization are in order: the Marche, Lazio and Umbria. In Southern Italy, Calabria and Apulia are penalized by regionalization. Finally as regards the Islands, there is a transfer of resources from Sicily to Sardinia. It is worth remembering that the overall data for administrative Regions and for Provinces conceals internal trends which can vary widely. There will be areas and farms that gain, or lose more or less than the average and areas and farms that lose, or gain more or less than the average.

Finally, in the case of Italy as a whole, one finds a much more significant redistribution of resources. Overall Southern and Northern Italy suffer a negative effect from regionalization. In fact, the first loses 11 per cent of historic resources and the second 5 per cent. From a glance at Table 30.1 one can see that the administrative Regions most penalized are Lombardy, followed by Apulia, Veneto and Calabria. The macro-regions that gain most are the Islands (+27 per cent) and Central Italy (+16 per cent). The administrative Regions that come out best are Sardinia Trentino Alto Adige, Tuscany and Sicily.

Table 30.1 Italy. Difference in absolute value (EUR) and percentage between the overall amount of support deriving from regionalization – in the hypothesis of a 'region' as macro-region and a 'region' as Italy as a whole, on the basis of 50 per cent regionalization – and the overall amount of historic support in 2006[1]

		50% regionalization						
	With special entitlements				Without special entitlements[2]			
	Macro-region		Italy		Macro-region		Italy	
Administrative Regions	Difference (EUR)	(%)	Difference (EUR)	(%)	Difference (EUR)	(%)	Difference (EUR)	(%)
Piedmont	22,153,789	7.6	4.327.627	1.5	16.685.515	5.7	2,648,893	0.9
Valle d'Aosta	8,122,562	268.1	7,097,304	234.2	7.762.435	256.2	6,955,130	229.6
Lombardy	-76,151,969	-16.0	-93,327,736	-19.6	-72,670,328	-15.3	-86,194,817	-18.1
Trentino Alto A.	53,948,403	226.4	46,942,022	197.0	58,188,975	244.2	52,672,032	221.1
Veneto	-55,095,950	-14.5	-69,396,235	-18.3	-50,328,294	-13.3	-61,588,579	-16.2
Friuli Venezia Giulia	1,167,349	1.6	-2,731,368	-3.8	-252,726	-0.4	-3,322,641	-4.7
Liguria	4,722,837	59.5	3,798,260	47.9	4,455,395	56.1	3,727,368	47.0
Emilia Romagna	41,132,980	16.0	23,107,479	9.0	36,159,028	14.1	21,965,444	8.6
Tuscany	14,488,895	9.6	42,425,913	28.1	14,004,624	9.3	39,521,864	26.2
Umbria	-4,992,248	-5.7	6,965,758	8.0	-5,328,474	-6.1	5,593,784	6.4
Marche	-12,535,937	-9.6	3,952,548	3.0	-12,982,123	-9.9	2,078,204	1.6
Lazio	-7,641,330	-4.7	15,184,487	9.4	-6,333,503	-3.9	14,515,245	9.0
Abruzzo	10,680,621	16.8	23,898,402	37.5	10,639,475	16.7	22,712,391	35.7

Northern Italy: Piedmont – Emilia Romagna

Central Italy: Tuscany – Abruzzo

	Molise	12,927,086	27.3	4,740,876	10.0	12,695,741	26.9	3,999,770	8.5
	Campania	19,895,002	12.2	-2,880,327	-1.8	20,336,177	12.4	-3,857,387	-2.4
Southern Italy	Apulia	-35,866,800	-7.2	-83,464,325	-16.8	-36,981,768	-7.4	-87,543,224	-17.6
	Basilicata	41,602,251	41.6	21,107,114	21.1	40,993,668	41.0	19,222,285	19.2
	Calabria	-38,557,539	-14.6	-59,489,013	-22.5	-37,043,819	-14.0	-59,278,711	-22.4
Islands	Sicily	-19,012,539	-7.6	37,602,732	15.1	-20,217,437	-8.1	35,573,803	14.3
	Sardinia	19,012,539	12.4	70,145,250	45.9	20,217,437	13.2	70,605,917	46.2

Note 1: The 2006 historic support is the value of the entitlements attributed before the application of the modulation

Note 2: Entitlements with the derogation to possess a number of eligible hectares equal to the number of entitlements. These include special entitlements, special entitlements to rent milk quotas and livestock lease special entitlements

Source: Elaborations on data from ISTAT and AGEA

With the increase in the size of the 'region' comes an increase in the resources transferred between areas. With a 50 per cent regionalization, in the case of a 'region' as an administrative Region the resources transferred between Provinces are equal to EUR212 million; this figure rises to EUR337 million in the case of a 'region' defined as macro-regions and EUR368 million in the case of Italy as a single 'region'.

From the analysis just carried out it emerges that, regardless of whether one opts for a 'region' defined in terms of macro-region, or Italy as a single 'region', and regardless of whether special entitlements are included or not, there are administrative Regions that, with regionalization, are better off in every case (Piedmont, Valle d'Aosta, Trentino Alto Adige, Liguria, Emilia Romagna, Abruzzo, Molise, Basilicata and Sardinia) and others that lose out in every case (Lombardy, Veneto, Apulia and Calabria). For Calabria and Apulia, for example, this happens because the average flat rate payment in the two hypotheses of regionalization is lower that average per hectare payment matured over the historic period, determined to a large extent by high payments per hectare for olive oil. This is equally the case for Veneto, where historic support is mainly determined by high payments for arable crops, especially maize, and for Lombardy, where a high level of historic support, beside arable crops, is generated by high milk premiums included in the SPS from 2006.

In other cases the position regarding the administrative Region is not so clear.

Expected effects of regionalization at farm level

The effects on farms are evaluated at aggregate level for Italy as a whole – what happens to the total of FADN farms in the different regionalization hypotheses.

From the comparison between historic support unaffected by the 2006 modulation and the support due to farms on the basis of the regionalization hypotheses considered, we have obtained the redistribution of farm for class of the variation (in percentage and absolute value) of support received according to the alternative hypotheses of 'region' and the percentage of regionalization. This enabled us to identify the critical areas of redistribution (how many farms gain, and how many lose out over a certain amount) and the crop land uses (Types of Farming – TF) most affected by the redistribution.

The number of the farms in the sample that see an increase in support of over 100 per cent grows with the increase in the percentage of regionalization (passing from 10 to 50 to 100 per cent) and the increase in size of the reference 'region' (moving from administrative Region, to macro-region, to Italy as a single 'region') (Table 30.2). This quota includes the new beneficiary farms, i.e. those that did not benefit from the historic support under the CAP direct payments scheme and now, thanks to regionalization, fall within the ambit of the SPS. This is confirmed by the fact that the farms that more than double their historic support, in all the scenarios, are mainly vineyards especially for the production of quality wine, fruit and citrus fruits and horticulture (flowers and market gardens), in other words farms with a crop land use that had not enjoyed support in the past. Let us now turn to the farms that find themselves worse off with regionalization, with particular reference to 50 per cent regionalization.

In the case of 50 per cent regionalization between 9 and 11 per cent of farms in the sample, depending on the 'region' considered, lose over 25 per cent of the historic support (Fig. 9.10). The losses are concentrated among dairy farms (especially in the scenario in which the 'region' is defined as administrative Region), olive farms (especially in the scenarios where Italy is a single 'region' and where 'region' is defined as macro-region), cattle farmers, who produce for the meat market, cereal farms, oilseed and protein crops and rice production.

Table 30.2 Italy. Gains/losses of farms (%) and farms (n) per class of % variation in support deriving from the regionalization in the three hypotheses of 'region' on the basis of 50% regionalization 2006[1]

| | 50% regionalization | | | | | |
| | Administrative Region | | Macro-region | | Italy | |
	Losses/gains (EUR)	Farms (n.)	Losses/gains (EUR)	Farms (n.)	Losses/gains (EUR)	Farms (n.)
-50%>	-	-	-	-	-	-
-50%≤ and <-25%	-20,656,105	1,556	-21,398,134	1,289	21702660	1,286
-25%≤ and <-10%	-6,474,441	2,084	-6,543,029	1,863	-5,409,891	1,584
-10%≤ and <0%	-965,904	1,371	-903,987	1,279	-1,129,990	1,526
≥0% and <10%	507,118	1,026	492,273	1,181	471,276	1,125
≥10% and <25%	1,827,201	1,173	1,525,272	1,175	1,517,727	1,148
≥25% and <50%	2,127,859	1,067	2,252,255	1,067	2,454,586	1,112
≥50% and <100%	2,415,229	888	3,113,025	1,059	3,619,264	1,051
≥100%	11,878,998	4,935	15,933,544	5,187	16,598,273	5,268
Total	**-9,340,045**	**14,1**	**-5,528,780**	**14,1**	**-3,581,415**	**14,1**

Note 1: Excluding Emilia Romagna

Source: Elaborations on data from FADN, ISTAT and AGEA

Overall, a little over 30 per cent of the farms in the sample transfer resources in favour of the remaining 70 per cent of farms, with varying differences depending on the size of the 'region' considered. In fact, in the case of 'region' defined as administrative Region the farms that suffer with regionalization make up 36 per cent of the total, which declines to 31 per cent both in the case where region' is defined as macro-region and where Italy is defined as a single 'region'.

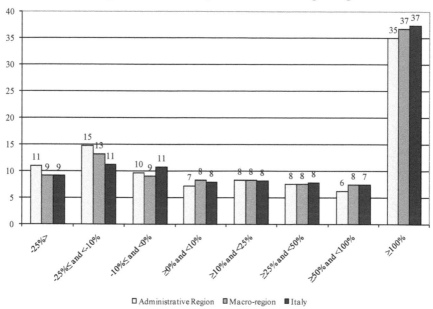

Figure 30.1   Italy. 50% regionalization. Farms (%) for class of % variation in support in the alternative hypotheses of 'region'- 2006[1]

1: Excluding Emilia Romagna

Source: Elaborations on data from FADN, ISTAT and AGEA

Now let us look at the distribution of farms per classes of variation of support (the difference between regionalized and historic support), in absolute values, (Table 30.3). With 50 per cent regionalization, it appears that the farms that are worse off, in the three hypotheses of 'region', are equally distributed in classes of losses below EUR5,000 and up to EUR500. At the opposite end, most of the better off farms are concentrated in the class with gains of between EUR500 and EUR1,000.

Table 30.3 Italy. Overall gains/losses of farms (%) and farms (n) per class of % variation in support deriving from the regionalization in the three hypotheses of 'region' on the basis of 50% regionalization – 2006[1]

	50% regionalization					
	Administrative Region		Macro-region		Italy	
	Losses/gains (EUR)	Farms (n)	Losses/gains (EUR)	Farms (n)	Losses/gains (EUR)	Farms (n)
-200,000>	-5,490,308	18	-5,814,928	18	-5,895,834	18
-200,000≤ and <-100,000	-3,020,253	23	-3,411,762	25	-3,462,023	25
-100,000≤ and <-50,000	-2,691,790	39	-2,989,950	44	-3,229,757	47
-50,000≤ and <-10,000	-9,195,650	437	-9,544,078	449	-8,957,862	417
-10,000≤ and <-5,000	-2,682,015	386	-2,912,505	413	-2,801,385	397
-5,000≤ and <-2,000	-2,847,572	912	-2,375,189	743	-2,125,650	670
-2,000≤ and <-1,000	-1,223,096	866	-1,027,680	717	-930,174	649
-1,000≤ and <-500	-631,006	860	-460,069	642	-525,68	727
-500≤ and <-300	-176,271	447	-175,244	445	-175,366	446
-300≤ and <-100	-118,349	604	-117,132	601	-121,334	623
-100≤ and <0	-20,14	419	-16,614	334	-17,475	377
≥0 and <100	57,113	1,102	41,697	772	42,614	773
≥100 and <300	309,952	1,597	320,802	1,633	322,387	1,625
≥300 and <500	421,614	1,059	509,689	1,277	483,721	1,224
≥500 and <1,000	1,284,296	1,779	1,445,862	1,991	1,451,704	2,02
≥1,000 and <2,000	2,204,462	1,549	2,438,009	1,72	2,296,355	1,629

≥2,000 and <5,000	3,639,250	1,181	3,836,136	1,243	4,084,616	1,313
≥5,000 and <10,000	3,306,588	476	3,832,841	552	3,945,181	566
≥10,000 and <50,000	6,209,017	330	8,850,458	453	10,188,451	527
>50,000	1,324,113	16	2,040,877	28	1,846,097	27
Total	**-9,340,045**	**14,1**	**-5,528,780**	**14,1**	**-3,581,415**	**14,1**

1: *Excluding Emilia Romagna*

Source: Elaborations on data from FADN, ISTAT and AGEA

From the analysis carried out so far, it emerges that the distribution of farms in the sample in classes of gains and losses *vis-à-vis* historic support expressed in percentages terms does not significantly change with an increase in the size of the 'region' considered. The picture changes considerably, on the other hand, when we consider the distribution of farms in the move from one percentage of regionalization to another.

Yet, if we consider farms from the point of view of the type of farming (TF) the analysis highlights the fact that specialist horticulture (TF 20), vineyards (TF 31) and fruit and citrus fruit (TF 32) farms in the sample all record a clear gain from regionalization (in over 90 per cent of cases, no matter which 'region' is chosen or percentage of regionalization is adopted). Specialized goat farms (TF 4430) and granivore farms (TF 50) both gain. Most sheep farms (TF 4410) gain, if support is set at macro-region level or if Italy is treated as a single 'region', but the results are more balanced if support is fixed at administrative Region level.

In a similar way, the analysis shows that over 90 per cent of tobacco farms (TF 1441) and nearly all rice growers (TF 1320) are penalized by the regionalization of support. Moreover, in roughly 60 per cent of cases, olives farms (TF 30) and dairy farms (TF 41) lose resources with regionalization. For specialist cattle-rearing and fattening farms (TF 42) the outcome varies depending on the scenario; nevertheless, we can say there is a certain balance between those that gain and those that lose out.

Conclusions

The analysis carried out here has allowed us to quantify the redistribution effects at territorial and farm level of alternative hypotheses of regionalization in Italy.

The territorial analysis has shown that also when the 'regions' are defined as administrative Regions, the redistribution of resources within may be quite high, and this is a function of the different systems of land crop use on which support was calculated in the historic reference period. As the size of the 'region' increases, and hence with the move from 'region' defined as administrative Region to that defined as macro-region, and then to Italy as a single 'region', the overall amount of resources to be redistributed between farms and areas (Provinces) grows. The loss or gain in each administrative Region, and the Provinces within the Region, will depend on the 'distance' between per hectare payment received in the past, in turn related to the crop land uses in the reference period used for the calculation of the single payment.

At a parity of 'region', the exclusion of special entitlements from regionalization does not generally lead to a significant added redistribution of resources within the 'region'.

The analysis at farm level was only considered regionalization on the assumption that special entitlements were included in the redistribution. The analysis has shown that the increase in size of the 'region' considered increases the percentage of farms in the sample that gain compared to historic support; this percentage is well over 60 per cent in all the cases considered.

Among the farms that gain more than 100 per cent are those vineyards specializing in the production of quality wine, fruit and citrus fruits farms, horticulture (flower and market gardens apart from fruit and vegetables for processing) that, in the past, had never benefited from direct support under the CAP. At the opposite extreme, the farms most damaged by regionalization are cattle farms especially dairy farms (above all, in the scenario where 'region' is defined as administrative Region) and olive farms (especially in scenarios where 'region' is defined as macro-region or as Italy as a whole), i.e. productions that benefited most in the past.

The analysis has shown how, for farms whose crop land uses were heavily subsidised under the CAP in the past (olive and rice growers, and cattle farms), the losses increase with the increase in the size of the reference 'region', because the internal crop land use become increasingly less uniform. The effect of redistribution, therefore, is greater the more diverse the crop land uses in the 'region' considered in the past and hence the variability of per hectare support currently received by farms.

The decisions on whether to apply the regionalization and how it should be applied heavily depend on the objectives of agricultural policy in Italy, on the perception that the current system of distribution is unfair and on decision makers' ability to find the more equitable and acceptable solution. In conclusion, this study has analysed the effects of redistribution on the basis of alternative hypotheses of regionalization, in an attempt to provide valuable information on which to base the choices that will have to be made in the near future.

In the evaluation of the results obtained it is necessary, however, to keep in mind certain implications of the assumptions on which they are based:

- the analysis is carried out with reference to 2006. Consequently, the historic date does not include the modifications stemming from the CMOs reform for fruit and vegetables and wine that came into force in 2008; the cut in the ceiling for tobacco that took place in 2010; the increase in the value of entitlements which beet producers will enjoy until 2010 for the progressive entry in force of the sugar CMO reform;
- the hypotheses of regionalization are carried out without taking into account the other decisions contained in the Health Check reform, that could influence the distribution of support;
- the results of the analysis at farm level for the Northern Italy macro-region and for Italy as a whole are affected by the lack of information in the FADN data on support received in 2006 for the farms in Emilia Romagna, that consequently were excluded from the simulations. This distorts, in the farm analysis, the redistributive effects in the ambit of these two 'regions' whose sign is difficult to foresee;
- the results of the analysis at farm level suffer from the fact that they are

based on a sample of farms (taken from the FADN data) that excludes the smallest producers (those under 4 ESU – Economic Size Unit). The extent to which the crop land uses of the smallest farms in the historic period were different from the larger one will be reflected in a distortion of the redistribution effects calculated in this work.

References

Anania, G. 2008. Il futuro dei pagamenti diretti nell'Health Check della Pac: regionalizzazione, condizionalità e disaccoppiamento, in *L'Health Check della* Pac. *Una valutazione delle prime proposte della Commissione*, edited by F. De Filippis, Quaderni Gruppo 2013, Coldiretti, Roma: Edizioni Tellus, 29-39.

Anania, G. and Tenuta, A. 2008. *Effetti della regionalizzazione degli aiuti nel Regime di pagamento unico sulla loro distribuzione spaziale in Italia*, Working paper n. 9, Coldiretti.

Chatellier, V. 2007. *La sensibilité des exploitations agricoles françaises à une modification des budgets et des instruments de soutien*, Report realized in the framework of the expertise Inra «Agriculture Prospective 2013».

Chatellier, V. 2008. *The financing and effectiveness of agricultural expenditure.* paper requested by the European Parliament's committee on Agriculture and Rural Development, IP/B/AGRI/IC/2008-013.

DEFRA (Department for Environment, Food and Rural Affairs). 2005. *Cap Single Payment Scheme: Basis for allocation of entitlement. Impacts of the Scheme to be adopted in England.* Agriculture in the United Kingdom' Seminar 2005, 10 May 2005.

DEFRA (Department for Environment, Food and Rural Affairs). 2007a. *Agricultural Change and Environment Observatory Programme. Annual Review*. DEFRA Agricultural Change and Environment Observatory.

DEFRA (Department for Environment, Food and Rural Affairs). 2007b. *Updated Projections Of The Distribution Of Single Payment Scheme Payments In 2012 By Farm Type, Size And Region*. DEFRA Agricultural Change and Environment Observatory.

Frascarelli, A. 2008. Le due vie della regionalizzazione. *Agrisole*, 6-12 giugno.

Sotte, F. 2005. La natura economica del PUA. *AGRIREGIONIEUROPA*, 3, anno 1. [Online]. Available at: http://agriregionieuropa.univpm.it/riviste/ agriregionieuropa_n3.pdf [accessed: 4 June 2010].

Swinbank, A., Tranter, R., Daniels, J. and Wooldridge, M. 2004. *An examination of various theoretical concepts behind decoupling and review of hypothetical and actual de-coupled schemes in some OECD countries*. Deliverable D1.1, GENEDEC Project. [Online]. Available at: http://www.grignon.inra.fr/ economie-publique/genedec/eng/enpub.htm [accessed: 4 June 2010].

Welsh Assembly Government. 2004. *The Economic and Distributional Impacts of the Implementation Options Available for CAP Reform*. Working document. [Online]. Available at: www.cefngwlad.cymru.gov.uk/fe/fileupload_getfile.asp?filePathPrefix=2152&fileLanguage=w.pdf. [accessed: 4 June 2008].

Chapter 31

Measuring Cross-Subsidisation of the Single Payment Scheme in England

Alan Renwick and Cesar Revoredo-Giha

Introduction

This chapter derives from a project for the UK Department of Environment, Food and Rural Affairs (Defra) 'Estimating the Environmental Impacts of Pillar I Reform and the Potential Implications for Axis II funding'. The purpose of the chapter is to estimate the extent that farmers are cross-subsidising the Single Payment Scheme (SPS) payments by applying them to productive activities as if they were coupled payments, and not selecting the most profitable ones at the new prices.

The motivation behind the chapter is to increase our understanding of the impact of the SPS by providing information on how farmers are utilising the proceeds of the SPS. There is a need for detailed analysis using real farm data as most of the available information about the use of the SPS is either anecdotal or simulated based on assumptions about the degree of coupling and without any empirical basis. Furthermore, understanding the behaviour of farmers in respect to the SPS is important because of the possible implications for future scenarios. For example, if farmers are using the SPS to cross-subsidise activities that are the not the most profitable then: (1) the SPS, despite what economic theory and policy makers may say, is having an impact on production.[1] (2) removal of the SPS (say by 2013) may have important implications for the level of production if farmers continue to cross-subsidise.

Section II briefly outlines the background to the implementation of the SPS. Section III outlines the empirical approach adopted for the study, whilst Section IV presents the results and discussion. The chapter concludes with a brief consideration of the need for further analysis.

Background

On 26 June 2003, EU farm ministers adopted a fundamental reform of the Common Agricultural Policy (CAP) and introduced a new Single Payment Scheme (SPS)

---

1   For instance, OECD (2006) considers how alternative indirect channels towards decoupled payments can affect production.

for direct subsidy payments to landowners. Although the SPS applies throughout the European Union according to rules agreed between the member states, the implementation details vary from country to country.

The intention of the SPS was to change the way the EU supports its farm sector by removing the link between subsidies and production of specific crops (e.g., area and headage payments). In this sense, the scheme replaced eleven previous subsidy schemes which were based on the production of crops and/or livestock e.g. suckler cow premium and arable area payments scheme. It should be noted that Member States have the choice to maintain a limited link between subsidy and production to avoid abandonment of particular production.

Member States had options in the way they calculated and made payments. The main difference lies in whether they calculated SPS on the basis of individual farmers' direct payments during a past reference period, thus producing a patchwork of different payments, or whether all payments are averaged out and paid uniformly over a region or state. Within the latter approach, payment levels may be varied between specific areas (e.g. disadvantaged and non-disadvantaged areas). An in-between system is also available which allows Member States either to operate a mixed historic/flat rate approach that stays the same over time ('static'); or they may choose a mix that alters over time ('dynamic'), usually so that the proportion of SPS based on historic references reduces as the flat rate element increases, offering a means to transit from the basic to the flat rate approach. For England, Defra decided to implement a dynamic flat rate approach.

The UK Government introduced the SPS in 2005. For the purposes of the SPS, the UK is divided into four regions: England, Northern Ireland, Scotland and Wales. England is further divided into three areas: (1) England outside the upland Severely Disadvantaged Area (SDA); (2) English upland SDA (other than moorland); and (3) English moorland within the upland SDA (Defra 2006).

The SPS is linked to meeting environmental, public, animal- and plant-health and animal welfare standards and the need to keep land in good agricultural and environmental condition. To gain funds from the SPS, the farmer has to cross comply – that is, to farm in an environmentally friendly way, with careful use of pesticides and fertilisers. Farmers also had to set aside 8 per cent of their productive land annually (although this has since been set to zero); in addition two metres on the perimeter of each field must be left uncropped to become overgrown.

Empirical Approach

Data

The information used in the chapter was extracted from Defra's Farm Accounts in England (Defra 2008), which is prepared from the results of the Farm Business Survey (FBS) in England. Nearly all farms in the FBS have accounting years ending between 31st December and 30th April, although on average, the accounts end in February (Defra 2007).

The data used covered the eight year period from 1998/99 to 2005/6 (the first year after implementation of the SPS). The information available was by Defra's robust farm type (i.e., cereals, dairy, general cropping, horticulture, LFA grazing livestock, lowland grazing livestock, mixed, pigs and poultry) and farm size (i.e., small, medium and large). This resulted in a balanced panel dataset of 192 observations. Table 31.1 provides information on the number of farms in England and by region and type that the data represents.

Variable costs were allocated to one of six groups: feed, livestock services, seed, fertilisers, crop protection and other goods and services. The outputs considered in the estimation were 19 (i.e., wheat, barley, other cereals, oilseed rape, potatoes, sugar beet, other crops, vegetables and fruits, by-prods, forage and cultivations, set aside, dairy cows and heifers in milk, beef cows, other cattle, ewes, other sheep, breeding sows, other pigs, hens and pullets in lay, other poultry).

The estimation of cost functions requires input prices. Defra's input price data for the United Kingdom were used for all the input categories. Output prices by Government Office Region were from Defra's Farm Accounts in England (Defra 2008).

Methodology

The approach adopted was to estimate farm level marginal cost functions by region, farm size and farm type and use them to predict the optimal output allocation, after the first year of the single payment scheme[2] given the prevailing input and output prices (i.e., it is an ex-post analysis). Comparison between the observed and predicted output is used to estimate whether cross-subsidisation is occurring. We concentrate the analysis on cereals, cattle and sheep for two main reasons. First, these enterprises were subject to coupled payments before the SPS (i.e., arable area payments and headage payments). Second, the fact that production was maintained at similar levels in the first year after decoupling was implemented, despite the prevailing low commodity prices implies that some degree of cross-subsidisation was occurring.

The starting point of the methodology was the estimation of a variable cost function considering terms by farm size and type. A multi-product cost function was chosen due to the fact that most of farms produce more than one output and also because itemised cost data by individual enterprise (which is now collected by Defra) was only available for the last two years of our eight year period.

From the aforementioned variable cost function, marginal cost functions were derived and calibrated for each Government Office Regions (i.e., East Midlands, East of England, North East, South East and London, South West, West Midlands and Yorkshire and Humber) using available output prices. It was assumed that

---

2   Data availability limited our analysis to the first year after implementation

each region was a separate market and therefore all producers in the region faced the same prices. It should be noted that Government Office Regions classification, although chosen because of data availability, does approximate quite well differences in natural resources (e.g. land quality) and production specialisation (e.g. the Eastern region for cereal production) across England.

The exercise of computing marginal cost functions by region effectively meant that for each region (denoted by the sub-index r), we constructed farm models (i.e., 'representative farms') which were disaggregated by farm type (denoted by the sub index t) and farm size (denoted by the sub index s). Therefore, a maximum of 24 supply relationships (i.e., three farm sizes multiplied by eight farm types) were possible in a region. An alternative way to view this is to consider a regional market comprising 24 different possible producers (large cereal farm or small LFA livestock) for each commodity.

Instead of using quantities produced (e.g. tonnes) in the estimation of the cost function, we used areas or average animal numbers. Whilst, perhaps unorthodox, this approach has two advantages for this study: first, the resultant profit maximisation situation subject to this cost function yields directly the area allocated to a crop and the average number of animals and; second, it avoids the problem of estimating a cost function where the regressors (i.e., crop outputs) are stochastic (since quantities produced are the multiplication of areas and yields and the latter are normally considered random terms).

Table 31.1 England – Number of businesses according to farm type, size (SLR) and region according to Census 2006

Type			East Midlands	East of England	North East	North West	South East and London	South West	West Midlands	Yorkshire Humber	Total
Cereals	Size	Small	4,256	5,188	1,161	538	2,798	1,999	1,560	661	18,161
		Medium	2,558	5,955	869	816	3,118	1,654	1,505	1,644	18,120
		Large	6,697	19,557	5,132	395	12,643	2,147	1,833	2,092	50,495
	Total		13,511	30,701	7,161	1,750	18,559	5,800	4,899	4,397	86,776
Dairy	Size	Small	406	258	190	745	265	784	335	376	3,358
		Medium	1,194	486	277	1,746	623	1,652	930	787	7,695
		Large	1,487	2,222	0	4,205	2,680	7,368	2,089	1,397	21,449
	Total		3,087	2,967	467	6,696	3,568	9,803	3,354	2,560	32,501
General Cropping	Size	Small	1,477	2,347	0	683	291	207	558	715	6,278
		Medium	878	1,724	157	0	120	233	448	1,459	5,018
		Large	7,105	16,188	1,688	758	2,144	2,073	2,361	1,774	34,090
	Total		9,460	20,259	1,845	1,441	2,555	2,513	3,366	3,947	45,386
Horticulture	Size	Small	0	82	0	56	155	28	15	1	336
		Medium	0	114	0	0	18	10	31	0	173
		Large	747	2,324	8	185	2,838	345	633	62	7,142
	Total		747	2,519	8	241	3,011	383	679	63	7,652

LFA Grazing Livestock	Size	Small	1,075	0	1,467	1,960	0	1,005	755	1,171	7,434
		Medium	641	0	2,943	3,372	0	1,188	397	2,870	11,411
		Large	1,671	0	5,400	13,422	0	3,177	580	6,059	30,308
	Total		3,387	0	9,810	18,754	0	5,369	1,732	10,099	49,153
Lowland Grazing Livestock	Size	Small	293	318	228	324	1,353	1,350	855	453	5,173
		Medium	325	1,148	517	268	929	1,295	339	111	4,932
		Large	1,249	2,594	987	1,140	1,692	2,135	694	83	10,574
	Total		1,867	4,060	1,731	1,732	3,974	4,781	1,888	646	20,680
Mixed	Size	Small	836	514	359	577	746	724	318	777	4,850
		Medium	287	689	254	336	539	831	562	439	3,939
		Large	3,016	2,360	3,056	1,447	4,950	4,976	3,656	525	23,986
	Total		4,139	3,563	3,669	2,361	6,235	6,531	4,536	1,742	32,775
Pigs and Poultry	Size	Small	44	91	0	0	136	29	58	114	471
		Medium	92	99	1	26	0	56	26	73	373
		Large	92	99	1	26	0	56	26	73	373
	Total		227	289	2	52	136	142	109	260	1,217
Totals by row											
Cereals			13,511	30,701	7,161	1,750	18,559	5,800	4,899	4,397	86,776
Dairy			3,087	2,967	467	6,696	3,568	9,803	3,354	2,560	32,501

General Cropping	9,460	20,259	1,845	1,441	2,555	2,513	3,366	3,947	45,386
Horticulture	747	2,519	8	241	3,011	383	679	63	7,652
LFA Grazing Livestock	3,387	0	9,810	18,754	0	5,369	1,732	10,099	49,153
Lowland Grazing Livestock	1,867	4,060	1,731	1,732	3,974	4,781	1,888	646	20,680
Mixed	4,139	3,563	3,669	2,361	6,235	6,531	4,536	1,742	32,775
Pigs and Poultry	227	289	2	52	136	142	109	260	1,217
Small	8,386	8,797	3,404	4,884	5,743	6,126	4,455	4,267	46,061
Medium	5,975	10,215	5,019	6,565	5,348	6,919	4,238	7,384	51,662
Large	22,064	45,344	16,272	21,578	26,947	22,276	11,871	12,064	178,417
Total	36,424	64,357	24,694	33,027	38,037	35,321	20,565	23,715	276,140

Source: Defra 2008

The functional form for the cost function was chosen due to its simplicity and adequacy for the task of estimating theoretically consistent marginal costs (i.e., supply relationships). The cost function omitting the sub-indices f,s,r for simplicity and also the specific dummies, is given by (where the sub-index t represents the time period, m is the number of outputs and n is the number of inputs):

$$C_t(W, A) = \left[\alpha_0 + \sum_{h=1}^{m} \alpha_h A_{ht} + \tfrac{1}{2} \sum_{h=1}^{m} \alpha_{hh} A_{ht}^2 \right] \cdot \exp\left[\beta_0 + \sum_{j=1}^{n} \beta_j \ln W_{jt} + \tfrac{1}{2} \sum_{j=1}^{n} \sum_{k=1}^{n} \beta_{jk} \ln W_{jt} \ln W_{kt} \right] + \varepsilon_t$$

It should be noted that the first part in brackets corresponds (excluding the parameter α_0) to the quadratic cost function frequently used in positive mathematical programming models, where separability amongst outputs (where the As in the formula represent the crop areas or average livestock numbers) is assumed. The second term in brackets corresponds to the input prices (Ws). This functional form can be deduced from the more general cost function presented in Pulley and Braunstein (1992).

The cost function was estimated with the inclusion of dummies for farm type and farm size and in addition a quadratic trend was included to try to capture any cost change over time. The results of the cost function estimation are presented in the annex. After the cost function was estimated, the parameters were adjusted to reproduce exactly the results of the season 2005/06, (i.e., the one year after the implementation of SPS).

The approach adopted to compute the degree of cross-subsidisation for a particular enterprise is highlighted diagrammatically in Figure 31.1. The cross-subsidy for one commodity for the farm is estimated as the difference between the implicit price ($P^{Implicit}$) at the level of observed production ($Q^{observed}$) minus the actual market price (P^{market}). The implicit price is computed using the estimated marginal cost function. Under the assumption that the cost function remains constant, if the market price is below the implicit price, the farmer is using part of his/her proceeds from the SPS to cross-subsidise the production of the commodity.

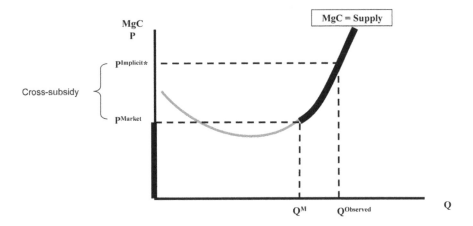

Figure 31.1   Estimation of the cross-subsidy

This approach therefore forms the basis for the results presented in the following section.

Results and discussion

Table 31.2 presents the results from the cross-subsidy estimation exercise. The results are presented as weighted averages (using production as the weighting variable) over size and farm type for the eight regions in England. As mentioned earlier, the analysis focuses on those crops and livestock that were receiving area or headage payments before decoupling was introduced.

The results highlight substantial levels of cross-subsidisation by commodity but with differences by regions. Although by no means universal, the results do reflect the process of specialisation that has occurred within England. That is, the level of cross subsidisation that is occurring at an enterprise level is less for those areas which tend to have a comparative advantage in production. For example, the East of England and East Midlands appear to have lower levels for cereal production and the South West for beef production. There are exceptions to this, but this may be a result of small levels of production skewing the results (for example cereals in the North West or beef cows in the East of England).

Table 31.2   England-Average weighted cross-subsidy by enterprise and region (£ per hectare or animal)

	East Midlands	East of England	North East	North West	South East and London	South West	West Midlands	Yorkshire and Humber
Wheat	272.4	312.0	453.6	288.2	n/s	393.0	307.4	367.6
Barley	384.7	264.4	296.3	n/s	378.4	273.7	308.4	269.8
Beef cows	155.8	72.5	154.6	109.6	155.8	45.7	173.2	231.5
Other cattle	134.9	174.8	125.9	133.6	133.5	50.6	6.0	117.7
Ewes	49.9	133.5	15.8	32.1	16.6	13.7	19.5	13.0
Other sheep	30.6	36.8	24.8	17.1	16.0	28.8	35.7	17.7

Note: n/s – marginal cost parameters were not statistically significant

The results clearly indicate that, in nearly all circumstances, the level of production found in 2005/6 was higher than that which would have been predicted under the prevailing market conditions. Of course there may be a number of reasons for this which do not, necessarily, involve a process of systematic cross-subsidisation. These include:

- the prices achieved in 2005/6 could have been lower than those expected at the time the level of production was decided
- the time lag associated with changing production levels (particularly for livestock) might infer that that any adjustments made may not be apparent within the first year of the SPS
- the fact that the policy change was so marked that farmers were just uncertain as to the impact and initially adopted a policy of maintaining the status quo in terms of production.

In terms of the first point above, it should be noted that prices in 2005/6 were in line with prices in the recent past and there was no general expectation that they would necessarily rise. The second and third points relate to the speed of the process of adjusting to the single payment. For example, recent research undertaken in Scotland based on more recent census data, does highlight that sheep numbers have declined markedly in the last couple of years as farmers seem to be adjusting stocking in response to the low market prices.

Another interesting feature of the degree of cross-subsidisation is that in many cases it appears higher than the single payment itself. This raises the question as to the extent that farmers are using other sources of income to support the farm business.

Conclusions and further research

The purpose of the chapter has been twofold: first to present a methodology to estimate the level of SPS that is used in the productive activities and; second, to analyse whether decoupled payments are truly decoupled. That is whether farmers are determining the allocation of their production simply according to market prices.

The results for the first year of application of the SPS 2005/06 indicate that for the key commodities that were under area or headage payments, farmers appear to have continued considering the SPS as coupled payments and therefore produced accordingly. Therefore, the SPS, despite what economic theory and policy makers might suggest may be having effects on production though a channel that is more direct than the ones pointed out by OECD (2006).

However, as mentioned in Section IV, it is important to mention that the obtained results might be due to some inertia in the production, associated for instance to rotation considerations or due to the fact that, as in the case of livestock, it takes time to restructure production. In this sense, it is worthwhile to repeat the exercise as more recent data becomes available, because this will provide a solid base to judge the ways farmers are restructuring their businesses in the presence of the SPS. This information is important if one needs to evaluate the impact of removing the SPS because if farmers do not become more market oriented (i.e., do not take their decisions based on market signals) the elimination of the SPS may have important productive implications in the future than those predicted by models that assume that farmers consider the SPS as a decoupled from production support.

The work of the chapter opens several possible paths for future research. The first is to use individual farm data from the FBS in the estimation of the cost functions. This would allow the computation of specific parameters for all regions. As more detailed cost data (at the individual enterprise level) become available a second line of research would be to compare the results obtained from multi-product cost functions with those by enterprise.

References

Caves, D.W., Christensen, L.R. and Tretheway, M.W. 1980. Flexible Cost Functions for Multiproduct Firms. *Review of Economics and Statistics*, 62, 477-81.

Chambers, R.G. 1988. *Applied production analysis: A dual approach.* New York, Cambridge University Press.

European Commission. 2006. Single Payment Scheme – the detail. Available at: http://ec.europa.eu/agriculture/markets/sfp/index_en.htm#capinfosheets

Organisation for Economic Co-operation and Development (OECD). 2006. Decoupling Agricultural Support from Production. OECD Policy Brief, November. Available at: http://www.oecd.org/dataoecd/5/54/37726496.pdf

Pulley L.B. and Braunstein Y.M. 1992. A Composite Cost Function for Multiproduct Firms with an Application to Economies of Scope in Banking. *Review of Economics and Statistics,* 74, 221-230.

UK-Department of Food, Environment and Rural Affairs. 2006. Single Payment Scheme

Handbook and Guidance for England 2006. Available at: http://www.rpa.gov.uk/rpa/index.nsf/15f3e119d8abcb5480256ef20049b53a/29723f07882706488025 712c0041199d/$FILE/SPS%20Handbook%20and%20Guidance%20for%20 England%202006%20edition.pdf

UK-Department of Food, Environment and Rural Affairs. 2007. Farm Business Survey – England. Available at: https://statistics.defra.gov.uk/esg/asd/fbs/default.htm

UK-Department of Food, Environment and Rural Affairs. 2008. Farm Accounts in England. Available at: https://statistics.defra.gov.uk/esg/publications/fab/default.asp

Annex

Correlation between estimated and observed endogenous variable: 0.99 Log likelihood: -1969.88		Standard error	t ratios
Variables	Coefficients		
Intercept-dummies for farm type			
Cereals	8.0381	0.3860	20.8260
Dairy	8.1215	0.3902	20.8160
General cropping	8.1899	0.3831	21.3760
Horticulture	15.3020	0.3860	39.6400
LFA grazing livestock	-19.8520	1.0000	-19.8520
Lowland grazing livestock	-0.2545	1.0000	-0.2545
Mixed	-15.1000	1.0000	-15.1000
Pigs and poultry	-16.2430	1.0233	-15.8730
Intercept-dummies for farm size			
Small	-5.2610	0.3833	-13.7260
Medium	-5.3912	0.3827	-14.0860
Large	-6.1467	0.3821	-16.0880
Intercepts associated to trend			
Trend	-0.0514	0.0271	-1.8969
Squared trend	0.0105	0.0029	3.6827
Input prices variables			
$\ln(W_1)$	0.3878	0.0255	15.2100
$\ln(W_1) \cdot \ln(W_1)$	0.2192	0.5602	0.3913
$\ln(W_1) \cdot \ln(W_2)$	-0.0325	0.0964	-0.3366
$\ln(W_1) \cdot \ln(W_3)$	-0.0912	0.3083	-0.2957
$\ln(W_1) \cdot \ln(W_4)$	0.0569	0.1441	0.3949
$\ln(W_1) \cdot \ln(W_5)$	-0.0080	0.2478	-0.0322
$\ln(W_1) \cdot \ln(W_6)$	-0.1445	0.2682	-0.5388
$\ln(W_2)$	0.0535	0.0054	9.9754
$\ln(W_2) \cdot \ln(W_1)$	-0.0325	0.0964	-0.3366
$\ln(W_2) \cdot \ln(W_2)$	0.0415	0.1131	0.3670
$\ln(W_2) \cdot \ln(W_3)$	-0.0296	0.0710	-0.4171
$\ln(W_2) \cdot \ln(W_4)$	0.0146	0.0380	0.3828
$\ln(W_2) \cdot \ln(W_5)$	0.0272	0.0752	0.3622
$\ln(W_2) \cdot \ln(W_6)$	-0.0212	0.1072	-0.1982
$\ln(W_3)$	0.1114	0.0150	7.4363
$\ln(W_3) \cdot \ln(W_1)$	-0.0912	0.3083	-0.2957
$\ln(W_3) \cdot \ln(W_2)$	-0.0296	0.0710	-0.4171
$\ln(W_3) \cdot \ln(W_3)$	0.0342	0.2269	0.1506
$\ln(W_3) \cdot \ln(W_4)$	-0.0240	0.0914	-0.2630
$\ln(W_3) \cdot \ln(W_5)$	0.0500	0.1710	0.2925
$\ln(W_3) \cdot \ln(W_6)$	0.0606	0.2174	0.2789

Correlation between estimated and observed endogenous variable: 0.99 Log likelihood: -1969.88		Standard error	t ratios
Variables	Coefficients		
$\ln(W_4)$	0.1274	0.0073	17.3760
$\ln(W_4) \cdot \ln(W_1)$	0.0569	0.1441	0.3949
$\ln(W_4) \cdot \ln(W_2)$	0.0146	0.0380	0.3828
$\ln(W_4) \cdot \ln(W_3)$	-0.0240	0.0914	-0.2630
$\ln(W_4) \cdot \ln(W_4)$	0.0424	0.0616	0.6886
$\ln(W_4) \cdot \ln(W_5)$	0.0299	0.1024	0.2921
$\ln(W_4) \cdot \ln(W_6)$	-0.1197	0.1186	-1.0099
$\ln(W_5)$	0.1010	0.0127	7.9546
$\ln(W_5) \cdot \ln(W_1)$	-0.0080	0.2478	-0.0322
$\ln(W_5) \cdot \ln(W_2)$	0.0272	0.0752	0.3622
$\ln(W_5) \cdot \ln(W_3)$	0.0500	0.1710	0.2925
$\ln(W_5) \cdot \ln(W_4)$	0.0299	0.1024	0.2921
$\ln(W_5) \cdot \ln(W_5)$	0.0643	0.2092	0.3076
$\ln(W_5) \cdot \ln(W_6)$	-0.1635	0.2234	-0.7320
$\ln(W_6)$	0.2189	0.0136	16.0354
$\ln(W_6) \cdot \ln(W_1)$	-0.1445	0.2682	-0.5388
$\ln(W_6) \cdot \ln(W_2)$	-0.0212	0.1072	-0.1982
$\ln(W_6) \cdot \ln(W_3)$	0.0606	0.2174	0.2789
$\ln(W_6) \cdot \ln(W_4)$	-0.1197	0.1186	-1.0099
$\ln(W_6) \cdot \ln(W_5)$	-0.1635	0.2234	-0.7320
$\ln(W_6) \cdot \ln(W_6)$	0.3884	0.3213	1.2087
Output related terms (linear and squared)			
Intercept	82.3860	6.8916	11.9540
Wheat	-146.9800	8.3536	-17.5950
Squared wheat	3.3141	0.2061	16.0790
Barley	-63.1990	6.5282	-9.6809
Squared barley	-8.5487	0.6174	-13.8450
Other cereals	-338.0600	4.8795	-69.2810
Squared other cereals	23.6890	1.0280	23.0440
Oilseed rape	136.8800	3.1569	43.3600
Squared oilseed rape	-3.2183	2.0864	-1.5426
Potatoes	-48.0730	1.1903	-40.3860
Squared potatoes	21.7130	1.0177	21.3350
Sugar beet	582.1600	2.6269	221.6200
Squared sugar beet	0.0000	2.5838	0.0000
Other crops	-61.5760	2.1558	-28.5620
Squared other crops	0.0000	3.1766	0.0000
Vegetables and fruits	-2.9495	0.7401	-3.9855
Squared vegetable and fruits	0.0000	0.1130	0.0000

Correlation between estimated and observed endogenous variable: 0.99 Log likelihood: -1969.88		Standard error	t ratios
Variables	Coefficients		
By prods., forage and cultivations	-105.6800	2.1876	-48.3090
Squared by prods., forage and cultivations	-11.2580	0.3751	-30.0140
Set-aside	-48.0890	4.1656	-11.5440
Squared set-aside	-5.6264	0.9948	-5.6556
Dairy cows and heifers in milk	-1656.4000	4.6599	-355.4500
Squared dairy cows and heifers in milk	-4.8653	0.2972	-16.3690
Beef cows	-923.4100	1.0003	-923.1600
Squared beef cows	0.0000	4.9363	0.0000
Other cattle	2518.7000	3.3184	759.0000
Squared other cattle	0.0000	1.0441	0.0000
Ewes	-906.1600	1.3706	-661.1200
Squared ewes	0.0000	0.5541	0.0000
Other sheep	-124.7400	1.3843	-90.1110
Squared other sheep	0.0000	0.5935	0.0000
Breeding sows	15767.0000	1.0000	15767.0000
Squared breeding sows	10.8770	2.6830	4.0542
Other pigs	-85.2900	1.9025	-44.8310
Squared other pigs	1.1233	0.3069	3.6600
Hen and pullets in lay	-64.8380	1.3335	-48.6220
Squared hen and pullets in lay	-0.8074	0.1147	-7.0397
Other poultry	-13.1780	29.5680	-0.4457
Squared other poultry	0.0000	0.9414	0.0000
Notes:			
W_1= Feed grown and purchased price			
W_2= Livestock services price			
W_3= Seeds (purchased and grown) price			
W_4= Fertilizers price			
W_5= Crop protection price			
W_6= Other goods and services price			

Index